Piero Picci • Marco Manfrini
Nicola Fabbri • Marco Gambarotti
Daniel Vanel
Editors

# Atlas of Musculoskeletal Tumors and Tumorlike Lesions

## The Rizzoli Case Archive



Springer

*Editors*
Piero Picci
Muscoloskeletal Oncology
Istituto Ortopedico Rizzoli
Bologna
Italy

Marco Manfrini
Department of Surgery
Istituto Ortopedico Rizzoli
Bologna
Italy

Nicola Fabbri
Department of Surgery
Memorial Sloan-Kettering Cancer Center
New York, NY
USA

Marco Gambarotti
Department of Anatomy and Pathology
Istituto Rizzoli
Bologna
Italy

Daniel Vanel
Department of Anatomy and Pathology
Istituto Rizzoli
Bologna
Italy

ISBN 978-3-319-37819-0 ISBN 978-3-319-01748-8 (eBook)
DOI 10.1007/978-3-319-01748-8
Springer Cham Heidelberg Dordrecht London New York


Softcover reprint of the hardcover 1st edition 2014


Printed on acid-free paper

Springer is part of Springer Science+Business Media (www.springer.com)

*The Syllabus is dedicated to*
*Prof. Mario Campanacci*
*Prof. Mario Mercuri*
*Dr. Gaetano Bacci*
*Dr. Marco Alberghini*
*who strongly supported the dissemination*
*of knowledge of musculoskeletal tumors.*

# Preface

The Rizzoli Orthopedic Institute has a proud and dynamic history, with several very famous orthopedic surgeons over the past 120 years. Dr. Mario Campanacci was especially concerned with the diagnosis and therapy of bone and soft tissue tumors, particularly the malignant bone tumors, two of which are very common and often lethal to children, namely osteosarcoma and Ewing's sarcoma. Beginning in the early 1970s, he pioneered the use of chemotherapy in Europe for these two tumors. He organized a team of surgeons and oncologists, who helped make an amazing reversal for most of these unfortunate patients dying of lung metastases within 2 years, to the majority being cured. The Rizzoli Orthopedic Institute has an independent Department dedicated to The Treatment of Malignant Bone Tumors and they now achieve, for osteosarcoma and Ewing's sarcoma, an over 70–75 % cure rate for their patients (by cure is meant the total eradication of the cancerous tumor).

Prior to the era of modern chemotherapy in the treatment of malignant tumors of bone, most patients died, even if they were diagnosed within days of arriving at a hospital, and even if an amputation was quickly applied as a desperate measure. The reason they died despite appropriate rapid diagnosis and local ablative surgery is that about 90 % of these patients already had microscopic lung metastases, impossible to see on their initial admission to the hospital by standard radiology studies. Months later, however, the tiny, microscopic seeds of metastases to the lungs would grow to grossly visible proportions, and within months, the patient would succumb to death by suffocation. Thanks to the Rizzoli and other Research Institutes throughout the world, not only are most patients with Ewing's and osteosarcoma now cured, but as amazingly the majority can be cured without even the need for amputation. It is now possible to maintain their limb simply by removing the tumor area en bloc, with a margin of uninvolved tissue, and using a prosthetic replacement, after a course of pre-operative chemotherapy. The Rizzoli Institute also stands at the cornerstone of these superb surgical advancements in prosthetic replacement techniques. As an effect of these outstanding results, the Rizzoli Institute Bone Tumor Treatment Department is now not only the premier treatment center for such tumors in Europe, treating some 80 % of all such malignant tumor patients in Italy, but it may well be the largest and most renowned such treatment center in the world.

Another of the main goals of the Rizzoli has always been centered around the education of young doctors to become some of the finest orthopedic surgeons, oncologists, pathologists and radiologists in the world with respect to all aspects of orthopedics, and especially for the specialty in the bone tumor field. Bone tumors are very rare, representing only about 1 % of all benign and malignant tumors. It is extremely important that major bone tumor centers such as the Rizzoli exist, where patients afflicted by very rare tumors can be sent and where the physicians who will, or are directly caring for these patients, obtain the necessary training and experience. A standard community hospital serving a local population of some 100,000 individuals will only see about 1 patient per year with a malignant bone tumor. In my opinion, to begin to understand how to accurately diagnose (with over 95 % accuracy) and treat these tumors adequately requires a physician to have personally seen at least 500 such patients. In a general hospital that could take 500 physician years, well beyond the lifetime of any ordinary person I know. But at the Rizzoli it is possible for diagnosticians, treating physicians and student doctors to be involved with some 500 bone tumor patients in 2–3 years.

Which now brings us to the "Rizzoli Syllabus" which you now hold in your hands. This is a truly remarkable primer for students and even trained physicians to study the essentials of bone tumor diagnosis and treatment of virtually all of the benign and malignant tumor entities of bone. And for the first time in my experience, this syllabus also includes considerations of basic biology concepts of giant cell tumor, chondrosarcoma, osteosarcoma and several other important basic science oriented topics. Many of these topics have been pioneered in the Rizzoli Research Institute under Dr. Piero Picci's direction. In addition, very important principals of staging and radiology have been added to the syllabus, vital to an overall understanding of the treatment and diagnosis of tumors of the bone.

I have been involved with writing my own syllabi for The UCLA orthopedic residents in years past, and I have seen a number of other syllabi over the years, but for a bone tumor syllabus, this is by far the best I have ever seen published. It is informative, it is accurate, it is concise, and it is beautifully illustrated. The authors are to be highly commended for their efforts and dedication to teaching, from which a new generation of highly competent bone tumor specialists will emerge.

Joseph M. Mirra, M.D.,
UCLA Professor of Pathology,
Emeritus

# Acknowledgements

Revision: Alba Balladelli, Laboratory of Experimental Oncology, IOR
Graphic work: Cristina Ghinelli, Laboratory of Experimental Oncology, IOR

# Contents

# Introduction

This book reflects the experience of the Rizzoli Orthopedic Institute in over 100 years of treatment of musculoskeletal tumor and tumorlike lesions. The first treated case dates September 28, 1900, and the archive contains the original material (clinical charts, imaging, paraffin blocks, and histological slides) of more than 40,000 cases (about 29,000 bone lesions and 11,000 soft tissue lesions). This syllabus briefly reports the most relevant entities.

Each single entity is presented multidisciplinarily, with the pertinent clinical, radiological, and histological correlations. Treatment is briefly reported for each entity. Other separate chapters analyze the more recent biomolecular findings useful for diagnosis, prognosis, and treatment.

The text reflects the improvements in knowledge of musculoskeletal tumors as presented during the yearly international course held at the Rizzoli Institute.

This course, promoted by Prof. Mario Campanacci since the 1970s, has seen the participation as guest professors of the major international experts in musculoskeletal lesions:

D. Dahlin (Rochester) 1974; 1984
W.F. Enneking (Gainesville) 1984; 1989–1990; 1992–1994; 1998; 2004
N. Jaffe (Houston) 1984
D. Springfield (Gainesville, New York, Boston) 1995; 1997; 2000; 2002–2007
J.M. Mirra (Los Angeles) 1996; 2008
H. Mankin (Boston) 1999
A.L. Schiller (Boston) 2002
D. Vanel (Villejuif-Bologna) 2003; 2005; 2007–2008
P.C.W. Hogendoorn (Leiden) 2006–2008
N. Athanasou (Oxford) 2008
M.C. Gebhardt (Boston) 2008–2013
F.H. Sim (Rochester) 2010–2011
M.I. O'Connor (Jacksonville) 2011–2013
J.M. Coindre (Bordeaux) 2011
M.J. Klein (New York) 2012–2013
J.H. Healey (New York) 2013
A.P. Dei Tos (Treviso) 2013

*Rizzoli collaborators to the Annual Course of the last years*:

| | |
|---|---|
| Patrizia Bacchini | Pathology |
| Maria Serena Benassi | Biology |
| Stefania Benini | Biology |
| Franco Bertoni | Pathology |
| Roberto Biagini | Orthopedics |
| Giuseppe Bianchi | Orthopedics |
| Stefano Boriani | Orthopedics |
| Laura Campanacci | Orthopedics |
| Roberto Casadei | Orthopedics |
| Massimiliano De Paolis | Orthopedics |
| Davide Donati | Orthopedics |
| Costantino Errani | Orthopedics |
| Nicola Fabbri | Orthopedics |
| Stefano Ferrari | Oncology |
| Andrea Ferraro | Orthopedics |
| Marco Gambarotti | Pathology |
| Alessandro Gasbarrini | Orthopedics |
| Marco Manfrini | Orthopedics |
| Emanuela Palmerini | Oncology |
| Piero Picci | Oncology |
| Alberto Righi | Pathology |
| Eugenio Rimondi | Radiology |
| Pietro Ruggieri | Orthopedics |
| Katia Scotlandi | Biology |
| Massimo Serra | Biology |
| Eric Staals | Orthopedics |
| Daniel Vanel | Radiology |
| Licciana Zanella | Biology |

# Staging

**(Updated by Pietro Ruggieri)**

Tumors are labeled and classified according to their clinical and histologic features. Such assessment however does not suffice to describe the actual behaviour of a given tumor in an individual patient. Moreover, to determine the anatomo-clinical diagnosis, staging of the tumor is necessary for each individual case. The staging system presently adopted, for benign and malignant primary tumors of bone and soft tissues, has been devised by Enneking (1980). According to Enneking (1983), "by compartment is meant an anatomical structure or space bounded by natural barriers to tumor extension". Natural barriers are the cortical bone, fascia and fascial septa, articular cartilage, joint capsule, tendons and tendon sheaths. Fat and interstitial areolar tissues outside these compartments are extra-compartmental, like the tissue around neuro-vascular bundles. Natural barriers as cortical bone or fascia can be breached especially along their vascular perforations. The most resistant barrier is the articular cartilage, having no vascular perforations and probably an intrinsic resistance to tumor. The growth plate functions as a relative barrier, depending on anatomical site and age (perforated by vessels in early infancy and again at puberal age). The periosteum, the synovial membrane (until it is ulcerated and bleeding, producing a hemarthrosis containing neoplastic cells), and the sheath of the major nerves (perinervium) can be considered relative although very thin barriers. Where the joint capsule and synovium, the ligaments and tendons insert into an epiphysis, apophysis or metaphysis, the only barrier is a thin bony cortex with vascular perforations, and therefore the tumor easily extends from the cancellous bone to these structures and vice versa.

The system is based on the three classic parameters, **G**, **T**, and **M**.

**G** is the grade of the tumor, mainly dictated by histology. G0 is benign, G1 is low-grade malignant, and G2 is high-grade malignant. When a four-grade classification of malignancy is used, histological grades 1 and 2 are low; grades 3 and 4 high.

**T** is the tumor anatomic extension. T0 means a benign tumor contained by a true capsule (intracapsular). T1 is a benign or malignant tumor not having a true capsule, yet confined within an anatomical compartment. T2 is a benign or malignant tumor without a true capsule, originating in an extra-compartmental space, or expanded extra-compartmentally by violating the natural barriers.

**M** are the metastases, either regional (skip, lymph nodes) or distant. M0 means absence, and M1 presence of metastases.

Benign tumors, of bone and soft tissues, are both staged 1 (latent, inactive), 2 (active), and 3 (aggressive).

Sarcomas (bone and soft tissues) are staged I (low-grade malignancy), II (high-grade malignancy) and III (with metastases). Each of the three stages is subdivided in A and B: whether the tumor is intra- or extra-compartmental in stage I and II, and if it is low- or high-grade in stage III.

**Surgical stages for benign musculoskeletal tumors**

| Stage | Grade | Site | Metastases | Definition |
|---|---|---|---|---|
| 1 | G0 | T0 | M0 | Latent or inactive |
| 2 | G0 | T0 | M0 | Active |
| 3 | G0 | T1–T2 | M0–1 | Aggressive |

**Surgical stages for malignant musculoskeletal tumors**

| Stage | Grade | Site | Metastases | Definition |
|---|---|---|---|---|
| IA | G1 | T1 | M0 | Low grade |
| IB | G1 | T2 | M0 | A Intra-compartmental |
| | | | | B Extra-compartmental |
| IIA | G2 | T1 | M0 | High grade |
| IIB | G2 | T2 | M0 | A Intra-compartmental |
| | | | | B Extra-compartmental |
| IIIA–B | G1–2 | T1–T2 | MI | Metastatic |
| | | | | Either grade (A or B) |
| | | | | Distant metastases |

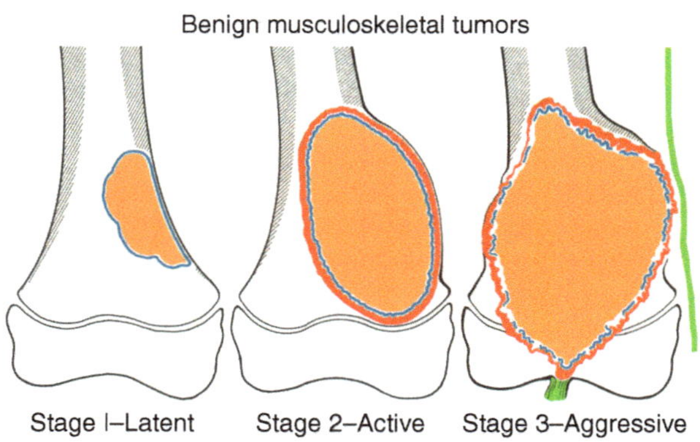

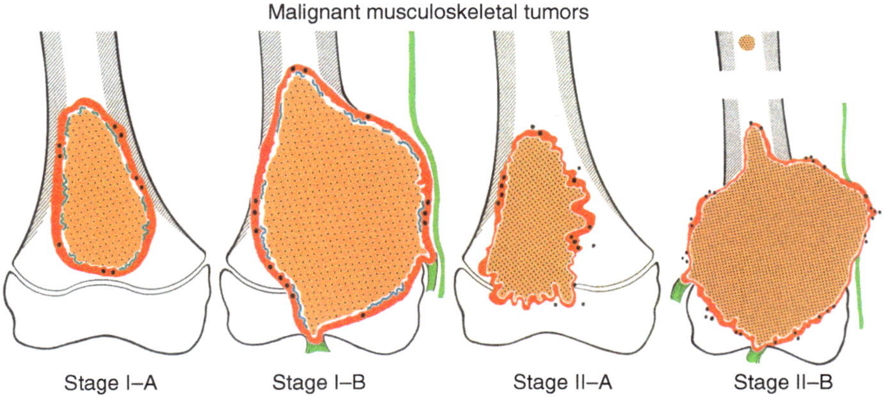


The Enneking system is best suited for well-documented sarcomas arising in the extremities. It does not include the type, size and depth of the tumor as separate parameters; its two-tier grading system is too narrow for the wide biological range of soft tissue tumors. Oncologists prefer the American Joint Committee system (AJCS) because it is applicable for any site. It is based on the TNMG system, and uses size and tumor extension (T), involvement of lymph nodes (N), presence of metastasis (M) and the type and grade of tumor (G).

**T1**: <5 cm; **T2**: 5 cm or greater; **T3**: involvement of bone, vessels, nerve. **N0**: no metastasis to regional lymph node; **N1**: lymph node metastases.
**M0**: no distant metastasis; **M1**: distant metastasis.
**G1**: low grade, well-differentiated; **G2**: moderate (moderately well-differentiated).
**G3**: high-grade, poorly differentiated.

| Stage | Grade | Site | Lymph node metastases | Distant metastases | Definition |
|---|---|---|---|---|---|
| IA | G1 | T1 | N0 | M0 | Low grade without metastases |
| IB | G1 | T2 | N0 | M0 | |
| IIA | G2 | T1 | N0 | M0 | Moderate grade without metastases |
| IIB | G2 | T2 | N0 | M0 | |
| IIIA | G3 | T1 | N0 | M0 | High grade without metastases |
| IIIB | G3 | T2 | N0 | M0 | Any tumor with lymph node metastases |
| IIIC | G1–3 | T1–2 | N1 | M0 | |
| IVA | G1–3 | T3 | N0–1 | M0 | Tumor involving bone, vessel, nerve with or without lymph node metastases |
| IVB | G1–3 | T1–3 | N0–1 | M1 | Any tumor with distant metastases |

## Selected Bibliography

Enneking WF, Spanier SS, Goodman MA (1980) A system for the surgical staging of musculoskeletal sarcoma. Clin Orthop Relat Res (153):106–120

Kotilingam D, Lev DC, Lazar AJ, Pollock RE (2006) Staging soft tissue sarcoma: evolution and change. CA Cancer J Clin 56(5):282–291; quiz 314–5. Review

Saddegh MK, Lindholm J, Lundberg A, Nilsonne U, Kreicbergs A (1992) Staging of soft-tissue sarcomas. Prognostic analysis of clinical and pathological features. J Bone Joint Surg Br 74(4):495–500

Wolf RE, Enneking WF (1996) The staging and surgery of musculoskeletal neoplasms. Orthop Clin North Am 27(3):473–481. Review

Wunder JS, Healey JH, Davis AM, Brennan MF (2000) A comparison of staging systems for localized extremity soft tissue sarcoma. Cancer 88(12):2721–2730

# Part I
# Bone Tumors

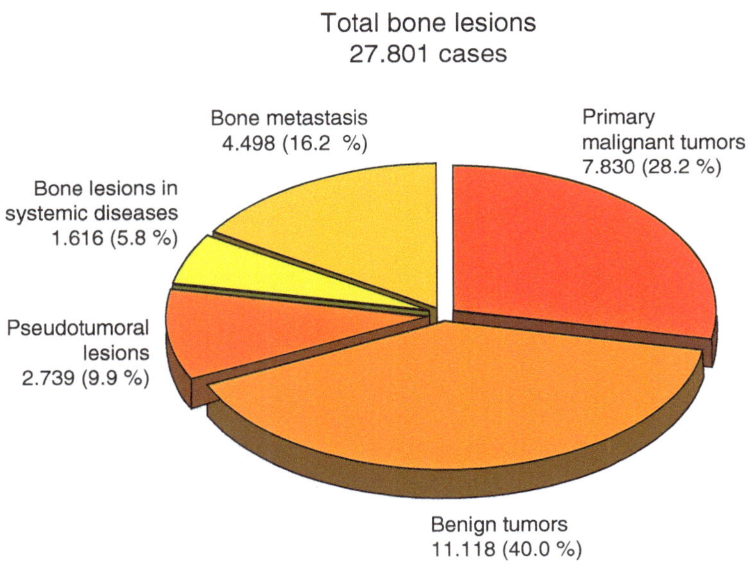


1900–2012 - Istituto Ortopedico Rizzoli - Laboratory of Experimental Oncology - Section of Epidemiology - Bologna - Italy

## Epidemiology

**(Updated by Piero Picci)**

There are no data regarding epidemiology of pseudotumoral and benign bone lesions. Most are incidental findings during examinations for other conditions. As for soft tissue lesions, it is generally accepted that benign conditions are 100 times more frequent than malignant primary bone tumors.

Figures from the Rizzoli archive are not representative of the true incidence, as they represent cases treated in a specialized center and therefore biased for more severe or complicated cases. More reliable are data regarding age, sex and sites of presentation.

## *Pseudotumoral Lesions*

The male to female ratio is 1.5:1. Typically, they affect children and teenagers with a median age of 12 years. In fact, the most represented entities are simple bone cyst and Langherans cells histiocytosis, which are typical of young age. Aneurysmal bone cyst, once included in pseudotumoral lesion, is now considered a benign entity, due to the detection of a specific translocation.

These two entities together represent more than 50 % of all pseudotumoral lesions. Preferred sites are metaphysis of long bones, especially proximally.

# Chapter 1
# Epidemiology

Piero Picci

There are no data regarding epidemiology of pseudotumoral and benign bone lesions. Most are incidental findings during examinations for other conditions. As for soft tissue lesions, it is generally accepted that benign conditions are 100 times more frequent than malignant primary bone tumors.

Figures from the Rizzoli archive are not representative of the true incidence, as they represent cases treated in a specialized center and therefore biased for more severe or complicated cases. More reliable are data regarding age, sex, and sites of presentation.

## 1.1 Pseudotumoral Lesions

The male/female ratio is 1.5:1. Typically, they affect children and teenagers with a median age of 12 years. In fact, the most represented entities are simple bone cyst and Langerhans cells histiocytosis which are typical of young age. Aneurysmal bone cyst, once included in pseudotumoral lesion, is now considered a benign entity, due to the detection of a specific translocation.

These two entities together represent more than 50 % of all pseudotumoral lesions. Preferred sites are metaphysis of long bones, especially proximally.

P. Picci, MD
Laboratory of Experimental Oncology, Musculoskeletal Oncology,
Istituto Ortopedico Rizzoli, Bologna, Italy
e-mail: piero.picci@ior.it

P. Picci et al. (eds.), *Atlas of Musculoskeletal Tumors and Tumorlike Lesions*,
DOI 10.1007/978-3-319-01748-8_1, © Springer International Publishing Switzerland 2014

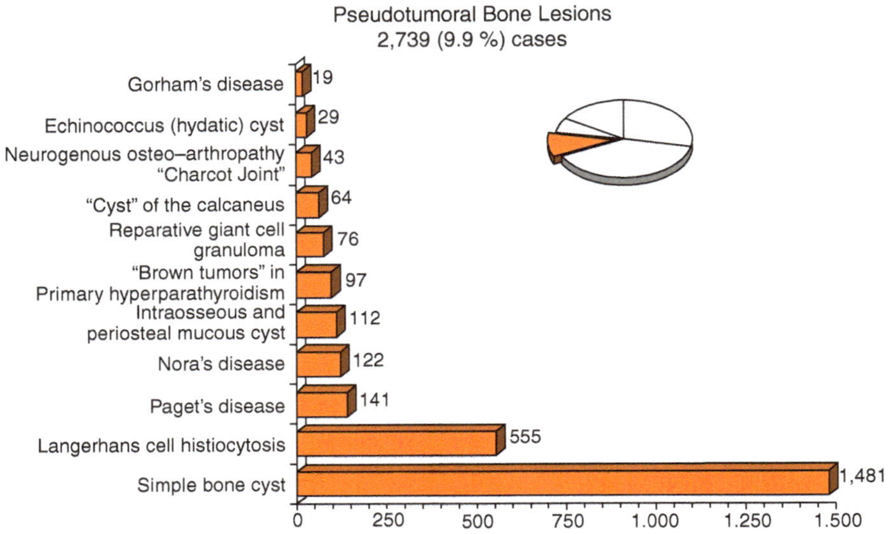


1900–2012 - Istituto Otopedico Rizzoli - Laboratory of Experimental Oncology - Section of Epidemiology - Bologna - Italy

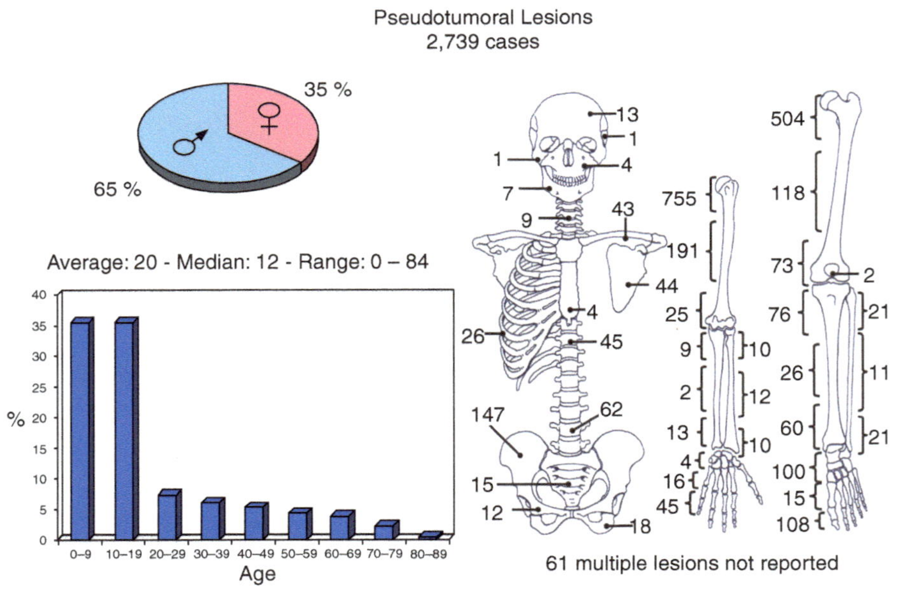


1900–2012 - Istituto Ortopedico Rizzoli - Laboratory of Experimental Oncology - Section of Epidemiology - Bologna - Italy

## 1.2 Benign Lesions

The male/female ratio is the same as for pseudotumoral lesions (1.5:1). Age comprises the first four decades of life, with a median age of 18 years. They are very rare in the elderly.

The most represented are benign chondroblastic lesions (osteochondromas and chondromas) and osteoid osteoma. Aneurysmal bone cyst is now considered a benign entity, due to the detection of a specific translocation. Preferred sites are in the bone around the knee, but practically all bones may be affected with localizations in flat bones that are not rare

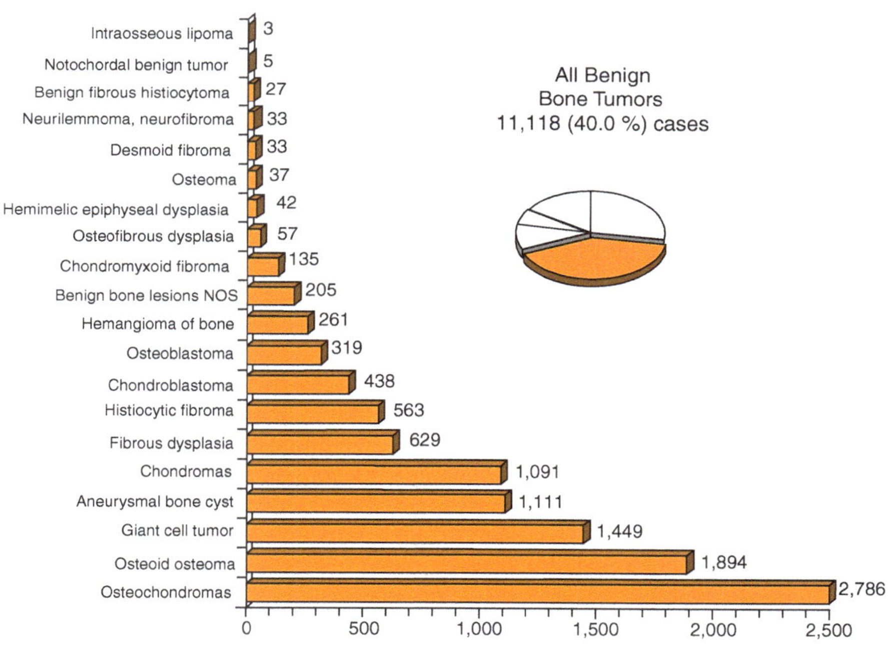


1900–2012 - Istituto Ortopedico Rizzoli - Laboratory of Experimental Oncology - Section of Epidemiology - Bologna - Italy

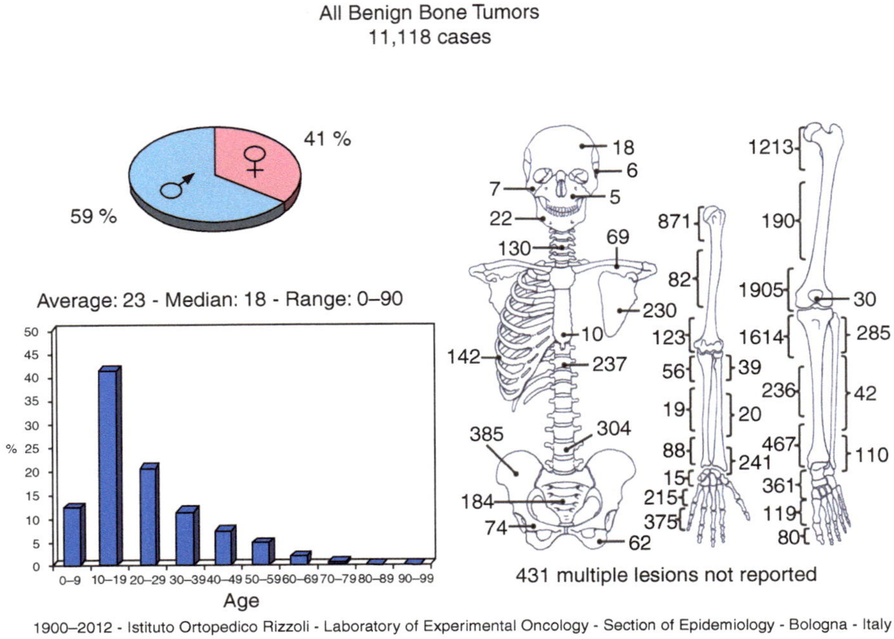


## 1.3 Primary Malignant Lesions

More reliable epidemiologic data are available for primary malignant bone tumors, or with a better definition bone sarcomas. Incidence usually is evaluated including soft tissue sarcomas. Global incidence reported is around 5 new cases/100,000 inhabitants/year. This incidence is similar in different countries with about 3,000 new cases per year in Great Britain and Italy (with 60 million inhabitants) and about 12,000 new cases in the USA (with about 300 million inhabitants).

This incidence is considered generally not more than 1 % of all malignant tumors.

Specifically regarding bone sarcomas, their incidence is considered 1/5–1/6 of all sarcomas, with an incidence of 0.8–1 new case/100,000 inhabitants/year.

The Rizzoli archive figures are more reliable in comparison to nonmalignant conditions. The only bias is related to the lack of registration for those localizations generally not treated at an orthopedic hospital as the trunk and skull.

The male/female ratio is again 1.5:1, as for the pseudotumoral and benign lesions. Median age is 25 years, with a peak in the second and third decades of life. Incidence is also evident and constant for all the adult age.

Osteosarcomas are the most frequent with an incidence of about 0.2 new cases/100,000 inhabitants/year.

A similar incidence is also reported for chondrosarcomas, followed by Ewing sarcoma with an incidence considered one-half in comparison to osteosarcomas. All other entities are very rare. Affected sites are the same as those of benign tumors.

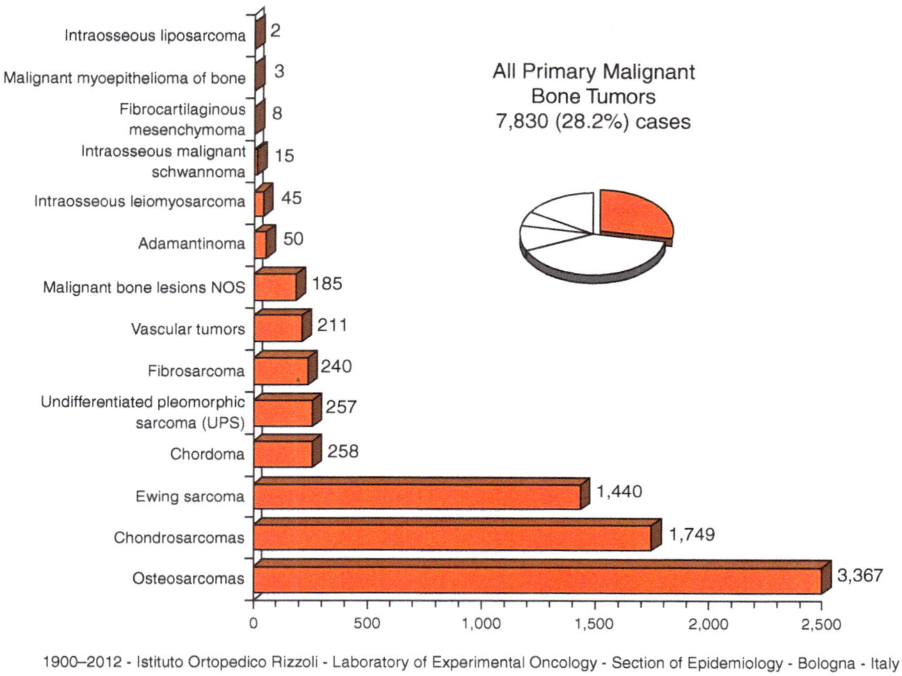
Intraosseous liposarcoma 2
Malignant myoepithelioma of bone 3
Fibrocartilaginous mesenchymoma 8
Intraosseous malignant schwannoma 15
Intraosseous leiomyosarcoma 45
Adamantinoma 50
Malignant bone lesions NOS 185
Vascular tumors 211
Fibrosarcoma 240
Undifferentiated pleomorphic sarcoma (UPS) 257
Chordoma 258
Ewing sarcoma 1,440
Chondrosarcomas 1,749
Osteosarcomas 3,367
0 500 1,000 1,500 2,000 2,500
All Primary Malignant Bone Tumors
7,830 (28.2%) cases
1900–2012 - Istituto Ortopedico Rizzoli - Laboratory of Experimental Oncology - Section of Epidemiology - Bologna - Italy

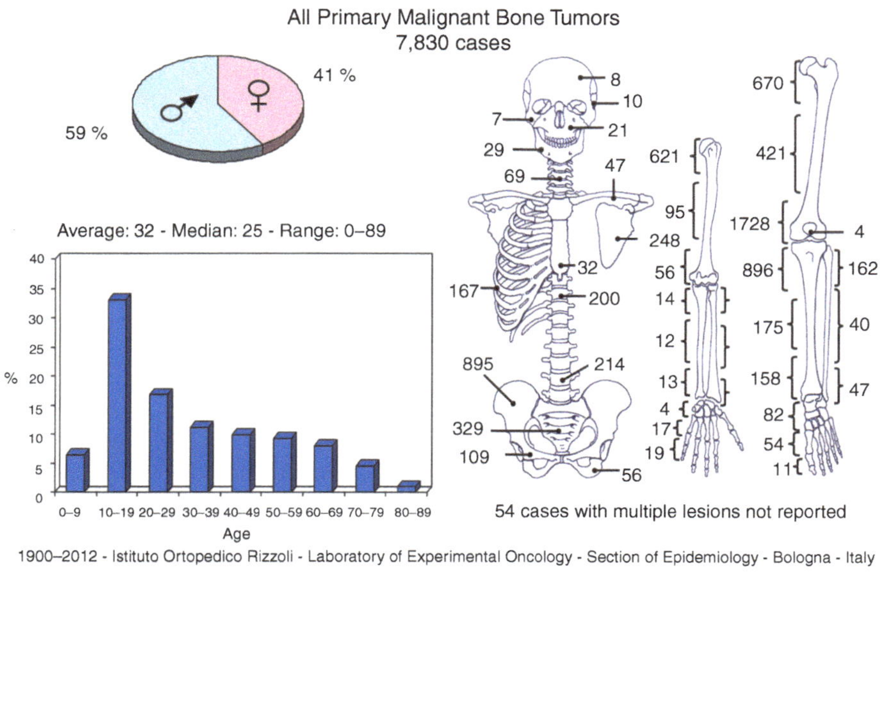
All Primary Malignant Bone Tumors
7,830 cases
59 %
41 %
Average: 32 - Median: 25 - Range: 0–89
%
Age
0–9 10–19 20–29 30–39 40–49 50–59 60–69 70–79 80–89
8 10 7 21 29 47 69 248 32 167 200 895 214 329 109 56
621 95 56 14 12 13 4 17 19
670 421 1728 4 896 162 175 40 158 47 82 54 11
54 cases with multiple lesions not reported
1900–2012 - Istituto Ortopedico Rizzoli - Laboratory of Experimental Oncology - Section of Epidemiology - Bologna - Italy

## Selected Bibliography

Jemal A, Murray T, Samuels A, Ghafoor A, Ward E, Thun MJ (2003) Cancer statistics, 2003. CA Cancer J Clin 3:5–26

Jemal A, Siegel R, Ward E, Murray T, Xu J, Thun MJ (2007) Cancer statistics, 2007. CA Cancer J Clin 57:43–66

Storm HH (1994) Cancers of the soft tissues. In: Doll R, Fraumeni JF Jr, Muir CS (eds) Trends in cancer incidence and mortality, cancer surveys, vol 19/20. Imperial Cancer Research Fund, Cold Spring Harbour Laboratory Press, Plainview, pp 197–217

Whelan J, McTiernan A, Cooper N, Wong YK, Francis M, Vernon S, Strauss SJ (2012) Incidence and survival of malignant bone sarcomas in England 1979–2007. Int J Cancer 131(4):E504–E517

# Chapter 2
# General Principles of Imaging

**Daniel Vanel**

Bone imaging has undergone significant changes during the last decades. Ultrasounds, CT, MR imaging, and PET have been added to arteriography, scintigraphy, and radiographs. At the same time, survival of primary malignant tumors has improved with chemotherapy, and local results have improved with conservative surgery. Imaging plays a major role in diagnosis, local and general staging, monitoring the treatment, and detecting recurrences.

## 2.1 Diagnosis

Patient age, location (bone or part of the bone involved), size (small is usually benign), calcifications and ossifications, limitation (the better seen, the slower the lesion growth), periosteal bone formations, and soft tissue involvement must be added to reach a reliable diagnostic probability. Combining clinical information, imaging, and histology leads to the most accurate results. Diagnosis of a bone tumor must be teamwork.

Radiographs remain the mandatory first step. They allow diagnosis of "leave-me-alone" lesions, and nothing more is usually added. If the lesion on radiographs is probably malignant, the next step should be immediate MR to stage the lesion.

Radiographs have limitations: superimpositions, partial cortex destruction could be overlooked, and flat and short bones and soft tissues are poorly analyzed.

CT is used in case of a diagnostic problem on radiographs. It allows a better study of the cortex, to detect and analyze small calcifications and thin periosteal bone formations. The nidus of an osteoid osteoma is well found (much better than on MR). Measuring tissue density can help characterize fat, fluid, blood, and

D. Vanel, MD
Department of Anatomy and Pathology, Istituto Ortopedico Rizzoli, Bologna, Italy
e-mail: daniel.vanel@ior.it

P. Picci et al. (eds.), *Atlas of Musculoskeletal Tumors and Tumorlike Lesions*,

DOI 10.1007/978-3-319-01748-8_2, 

calcification. After injection of contrast medium, soft tissue extension and vessels are well located.

The contrast remains however much lower than on MRI, which is the main modality for local staging. Its diagnostic role is limited, as calcifications and periosteal bone formations are more difficult to analyze (a black signal on MRI may be a calcification, but also fibrous tissue or chronic bleeding with hemosiderin). Fluid-fluid levels are better depicted than on CT, because of higher contrast and longer examination time. They are frequently seen in aneurysmal bone cysts but are nonspecific.

## 2.2 Local Staging

On MRI, the precise location of the tumor is well analyzed.

Intramedullary extension (and the level of surgical resection), skip metastases, soft tissue involvement, and extension to vessels and nerves are easily and reliably detected. The main limitation is articular extension, which could change the surgical technique: if the tumor abuts on the cartilage, we cannot reliably predict the joint involvement.

In case of contraindications (pacemakers and metallic ocular foreign bodies), CT is used but has a lower accuracy.

## 2.3 Distant Spread

Bone metastases and multiple lesions are detected on scintigraphy. Total body MRI is more sensitive without irradiation. Pulmonary metastases are sought by chest CT. We must keep in mind that although sensitivity is good, specificity is poor (or if we detect a lesion, we are not sure that it is actually a metastasis).

PET, now combined with CT, allows a global study of the patient, studying the tumor metabolic activity as well as the distal extent. Its spatial resolution is limited (lesions of less than 5 mm may be overlooked), and some malignant lesions are not very active metabolically (such as sclerotic metastases).

## 2.4 Evaluation of Treatment Effectiveness

Most primary malignant tumors are treated with preoperative chemotherapy. Decrease of lesion size, ossification, and decrease of early contrast medium uptake on scintigraphy and most of all dynamic MRI are signs of an efficient treatment. They become reliable only late after the beginning of chemotherapy. The initial results of PET are not (yet?) much better.

## 2.5 Detection of Local Recurrences

In case of suspicion, a local MR can be performed if the prothesis is non-paramagnetic (that is in titanium).

## 2.6 Summary

Radiographs remain the first step to image a bone tumor. In case of diagnostic problems, the next step is CT. MR is the main imaging modality for local staging, treatment evaluation, and detection of recurrences. PET is still under evaluation.

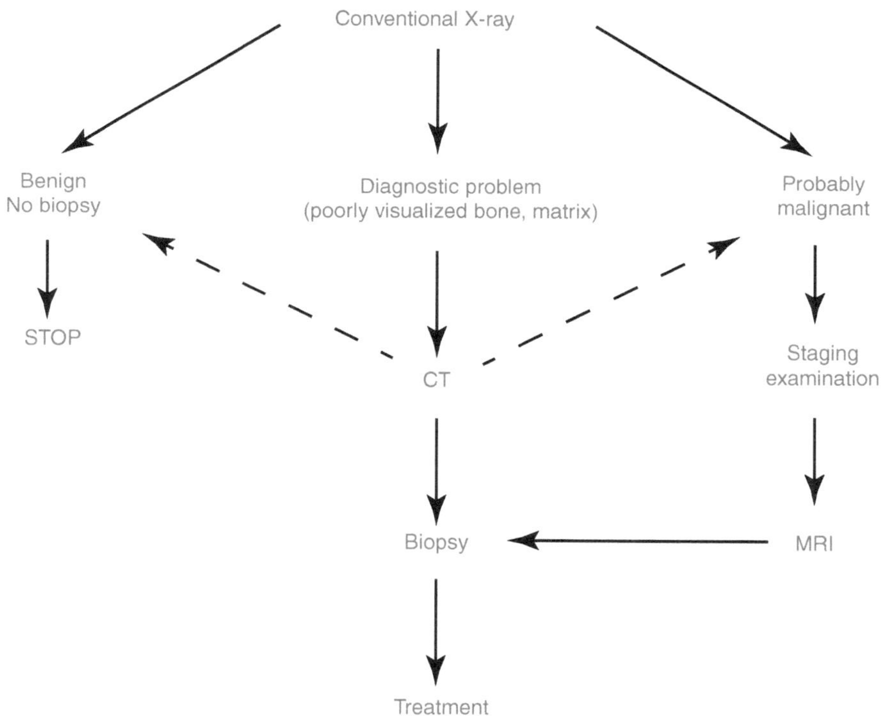


## Selected Bibliography

Azouz EM (2002) Magnetic resonance imaging of benign bone lesions: cysts and tumors. Top Magn Reson Imaging 13(4):219–229. Review
Miller SL, Hoffer FA (2001) Malignant and benign bone tumors. Radiol Clin North Am 39(4):673–699. Review

Murphey MD (2007) World Health Organization classification of bone and soft tissue tumors: modifications and implications for radiologists. Semin Musculoskelet Radiol 11(3):201–214. Review

Rodallec MH, Feydy A, Larousserie F, Anract P, Campagna R, Babinet A, Zins M, Drapé JL (2008) Diagnostic imaging of solitary tumors of the spine: what to do and say. Radiographics 28(4):1019–1041. Review

Woertler K (2003) Benign bone tumors and tumor-like lesions: value of cross-sectional imaging. Eur Radiol 13(8):1820–1835. Review

# Chapter 3
# General Principles of Bone Pathology

**Marco Gambarotti**

Bone tumors are among the rarest neoplasms in humans. They account for about 0.8–1 % of all malignancies arising in the body. If we consider that more than 40 bone malignant neoplasms have been described, it is reasonable to think that only large specialized centers may have enough experience with some of them. The peculiar multidisciplinary approach is mandatory in bone tumors, in order to avoid dramatic mistakes. The pathologist dealing with bone must follow a diagnostic flowchart that starts from the accurate collection of clinical information, followed by the careful examination of the imaging, then the decision about the kind of diagnostic procedure to apply, and finally the histological diagnosis. All these steps must be shared with colleagues of the team, the orthopedic surgeons, the radiologists, and the oncologists that will establish the proper treatment after diagnosis. In the last years, pathologists have started to collect biologic samples of fresh tumor tissue to store in biobanks that are necessary for the study of these rare tumors, because of the possibility to perform molecular analyses and to share tissue samples with other institutions in the context of large international scientific projects. Examining in detail every single step of the diagnostic approach, analysis of clinical features, such as patient age, symptoms, and anatomic location of the lesion, is necessary for a preliminary assessment of the lesion. Bone tumors like Ewing sarcoma and osteosarcoma usually occur in young patients. Tumors like chordoma, myeloma, and chondrosarcoma are typical of adults or elderly patients. When osteosarcomas occur in patients older than 50, they are frequently secondary to preexisting bone conditions, such as Paget's disease and bone infarcts, or arising after radiation therapy. Symptoms and features are frequently of great clue for diagnosis. Pain during the night that can be treated with salicylates is typical of osteoid osteoma; the presence of fever favors for the diagnosis of Ewing sarcoma rather than lymphoma.

M. Gambarotti, MD
Department of Anatomy and Pathology, Istituto Rizzoli, Bologna, Italy
e-mail: marco.gambarotti@ior.it

P. Picci et al. (eds.), *Atlas of Musculoskeletal Tumors and Tumorlike Lesions*,
DOI 10.1007/978-3-319-01748-8_3, 

An accurate collection of clinical information is mandatory for detection of a primary tumor in case of metastasis to the bone; the blood levels of parathormone are the key feature for diagnosis of hyperparathyroidism. The site of the tumor within the bone and the specific bone segment are very important. Some tumors occur usually in the epiphysis, such as giant cell tumor, chondroblastoma, and clear cell chondrosarcoma; most tumors are centrally located, while others are eccentrically located in the bone cortex; others, such as adamantinoma, occur almost exclusively in the tibial diaphysis. In low-grade chondroid lesions, the site of the lesion is very important for a correct interpretation of histology: if the lesion is in the small bones of the hands and feet, it is usually benign, while, with the same histological features, it is usually malignant if located in the ribs and sternum. Lesions that arise in the periosteum are generally clinically less aggressive than the medullary counterparts. The radiographic features of the lesion as well are very important for the pathologist: they have to be considered like a negative image of the macroscopic appearance of the neoplasm. Bone lesions can cause osteolysis or reactive bone production (osteosclerosis). Combinations of these two processes give three typical patterns of bone destruction:

1. The geographic bone destruction pattern, typical of slow-growing lesions.
2. The "moth-eaten" pattern that is typical of more aggressive lesions that usually have a faster growth rate, and for this reason the osteosclerosis in this situation is less evident.
3. The permeative growth that can be observed in the most aggressive lesions, such as lymphomas and Ewing sarcoma.

Finally, periosteal reaction gives a great clue for interpretation of biological features of a bone neoplasm. Fast-growing lesions do not cause a periosteal reaction that usually requires about 2 weeks to be detectable on radiographs. Some kinds of periosteal reactions suggest a specific diagnosis (onionskin reaction is frequently present in Ewing sarcoma). After a careful examination of all these aspects, the pathologist can correctly interpret the histology of a given bone lesion.

From a histological point of view, the 2013 WHO classification of bone tumors is based on cytologic and histogenetic criteria and on the kind of matrix produced by the tumor. Table 4.1 represents the classification used today. Bone-producing lesions, together with cartilaginous lesions and Ewing sarcoma, account for about 80 % of all bone tumors; the remaining entities are by far rarer and all together account for about 20 %. The use of ancillary techniques, such as immunohistochemistry, is very important for the assessment of the possible origin of a bone metastasis but also in some primary bone tumors (i.e., chordoma and adamantinoma). Molecular techniques have some technical limitations that are due to the use of decalcifying solutions in bone tumors but are widely used in the validation of the diagnosis of Ewing sarcoma and in other situations always more frequently described (primary aneurysmal bone cyst, fibrous dysplasia). The grading system for bone sarcomas used at the Rizzoli Institute is a four-grading system according to Broders (grades 1–2, low grade; grades 3–4, high grade).

The Rizzoli's syllabus is based on the study of the most numerically important series in the world, and the use of a schematic approach for every single entity gives the reader a useful diagnostic tool, very practical for such rare diseases.

## Selected Bibliography

Antonescu CR (2008) Molecular profiling in the diagnosis and treatment of high grade sarcomas. Ultrastruct Pathol 32(2):37–42. Review

Bovée JV, Hogendoorn PC (2010) Molecular pathology of sarcomas: concepts and clinical implications. Virchows Arch 456(2):193–199. Review

Broders AC (1920) Squamous-cell epithelioma of the lip (a study of five hundred and thirty-seven cases). JAMA 6:656–664

Broders AC (1925) The grading of carcinoma. Minn Med 8:726–730

Fletcher CDM, Bridge JA, Hogendoorn PCW, Mertens F (eds) (2013) WHO classification of tumors of soft tissue and bone, 4th edn. International Agency for Research on Cancer, Lyon

Unni K et al (2005) Afip atlas of tumor pathology - Series 4 - Tumors of the bones and joints. Arp Press, Silver Spring

# Chapter 4
# Classification of Primary Bone Tumors

**Piero Picci**

Primary bone tumors are named and classified based upon their histologic differentiation, as reported in the following Table 4.1. The term "tumor" refers in general to a growing mass and is also used by clinicians for lesions of nonneoplastic nature, such as histiocytic fibroma and osteochondroma, because their basic tissue expresses a recognizable differentiation, such as fibrohistiocytic and cartilaginous. Other growing lesions are usually defined as "reactive" and do not show a recognizable pattern of histologic differentiation. Lymphomas and myeloma are not considered primary bone tumors, deriving from hematopoietic system.

P. Picci, MD
Laboratory of Experimental Oncology, Musculoskeletal Oncology,
Istituto Ortopedico Rizzoli, Bologna, Italy
e-mail: piero.picci@ior.it

P. Picci et al. (eds.), *Atlas of Musculoskeletal Tumors and Tumorlike Lesions*,
DOI 10.1007/978-3-319-01748-8_4, 

**Table 4.1** Classification of primary bone tumors

| Histogenesis | Benign | Intermediate (locally aggressive and/or rarely metastasizing) | Malignant *[(*) low-grade lesions if not dedifferentiated]* |
|---|---|---|---|
| Fibrogenic and fibrohistiocytic | Histiocytic fibroma | Desmoplastic fibroma | Fibrosarcoma |
| | Benign fibrous histiocytoma | | |
| Chondrogenic | Osteochondroma | Chondromyxoid fibroma | Chondrosarcoma grade 2 |
| | Hemimelic epiphyseal dysplasia | | Chondrosarcoma grade 3 |
| | Enchondroma | Chondrosarcoma grade 1 | Dedifferentiated chondrosarcoma |
| | Periosteal chondroma | | Mesenchymal chondrosarcoma |
| | Osteochondromyxoma | Chondroblastoma | Peripheral chondrosarcoma (*) |
| | Subungual exostosis | | Periosteal chondrosarcoma (*) |
| | Bizarre parosteal osteochondromatous proliferation | Fibrocartilaginous mesenchymoma | Clear cell chondrosarcoma (*) |
| | Synovial chondromatosis | | |
| Osteogenic | Osteoma | Osteoblastoma | Low-grade central osteosarcoma (*) |
| | | | Conventional osteosarcoma |
| | | | Telangiectatic osteosarcoma |
| | | | Small cell osteosarcoma |
| | Osteoid osteoma | | Secondary osteosarcoma |
| | | | Parosteal osteosarcoma (*) |
| | | | Periosteal osteosarcoma (*) |
| | | | High-grade surface osteosarcoma |
| Osteoclastic | Giant cell reparative granuloma | Giant cell tumor of bone | Malignancy in giant cell tumor |
| Giant cell rich | | | |
| Vascular | Hemangioma | Epithelioid hemangioma | Hemangioendothelioma (*) |
| | Lymphangioma | | Epithelioid hemangioendothelioma (*) |
| | | | Angiosarcoma |

| | | | |
|---|---|---|---|
| Nervous | Schwannoma | | MPNST |
| | Neurofibroma | | |
| Lipogenic | Lipoma | | Liposarcoma |
| Myogenic | Leiomyoma | | Leiomyosarcoma |
| Notochordal | Benign notochordal cell tumor (BNCT) | | Chordoma (*) |
| Tumors of undefined neoplastic nature/miscellaneous tumors | Simple bone cyst | Aneurysmal bone cyst | Adamantinoma (*) |
| | Fibrous dysplasia | Langerhans cell histiocytosis | Ewing sarcoma |
| | Osteofibrous dysplasia | Erdheim-Chester disease | Undifferentiated pleomorphic sarcoma |
| | Chondromesenchymal hamartoma | | |
| | Rosai-Dorfman disease | | |

# Chapter 5
# Histiocytic Fibroma

**Pietro Ruggieri**

**Synonyms**: Non-ossifying fibroma, Metaphyseal fibrous defect, Fibrous cortical defect, Fibrous xanthoma
**Definition:** It is a common benign lesion essentially consisting of histio-fibroblastic tissue, most frequently originating eccentrically, in the metaphyseal portion of a long bone, in a skeletally immature individual.

The tendency to remain asymptomatic and spontaneously resolve in most instances and the common relationship with tendon and ligament insertions in proximity of the growth plate suggest a posttraumatic or developmental defect producing faulty ossification rather than a true neoplasm. It is not infrequent to see two or three lesions in either one or both lower extremities of the same patient. A different and very rare clinical scenario is the association of numerous multiple histiocytic fibromas, more extensively involving the skeleton, with extraskeletal abnormalities. This condition known as Jaffe-Campanacci syndrome will be discussed separately.
**Epidemiology:** Overall, it is a very frequent lesion, the incidence of which has been estimated around 30 %. However, patients are in most instances asymptomatic, and the true incidence is therefore underestimated. There is slight male sex predominance. It is typical in childhood and adolescence, being rare before 5 and after 20 years of age.

P. Ruggieri, MD, PhD
2nd Orthopaedic and Traumatologic Clinic, Istituto Ortopedico Rizzoli, Bologna, Italy
e-mail: pietro.ruggieri@ior.it

P. Picci et al. (eds.), *Atlas of Musculoskeletal Tumors and Tumorlike Lesions*,
DOI 10.1007/978-3-319-01748-8_5, 

Histiocytic Fibroma
563 cases (including 18 multicentric cases)

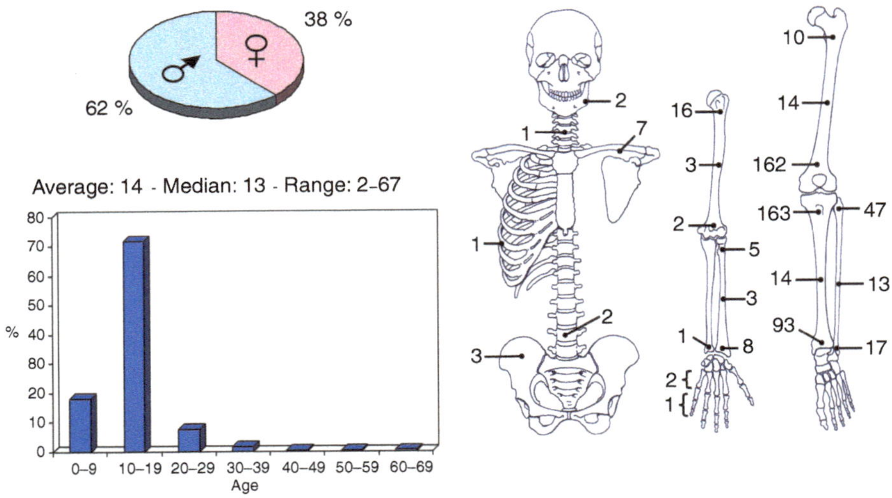


1900–2012 - Istituto Ortopedico Rizzoli - Laboratory of Experimental Oncology - Section of Epidemiology - Bologna - Italy

**Localization:** It is generally localized in the metaphysis at first and then displaced toward the diaphysis. It starts intracortically/subperiosteally. Most of the lesions are located around the knee and ankle. It is rare in the proximal femur and upper limb, exceptional in the trunk, hand, and foot. Two or three lesions may appear in the same or both lower limbs (e.g., distal femur and proximal tibia).
**Clinical Aspects:** Diagnosis is usually based on a radiogram obtained for unrelated reasons (usually trauma), histiocytic fibroma being the most classic and common lesion diagnosed as incidental finding. Rarely, pathologic fractures may occur in larger lesions (more than 1/2–2/3 of the bone cross section, particularly for the distal tibia and fibula).
**Imaging:** Standard X-rays are usually diagnostic. The defect is metaphyseal, intracortical, and/or subperiosteal. Generally lobulated, its inner boundaries are surrounded by a rim of bone sclerosis. The osteolytic image may appear multilocular due to corrugations of the bony wall. The cortex is sometimes attenuated, rarely slightly expanded due to chronic periosteal reaction.
**Pathology:** Tissue is compact, rubbery soft, tan-brown in color, sometimes containing yellow (foam cells) or dark (hemosiderin) areas. Dense network of cellular typically plump spindle cells with a prominent storiform pattern and with scattered

multinucleated giant cells are evident. Intra- and extra-cytoplasmatic hemosiderin and lipid-loaded foam cells are common findings.

**Course and Staging:** Tumor growth stops at skeletal maturity and very frequently even before. The lesion then tends to ossify slowly. Stage is initially 1 or 2 and then constantly becomes 1.

**Treatment and Prognosis:** The majority of histiocytic fibromas do not require treatment; diagnosis can usually be made on clinico-radiographic features avoiding biopsy. Occurrence of a pathologic fracture per se is not an indication for surgery, as fractures heal essentially in a normal fashion. Curettage and bone grafting are rarely needed for large lesions, possibly associated with internal fixation for displaced fractures.

**Key Points**

| | |
|---|---|
| Clinical | Incidental findings |
| Radiological | Eccentric, lobulated, subperiosteal, or intracortical |
| Histological | Bland spindle cells without atypia and with foam cells |
| Differential diagnosis | Chondromyxoid fibroma (at imaging) |

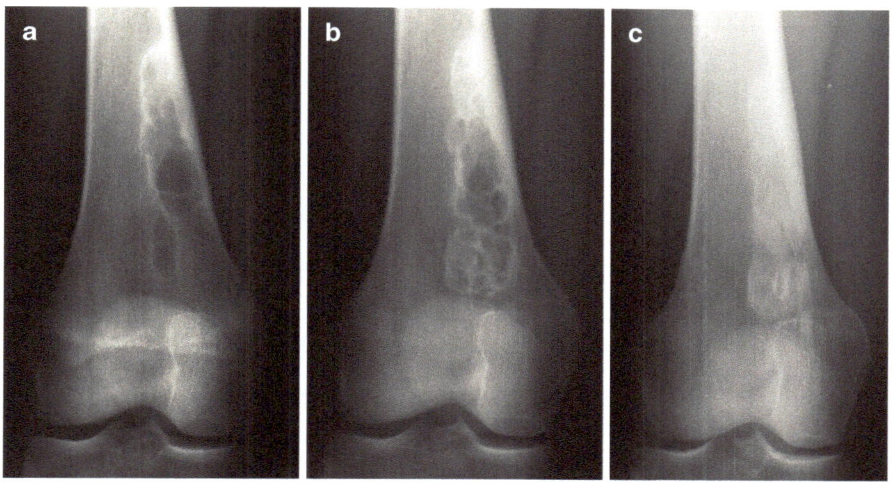


Radiographs of the knee. The lesion is lytic, metadiaphyseal, centered on the cortex, and limited by a sclerotic line. As it was painless, diagnosis of histiocytic fibroma was sure, and biopsy was not performed. With time ossification that slowly completely regressed appeared in the proximal (older part) lesion. (**a**) 1997 (**b**) 2000 (**c**) 2003

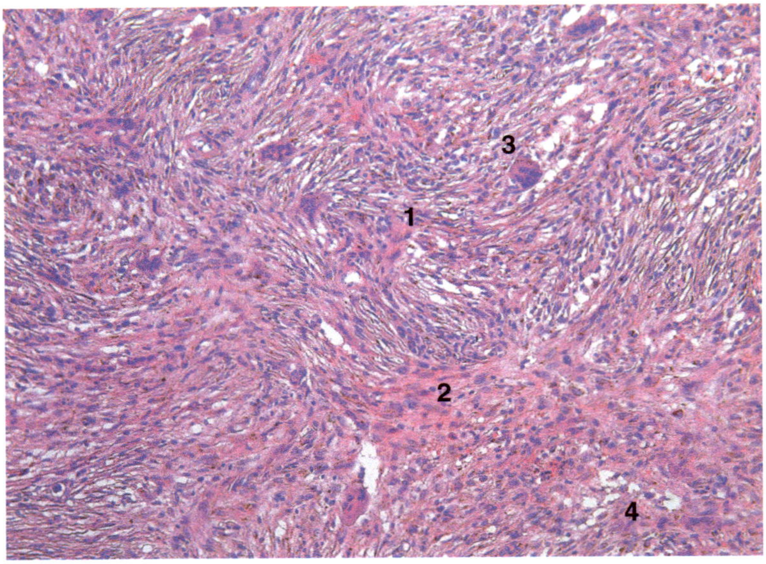


(1) Compact histio-fibroblastic tissue arranged in imbricated and whorled bundles (storiform pattern). (2) Great number of slightly plump and well-stained nuclei; rare and normal mitotic figures (active areas of the lesion). (3) Small and sparse giant cells, less numerous cells and more abundant collagenous bundles (florid stage). (4) Foam cells (regressive phenomena of the tumor)

## Selected Bibliography


Betsy M, Kupersmith LM, Springfield DS (2004) Metaphyseal fibrous defects. J Am Acad Orthop Surg 12:89–95

Jee WH, Choe BY, Kang HS et al (1998) Nonossifying fibroma: characteristics at MR imaging with pathologic correlation. Radiology 209:197–202

Levine SM, Lambiase RE, Petchprapa CN (2003) Cortical lesions of the tibia: characteristic appearances at conventional radiography. Radiographics 23:157–177

Mankin HJ, Trahan CA, Fondren G, Mankin CJ (2009) Non-ossifying fibroma, fibrous cortical defect and Jaffe-Campanacci syndrome: a biologic and clinical review. Chir Organi Mov 93(1):1–7. Review

Vanel D, Ruggieri P, Ferrari S, Picci P, Gambarotti M, Staals E, Alberghini M (2009) The incidental skeletal lesion: ignore or explore? Cancer Imaging 9(Spec No A):S38–S43. Review

# Chapter 6
# Multiple Histiocytic Fibromas with Extraskeletal Abnormalities (Jaffe-Campanacci's Syndrome)

**Pietro Ruggieri**

It is a very rare condition possibly linked to neurofibromatosis. Multiple large histiocytic fibromas extend to the long bones of one or both lower limbs or to the four limbs with prevalence in one side of the body and even including the pelvis. Café au lait spots are usually seen. Occasionally, mental retardation, hypogonadism or cryptorchidism, ocular and cardiovascular anomalies, and other skin alterations suggestive of neurofibromatosis are associated.

More frequently symptomatic, as compared to the usual histiocytic fibroma, they cause slight expansion of the bone, stress or pathologic fractures, and sometimes deformity or limb length discrepancy.

**Imaging**: Lesions are rather extensive and, from the metaphysis, tend to involve the diaphysis. Osteolyses are mostly intracortical or eccentric. The cortex may be very thin or absent. Differential diagnosis includes multifocal fibrous dysplasia, but imaging and histopathology are quite different in the two lesions.

**Pathology:** Same as conventional histiocytic fibroma.

**Course**: Similar to conventional histiocytic fibroma. The lesions do not expand after skeletal maturity and then tend to be replaced by sclerotic bone.

**Treatment:** Aimed to prevent pathologic fractures and address deformities.

P. Ruggieri, MD, PhD
2nd Orthopaedic and Traumatologic Clinic, Istituto Ortopedico Rizzoli, Bologna, Italy
e-mail: pietro.ruggieri@ior.it

P. Picci et al. (eds.), *Atlas of Musculoskeletal Tumors and Tumorlike Lesions*,
DOI 10.1007/978-3-319-01748-8_6, © Springer International Publishing Switzerland 2014

Radiograph of the knee, lateral view. Multiple histiocytic fibromas. The lesion of the tibia is broken. The one on the fibula appears centered, and not cortical. This pattern is frequent in thin bones

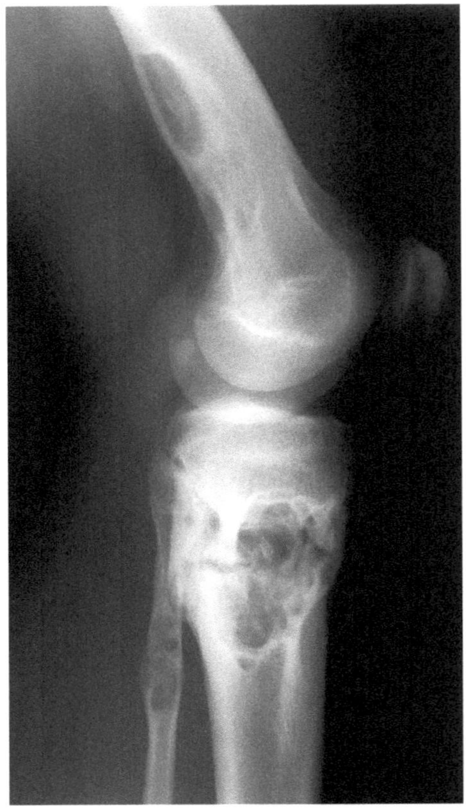

## Selected Bibliography

Mankin HJ, Trahan CA, Fondren G, Mankin CJ (2009) Non-ossifying fibroma, fibrous cortical defect and Jaffe-Campanacci syndrome: a biologic and clinical review. Chir Organi Mov 93(1):1–7. doi:10.1007/s12306-009-0016-4. Epub 2009 Apr 29. Review

Mirra JM, Gold RH, Rand F (1982) Disseminated nonossifying fibromas in association with café-au-lait spots (Jaffe-Campanacci syndrome). Clin Orthop Relat Res 168:192–205

# Chapter 7
# Fibrous Dysplasia

**Pietro Ruggieri**

## Related Conditions: McCune-Albright's Syndrome, Mazabraud's Syndrome

**Definition:** Intramedullary hamartoma consisting of a peculiar fibro-osseous tissue, it may be either monostotic or polyostotic.
**Epidemiology:** Monostotic fibrous dysplasia is frequent, polyostotic is uncommon, McCune-Albright's syndrome is rare. Difficult to assess true incidence because often asymptomatic. Slight female predominance. Usually diagnosed between age 10 and 30. When asymptomatic, it can be discovered at any age. The polyostotic forms and McCune-Albright's syndrome manifest in early childhood.

P. Ruggieri, MD, PhD
2nd Orthopaedic and Traumatologic Clinic, Istituto Ortopedico Rizzoli, Bologna, Italy
e-mail: pietro.ruggieri@ior.it

P. Picci et al. (eds.), *Atlas of Musculoskeletal Tumors and Tumorlike Lesions*,
DOI 10.1007/978-3-319-01748-8_7, 

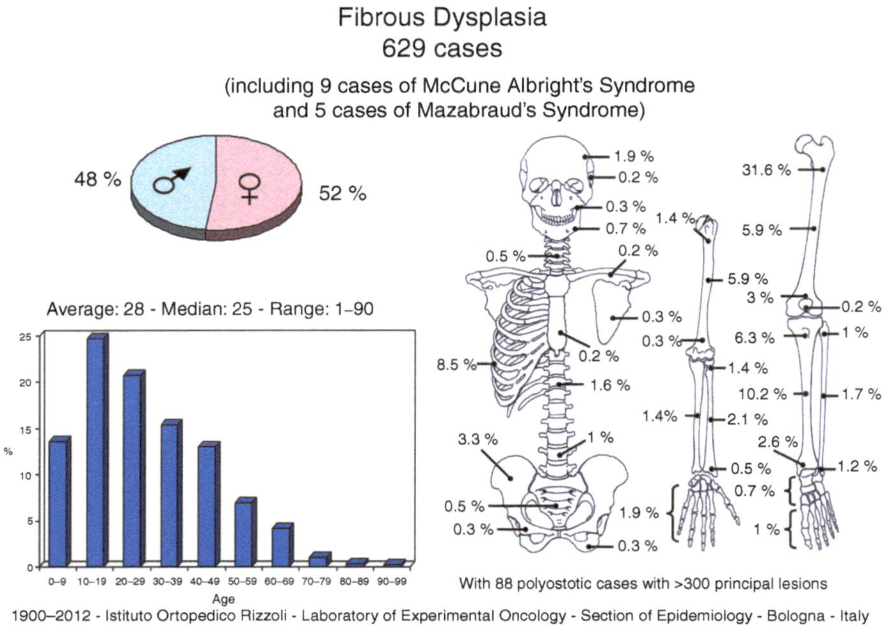


**Location:** Femur (proximal), tibia, craniofacial bones, ribs; then humerus, forearm, pelvis. Same sites in multicentric and polyostotic disease. Frequently more areas in the same long bone, or two to three adjacent bones affected. Lower limb more frequent than upper limb. Hand, foot involved almost only in extensive polyostotic forms. Spine, scapula, clavicle rarely affected. Poliostotic type is usually prevalent in one body side.

**Clinical:** Monostotic is usually asymptomatic representing an incidental finding. Polyostotic: discontinuous pain (fatigue fractures), bony expansion in superficial bone, pathologic fracture, deformity and lower limb-length discrepancy. In polyostotic forms also cafè au lait spots ("coast of Maine"), multiple endocrine abnormalities (McCune-Albright's syndrome), intramuscular mixomas (Mazabraud's syndrome).

**Imaging:** Standard X-rays show defined defect involving cortical and cancellous areas. Margins well defined, sometimes marked by a rind of bone sclerosis. The cortex is sometimes thinned and expanded but continuous. No periosteal reaction. Radiolucency of "ground glass" appearance is depending upon the amount of intratumoral trabeculae of woven bone. Severe "shepherd's crook" deformity of the proximal femur, usual in polyostotic form. Isotope scan: rather hot (diffuse dysplastic bone formation) and corresponding to radiographic extent. CT: homogeneity of ground glass radiolucency; cystic cavities and cartilaginous areas (sometimes calcified) when present. MRI: fairly homogenous low signal in T1.

**Pathology:** Periosteum not involved, underlying cortex is regularly smooth but thin. Lesional tissue, well defined from surrounding bone, whitish to pink, from fibrous to gritty, to hard bony. Sometimes, hemorrhagic areas or cystic spaces with serohematic content are present. Rarely, sparse lobules of hyaline cartilage are embedded in the above described tissue. Histology: mixture of benign proliferating fibroblastic cells and islands of woven bone. The bony trabeculae are arranged in a "Chinese letters" fashion. Usually, the bony trabeculae show no clear-cut osteoblastic rimming. Benign giant cells and foam cells are commonly found. No mitotic activity, no atypia. Islands of cartilage may dominate the histologic appearance. May have ABC-like areas.

**Course and Staging:** Lesions are usually stage 2 in children and adolescents and stage 1 in adults. If a lesion expands and becomes symptomatic in an adult, this may be due to hemorrhage (during pregnancy). Rarely (less than 0.5 % of the reported cases) a sarcoma may develop on a fibrous dysplasia. It usually occurs in adult-advanced age, both in monostotic and polyostotic fibrous dysplasias, more frequent after radiation therapy.

**Treatment:** Frequently not needed. Curettage and grafting of active lesions should be avoided. Deformities may need corrective osteotomies and internal fixation; preferably by intramedullary devices.

### Key Points

| | |
|---|---|
| Clinical | Incidental findings |
| Radiological | "Ground glass" appearance |
| Histological | Benign fibroblastic lesion producing immature woven bone. No osteoblasts |
| Differential diagnosis | Low-grade central osteosarcoma |

### Chromosomal Translocations

| | | |
|---|---|---|
| Point mutations GNAS1 | 20q13.32 | 93 % |

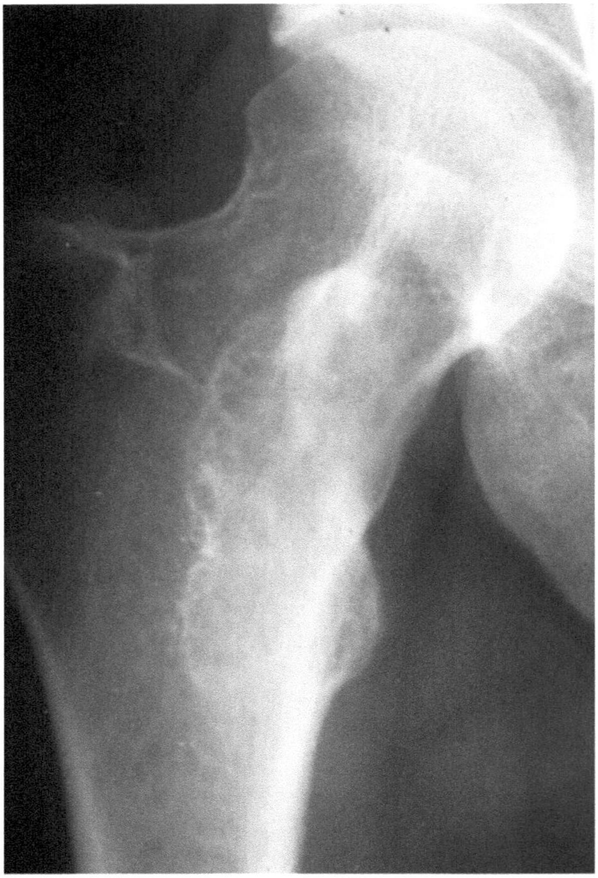

Radiograph of the proximal femur. The lesion is metadiaphyseal, eccentric, sclerotic, and heterogeneous. It is well limited by a sclerotic ring

Radiograph of the hip. Lytic and sclerotic well-limited lesions. Typical femoral neck deformity (Shepherd's crook)

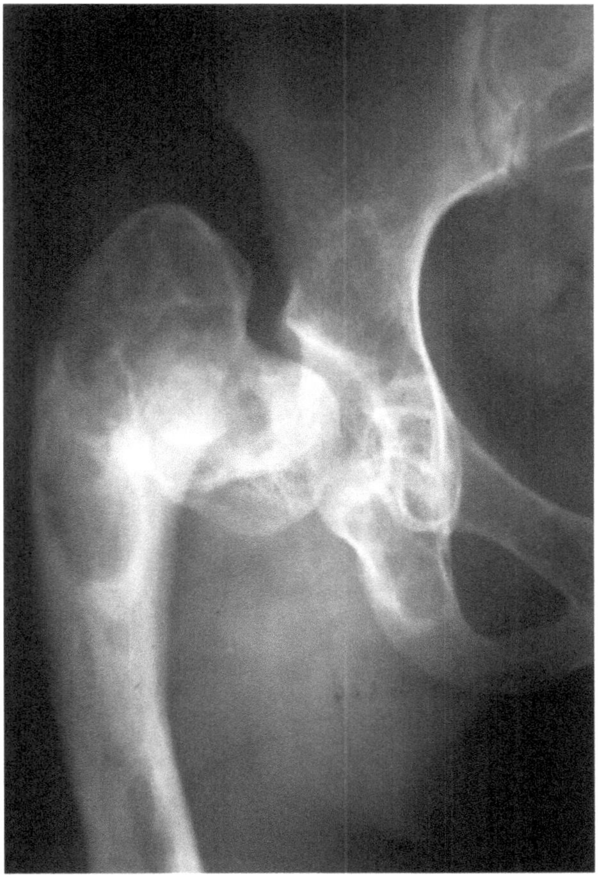

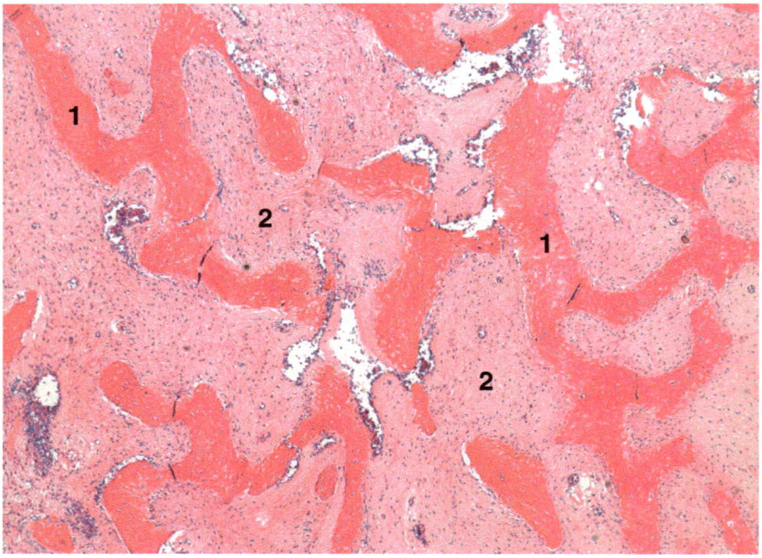


Histopathologic features are represented by immature bone trabeculae enmeshed in immature histio-fibroblastic tissue. (1) The dysplastic bone trabeculae are generally small and shaped like Chinese ideograms; they are usually not bordered by rows of osteoblasts. Trabeculae have a woven structure. (2) Undifferentiated fibrous connective tissue surrounding the trabeculae. The histio-fibroblasts are numerous and plump, with rare mitotic figures

## Selected Bibliography

Case DB, Chapman CN Jr, Freeman JK, Polga JP (2010) Best cases from the AFIP: atypical presentation of polyostotic fibrous dysplasia with myxoma (Mazabraud syndrome). Radiographics 30(3):827–832

Dorfman HD (2010) New knowledge of fibro-osseous lesions of bone. Int J Surg Pathol 18(3 Suppl):62S–65S. Review

Dujardin F, Binh MB, Bouvier C, Gomez-Brouchet A, Larousserie F, Muret AD, Louis-Brennetot C, Aurias A, Coindre JM, Guillou L, Pedeutour F, Duval H, Collin C, de Pinieux G (2011) MDM2 and CDK4 immunohistochemistry is a valuable tool in the differential diagnosis of low-grade osteosarcomas and other primary fibro-osseous lesions of the bone. Mod Pathol 24:624–637

Dumitrescu CE, Collins MT (2008) McCune-Albright syndrome. Orphanet J Rare Dis 3:12

Ruggieri P, Sim FH, Bond JR, Unni KK (1994) Malignancies in fibrous dysplasia. Cancer 73(5):1411–1424

Szuhai K, Cleton-Jansen AM, Hogendoorn PC, Bovée JV (2012) Molecular pathology and its diagnostic use in bone tumors. Cancer Genet 205(5):193–204. doi:10.1016/j.cancergen.2012.04.001

Yoshida A, Ushiku T, Motoi T, Shibata T, Beppu Y, Fukayama M, Tsuda H (2010) Immunohistochemical analysis of MDM2 and CDK4 distinguishes low-grade osteosarcoma from benign mimics. Mod Pathol 23(9):1279–1288

# Chapter 8
# Osteofibrous Dysplasia of Long Bones

**Pietro Ruggieri**

**Epidemiology:** It is rare, prevalent in males, and usually observed in infantile age, below 5 or 10. Rare cases are congenital.

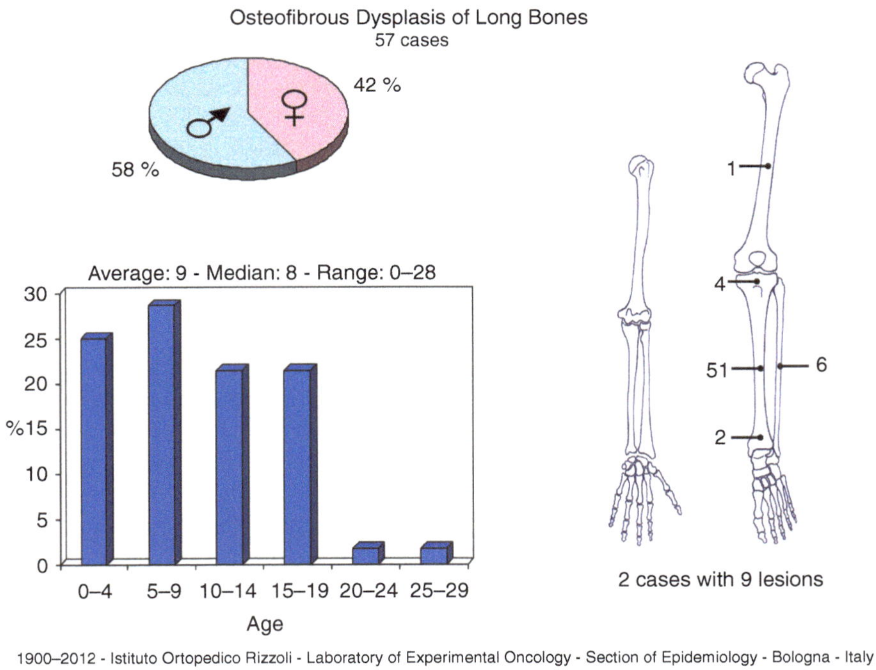


P. Ruggieri, MD, PhD
2nd Orthopaedic and Traumatologic Clinic, Istituto Ortopedico Rizzoli, Bologna, Italy
e-mail: pietro.ruggieri@ior.it

P. Picci et al. (eds.), *Atlas of Musculoskeletal Tumors and Tumorlike Lesions*,
DOI 10.1007/978-3-319-01748-8_8, © Springer International Publishing Switzerland 2014

**Localization:** The lesion is almost exclusively localized in the shaft of the tibia, occasionally involving the ipsilateral fibula, too. Exceptionally it is bilateral in tibias and fibulas.

**Clinical:** The symptoms are moderate expansion of the bone, anterior or valgus-varus bowing of the diaphysis. The lesion is painless. Stress and pathologic fractures are not uncommon, but pseudoarthrosis is exceptional.

**Imaging:** It shows a typically intracortical osteolysis, more frequently involving the anterior (occasionally valgus-varus) cortex, and often associated to an anterior bowing of the diaphysis. The cortex is regularly and moderately expanded, very thin, and with a smooth surface. Toward the medulla, instead, the osteolysis is bordered by a rim of sclerotic bone, so that the medullary canal is frequently narrowed or closed. The osteolysis can be multiple and having a multilocular or bubbly appearance. Some of the osteolyses are veiled by a tenuous ground-glass radiodensity.

**Pathology:** Histologically, it is characterized by a fibrous tissue, embedding small trabeculae of woven bone, rimmed by cubic osteoblasts. The fibrous component is less cellular as compared to active fibrous dysplasia, with delicate collagen, occasionally with a storiform pattern, or focally fibro-myxoid. The bone trabeculae tend to be sparse, thin, and woven in the center of the lesion, and to gradually become more abundant, anastomosing, and lamellar toward the periphery, where both the trabeculae and the fibrous tissue blend with the surrounding host bone (zonal architecture). Exceptionally, small foci of hyaline cartilage have been found. The immunohistochemical studies may show isolated cells, small nests, or microscopic tubes stained with the keratin antibodies. These cells are not obvious in the hematoxylin-eosin stain.

**Course and Staging:** The lesion may progress considerably in the first years of life, and thereafter stabilizes and, after 10–12 years of age, slowly repairs through peripheral bone thickening. The pathologic fractures do regularly united. Congenital pseudoarthrosis is not a feature of this disease.

**Treatment and Prognosis:** Attempts to intralesional excision before the lesion has stabilized are regularly followed by local recurrence. Surgery (not even biopsy) is usually not indicated. The child should be protected, when indicated, by bracing, and monitored, until spontaneous repair occurs. The prognosis is good, except for the more severe cases ending up with leg deformity and shortening. A small subset of these patients can develop into adamantinoma. This exceptionally occurs in children, rarely in adolescent and young adults.

**Key Points**

| | |
|---|---|
| Clinical | Tibia (or fibula) in children, moderate expansion of the bone |
| Radiological | Intracortical expansion with deformity |
| Histological | Benign fibroblastic stroma with immature woven bone bordered by osteoblasts |
| Differential diagnosis | Adamantinoma |

**Immunohistochemical Panel**

| | |
|---|---|
| CK | Rare positive cells |

## Selected Bibliography

Benassi MS, Campanacci L, Gamberi G, Ferrari C, Picci P, Sangiorgi L, Campanacci M (1994) Cytokeratin expression and distribution in adamantinoma of the long bones and osteofibrous dysplasia of tibia and fibula. An immunohistochemical study correlated to histogenesis. Histopathology 25(1):71–76

Campanacci M, Laus M (1981) Osteofibrous dysplasia of the tibia and fibula. J Bone Joint Surg Am 63(3):367–375

Gleason BC, Liegl-Atzwanger B, Kozakewich HP, Connolly S, Gebhardt MC, Fletcher JA, Perez-Atayde AR (2008) Osteofibrous dysplasia and adamantinoma in children and adolescents: a clinicopathologic reappraisal. Am J Surg Pathol 32(3):363–376

Khanna M, Delaney D, Tirabosco R, Saifuddin A (2008) Osteofibrous dysplasia, osteofibrous dysplasia-like adamantinoma and adamantinoma: correlation of radiological imaging features with surgical histology and assessment of the use of radiology in contributing to needle biopsy diagnosis. Skeletal Radiol 37(12):1077–1084

Most MJ, Sim FH, Inwards CY (2010) Osteofibrous dysplasia and adamantinoma. J Am Acad Orthop Surg 18(6):358–366. Review

Taylor RM, Kashima TG, Ferguson DJ, Szuhai K, Hogendoorn PC, Athanasou NA (2012) Analysis of stromal cells in osteofibrous dysplasia and adamantinoma of long bones. Mod Pathol 25:56–64

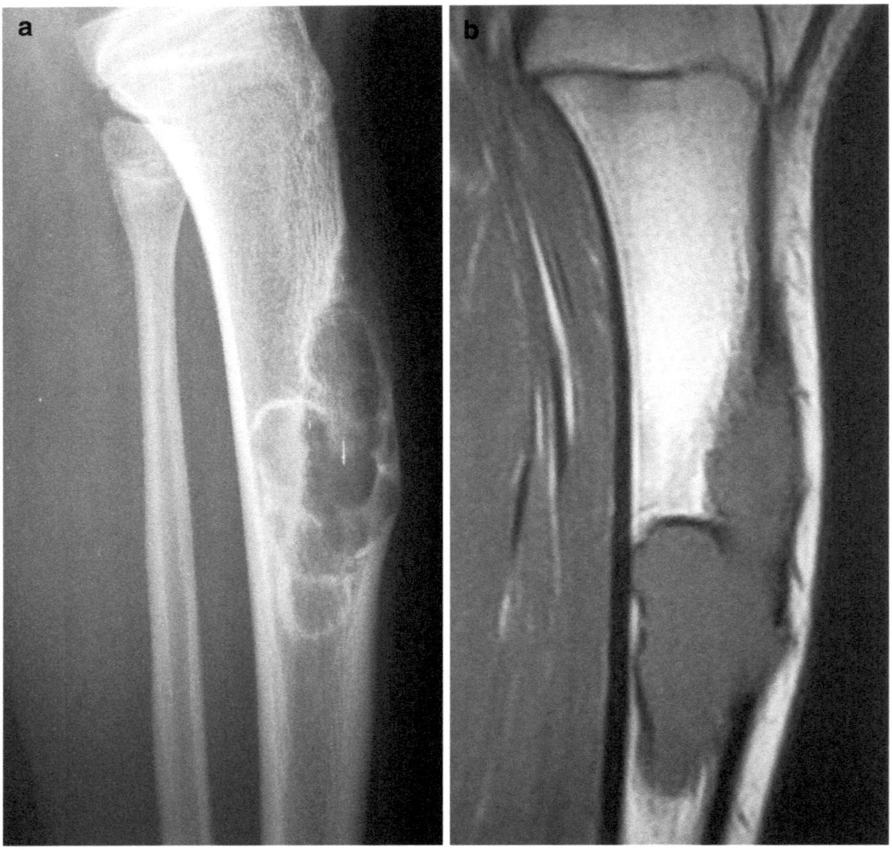


Lateral radiograph and sagittal T1 MR image. There is bowing of the tibia. The lesion is centered on the anterior cortex of the shaft of the tibia, lytic, and well limited

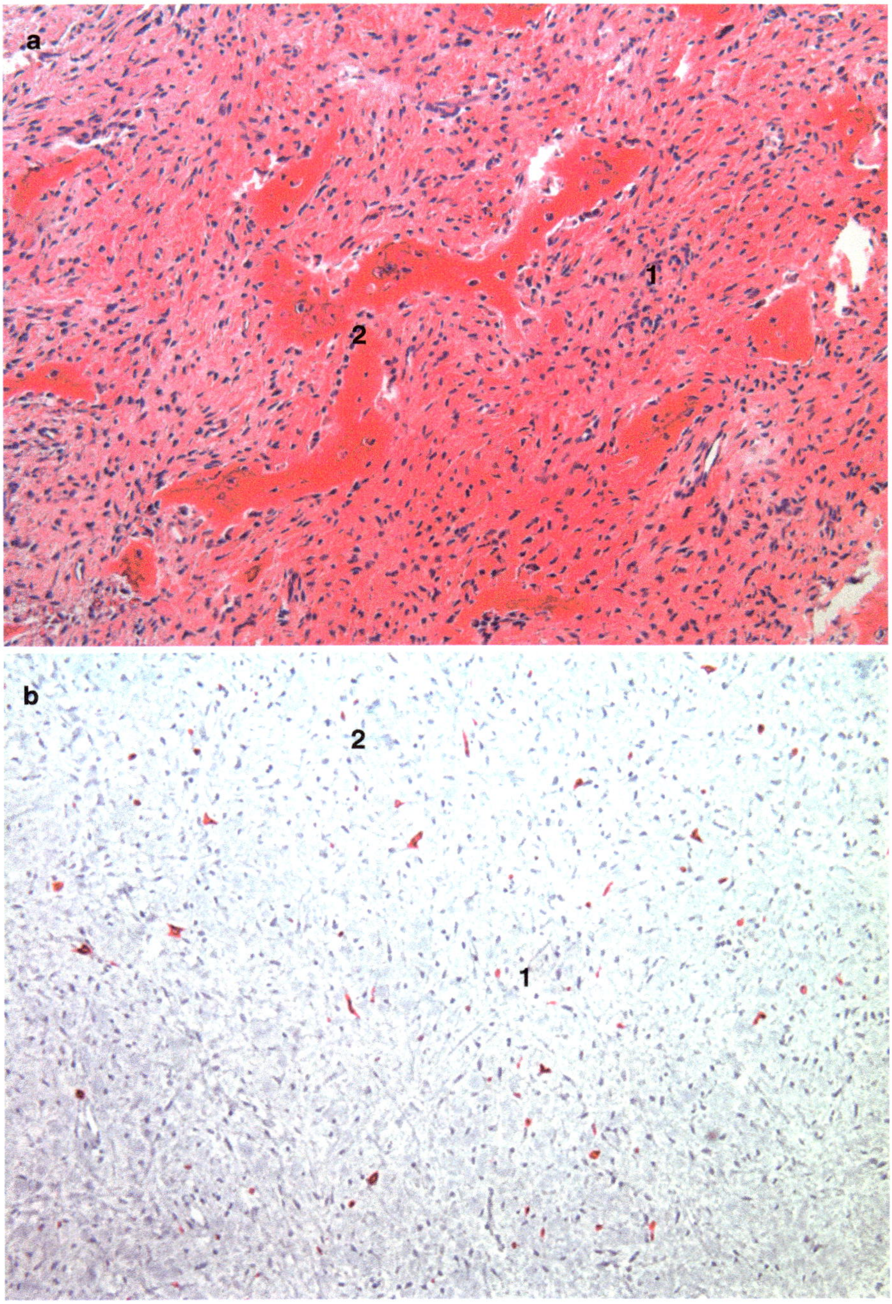


(**a**) Fibrous tissue embedding osseous trabeculae bordered by osteoblasts and "zonal" architecture is the histological characteristic of this lesion. (*1*) Fibrous tissue: well-differentiated fibroblasts with a storiform structure and collagen fibers. (*2*) Newly formed bony trabeculae with woven structure rimmed by osteoblasts. These are usually located in the center of the lesion and are more numerous and thick at the periphery. (**b**) Immunohistochemical pattern for cytokeratin staining in osteofibrous dysplasia. (*1*) Focal cytokeratin positivity. (*2*) Cytokeratin-negative fibrous stroma

# Chapter 9
# Osteofibrous Dysplasia and Adamantinoma

**Pietro Ruggieri**

These two lesions are so related as to deserve to be described under the same heading. Both are rare, but adamantinoma is three times rarer than osteofibrous dysplasia. Osteofibrous dysplasia is a benign even spontaneously regressing lesion, while adamantinoma is a low-grade malignant tumor also capable of metastasizing. There are, however, a few cases where an osteofibrous dysplasia changed into adamantinoma. Osteofibrous dysplasia is typical of children; adamantinoma occurs in adults or adolescents. There are, however, exceptional cases of adamantinoma in children. Osteofibrous dysplasia is almost exclusive of the tibia (and fibula); adamantinoma is prevalent in the tibia but can also rarely occur in other long bones and even in soft tissues. The clinico-imaging presentation of the two lesions can be different, but frequently it is quite similar or identical. Histologically, osteofibrous dysplasia is characterized by a fibro-osseous pattern similar to fibrous dysplasia but containing tiny foci of single epithelioid cells, which can only be seen well using a keratin stain. In an adamantinoma or in an osteofibrous dysplasia transforming into an adamantinoma, small to large nests of epithelioid cells can even be seen on routine H&E staining and can be fully corroborated as epithelioid cells by keratin immunostaining.

## Selected Bibliography

See Osteofibrous Dysplasia and Adamantinoma

P. Ruggieri, MD, PhD
2nd Orthopaedic and Traumatologic Clinic, Istituto Ortopedico Rizzoli, Bologna, Italy
e-mail: pietro.ruggieri@ior.it

P. Picci et al. (eds.), *Atlas of Musculoskeletal Tumors and Tumorlike Lesions*,
DOI 10.1007/978-3-319-01748-8_9, 

# Chapter 10
# Simple Bone Cyst (Unicameral Bone Cyst)

**Pietro Ruggieri**

**Definition:** Primary solitary, clear fluid-filled cystic lesion. It is the only true cyst of bone that conforms to the pathologic definition of cyst. Simple bone cyst cavity will only contain blood if associated with a pathologic fracture.
**Epidemiology:** The simple bone cyst represents one of the most frequent osseous lesions. Males are more affected than females (2:1). It usually occurs in infancy and teenage (highest incidence between 5 and 15 years of age).

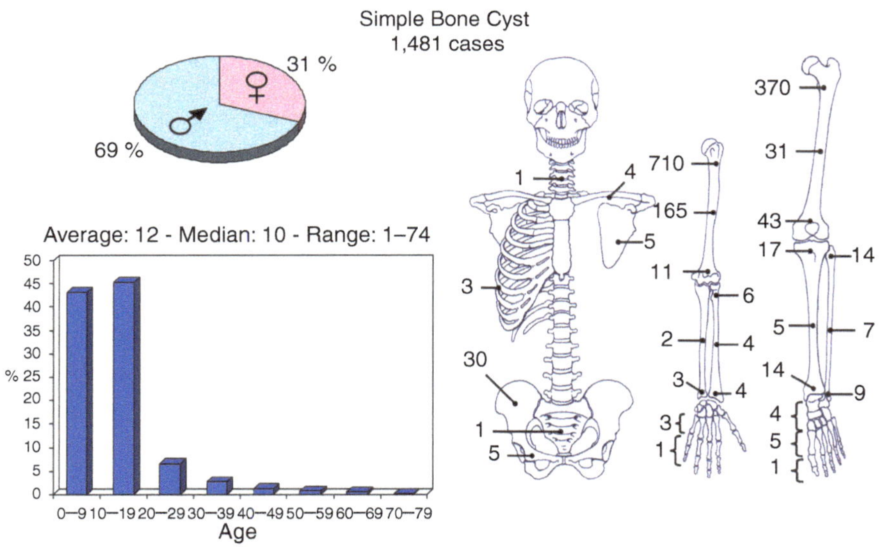


1900–2012 - Istituto Ortopedico Rizzoli - Laboratory of Experimental Oncology - Section of Epidemiology - Bologna - Italy

P. Ruggieri, MD, PhD
2nd Orthopaedic and Traumatologic Clinic, Istituto Ortopedico Rizzoli, Bologna, Italy
e-mail: pietro.ruggieri@ior.it

P. Picci et al. (eds.), *Atlas of Musculoskeletal Tumors and Tumorlike Lesions*,
DOI 10.1007/978-3-319-01748-8_10, © Springer International Publishing Switzerland 2014

**Localization:** It usually arises in the metaphyses of the long bones, beneath the growth plate, but along with patient growth, the cyst migrates to the diaphysis. The most common location is the proximal humerus, followed by proximal femur.
**Clinical:** Per se asymptomatic, the simple bone cyst becomes painful due to frequent pathologic fractures. Sometimes it presents as an incidental finding.
**Imaging:** The characteristic radiographic appearance is that one of a pure osteolysis, thinning, and "inflating" the cortex. Bone septa, usually present, give the appearance of a multiloculated cyst. When fracture occurs, a small fragment of the wall may be seen on X-rays in the cavity, also known as "fallen leaf" sign.
**Pathology:** Grossly it appears as a cavity filled by serous fluid similar to synovial fluid. When the cyst is growing, a thin membrane made of flattened connective cells with few osteoclasts lines the wall. When the lesion is inactive, cuboidal cells line the cavity, while cholesterol slits, hemosiderin deposits, scattered giant cells, and osteoblasts are present along the wall. About 20 % of the cyst walls contain clumps of pink, granular amorphous material that resembles cementum or fibrinoid material that is quite diagnostic of SBC when seen.
**Course and Staging:** Bone cysts are staged according to their activity. Lesions abutting to the growth plate usually have an internal pressure >30 cm H2O, a shiny membrane, and overall clinical features of activity, being considered stage 2. Lesions located far from the physis usually have an internal lower pressure, a thicker membrane, and overall features of inactivity, being considered stage 1. Occasionally the fracture may induce spontaneous healing of the cyst.
**Treatment and Prognosis:** Percutaneous methylprednisolone injections appear to have a reasonable success rate in the majority of active cysts. Effectiveness of injection in latent lesions is debatable and case-by-case approach suggested. Curettage and bone grafting +/− internal fixation in fracture or risk of fracture.

**Key Points**

| | |
|---|---|
| Clinical | Incidental findings or pain if fractured |
| Radiological | Pure lytic, central, metaphyseal |
| | After fracture, "fallen leaf" sign |
| Histological | Pure cavity, lined by a thin connective tissue |
| Differential diagnosis | None |

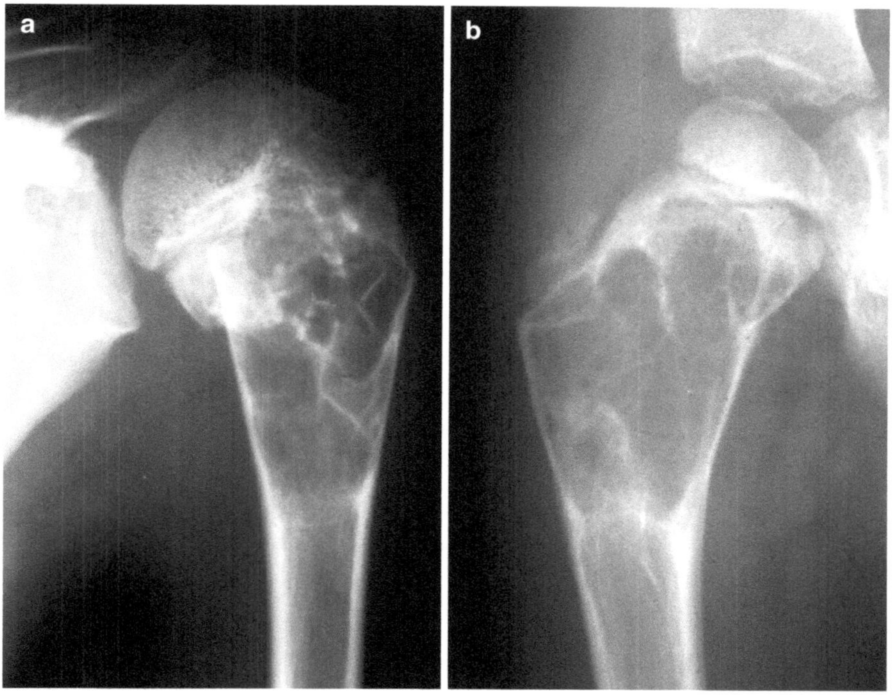


Radiographs of the proximal humerus and femur in two patients. Lesions are metaphyseal, start from the epiphyseal plate, are purely lytic, and are centered in the middle of the bone. The cortex is thin. The humeral lesion is broken

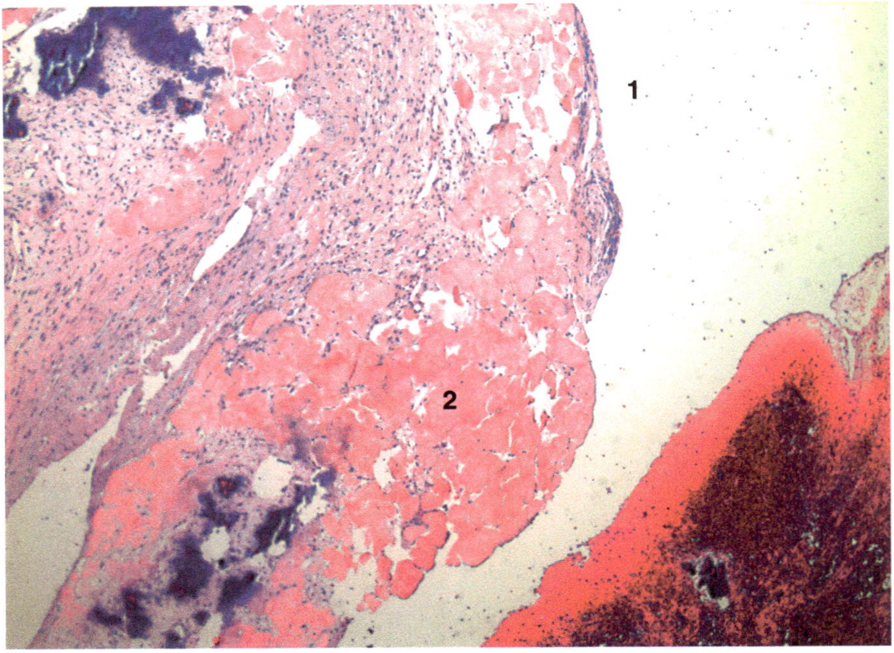


(*1*) Thin and shiny membrane made up of aspecific loose collagen. At the surface of this connective tissue, cells may assume a cuboidal, vaguely endotheliomorphous appearance; deep to the border of cuboidal cells, there is a zone of loose mesenchymal cells, where hemosiderin, macrophages, multinucleated giant cells, and scattered lymphomonocytary infiltrates are observed. (*2*) Trabecular masses of pseudofibrin, cementum-like material, acellular, and not bordered by osteoblasts. Common opinion is that this is coagulated fibrin material. At times, it calcifies; at times, it gradually merges with osteoid trabeculae; and at times osteoid-osteoblastic rims are deposited on its borders

## Selected Bibliography

Baschang A, von Laer L (1991) Indication and procedure of the operative treatment of benign bone cysts in children and adolescents. Eur J Pediatr Surg 1(4):207–209

Bensahel H, Jehanno P, Desgrippes Y, Pennecot GF (1998) Solitary bone cyst: controversies and treatment. J Pediatr Orthop B 7(4):257–261

Capanna R, Albisinni U, Caroli GC, Campanacci M (1984) Contrast examination as a prognostic factor in the treatment of solitary bone cyst by cortisone injection. Skeletal Radiol 12(2):97–102

Di Bella C, Dozza B, Frisoni T, Cevolani L, Donati D (2010) Injection of demineralized bone matrix with bone marrow concentrate improves healing in unicameral bone cyst. Clin Orthop Relat Res 468(11):3047–3055

Margau R, Babyn P, Cole W, Smith C, Lee F (2000) MR imaging of simple bone cysts in children: not so simple. Pediatr Radiol 30(8):551–557

# Chapter 11
# Langerhans Cell Histiocytosis

**Pietro Ruggieri**

**Definition:** A histiocytary proliferation of granuloma-like aspect with unknown etiology that may involve bone marrow, internal organs, skin, and mucosae.

(a) Localized in the skeleton: solitary or multiple eosinophilic granuloma
(b) Chronic disseminated: including Hand-Schuller-Christian disease
(c) Acute/subacute diffused: including Letterer-Siwe disease

## 11.1 Eosinophilic Granuloma

**Epidemiology:** More frequently solitary, multiple <10 %. More frequently in males. 5–10 years of age.

P. Ruggieri, MD, PhD
2nd Orthopaedic and Traumatologic Clinic, Istituto Ortopedico Rizzoli, Bologna, Italy
e-mail: pietro.ruggieri@ior.it

P. Picci et al. (eds.), *Atlas of Musculoskeletal Tumors and Tumorlike Lesions*,
DOI 10.1007/978-3-319-01748-8_11, © Springer International Publishing Switzerland 2014

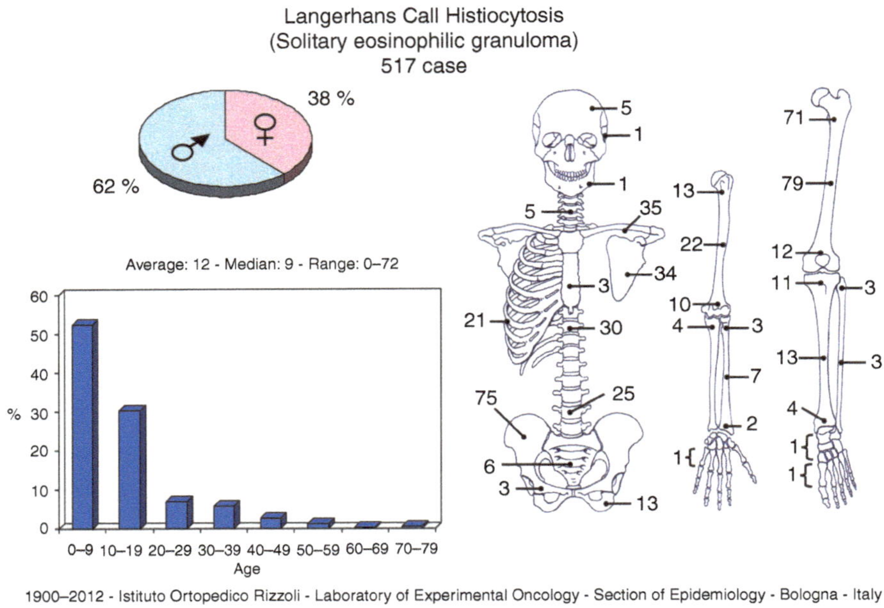


**Location:** Flat and short bones of the trunk: skull (parietal, frontal site), ribs, pelvis, vertebral body, clavicle, and scapula. Among long bones: proximal half of the femur, humerus, and tibia (diaphyseal localization is typical). Very rare in the hand and foot.

**Clinical:** Pain, wide swelling in superficial bones, rare pathologic fracture, radicular pain, rarely signs of medullary compression, and deformities of the column in vertebral lesions. Rare mild increase in sedimentation rate and mild eosinophilia.

**Imaging:** Standard X-rays show an osteolytic lesion with variable features. Sometimes rounded, often polycyclic, well-defined margins, thin sclerotic rim, and overall benign-looking appearance. Sometimes "moth-eaten" pattern with ill-defined margins, no sclerotic rim, and "onionskin" periosteal reaction mimicking a malignant process (e.g., Ewing sarcoma). Uniform, rapid flattening of the vertebral body reduced to a thin bony lamina (vertebra plana) is typical. Bone scan sometimes shows multiple lesions. On MRI light gray, intermediate low signal on T1, and a higher than fat, bright signal on T2.

**Pathology:** Soft, semiliquid, yellowish gray, with areas of hemorrhage or necrosis. Background of large pale-staining cells (histiocytes) with infiltration of leukocytes, without intercellular matrix, punctuated by nodules of small eosinophilic cells. Histiocytes have wide cytoplasm, with ill-defined membrane, with reniform, indented, pale nucleus, small nucleolus, and may be collected in nests or nodules or form a more rarefied background to the infiltration of eosinophiles with small lobate, dark nucleus, and cytoplasm stuffed with bright red granules. Few neutrophiles and lymphocytes. Abundant reticulum surrounding small group of cells. Sporadic giant cells, foam cells, mitotic figures.

**Course and Staging:** Rapid growth, spontaneously self-limiting and tendency to healing with at least partial bone repair, rare evolution into multiple type, exceptional transformation into chronic diffused type. Usually, stage 2, rarely stage 3.
**Treatment:** Needle biopsy, frozen section, and steroid injection treatment of choice. Clinically very successful with complete or almost complete repair in 2 years. Bracing or casting required in the spine. Radiotherapy may be used (2–3,000 r). Systemic cortisone and chemotherapy are used in multiple lesions.

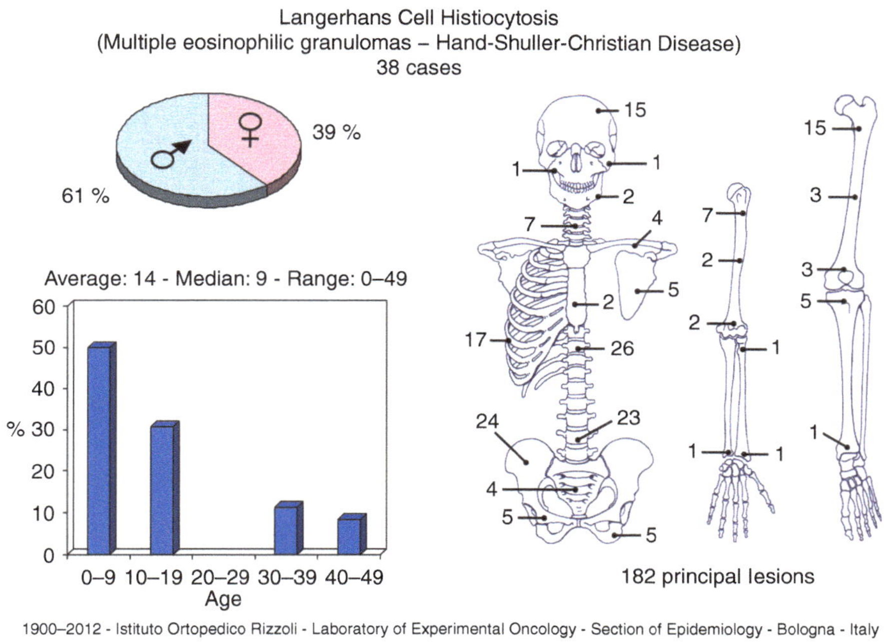


## Key Points

| | |
|---|---|
| Clinical | Pain, swelling, rapidly increasing |
| Radiological | Variable aspects, from benign to aggressive aspects |
| Histological | Histiocytes, leukocytes, eosinophilic cells |
| Differential diagnosis | Osteomyelitis, Ewing sarcoma |

## Immunohistochemical Panel

| | |
|---|---|
| CDla | + |
| S100 | + |
| Langerin | + |

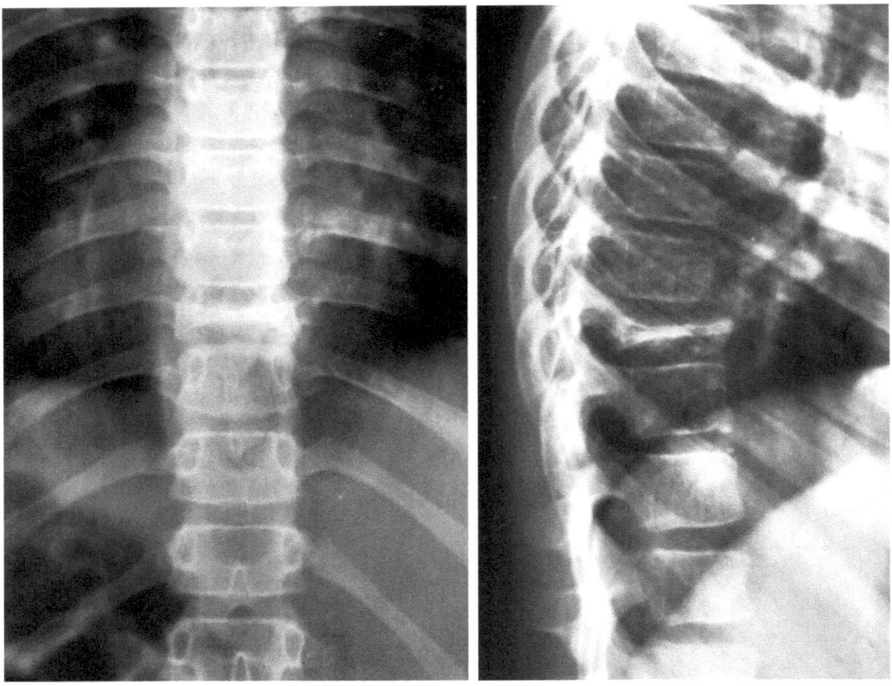

AP and lateral radiographs of the spine: complete vertebral collapse of T9. No soft tissue mass

Radiograph of the fibula: poorly limited lytic lesion of the shaft, with irregular periosteal bone formation. Differential diagnosis with Ewing sarcoma or an acute infection may be very difficult at the onset of disease

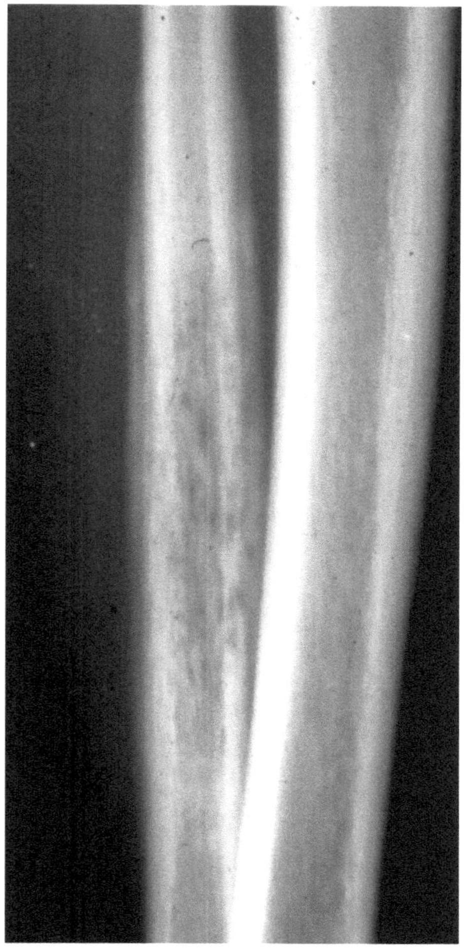

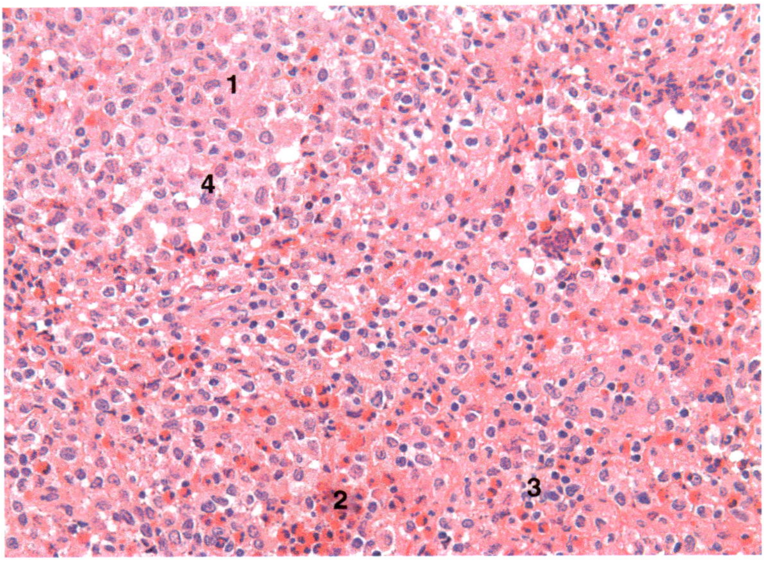


(1) Background of large Langerhans cells, generally organized in a more or less loose net, rarely collected in nests or nodules. (2) Infiltration of eosinophilic granulocytes (active areas of recent lesions). (3) Mixed inflammatory cells (eosinophiles, neutrophiles, and lymphocytes). (4) Coffee bean nuclei, with indentations in the nuclear membrane, typical and diagnostic feature of Langerhans cells

## Selected Bibliography

Ghanem I, Tolo VT, D'Ambra P, Malogalowkin MH (2003) Langerhans cell histiocytosis of bone in children and adolescents. J Pediatr Orthop 23(1):124–130

Han I, Suh ES, Lee SH, Cho HS, Oh JH, Kim HS (2009) Management of eosinophilic granuloma occurring in the appendicular skeleton in children. Clin Orthop Surg 1(2):63–67

Howarth DM, Gilchrist GS, Mullan BP, Wiseman GA, Edmonson JH, Schomberg PJ (1999) Langerhans cell histiocytosis: diagnosis, natural history, management, and outcome. Cancer 85(10):2278–2290

Kilpatrick SE, Wenger DE, Gilchrist GS, Shives TC, Wollan PC, Unni KK (1995) Langerhans' cell histiocytosis (histiocytosis X) of bone. A clinicopathologic analysis of 263 pediatric and adult cases. Cancer 76(12):2471–2484

Plasschaert F, Craig C, Bell R, Cole WG, Wunder JS, Alman BA (2002) Eosinophilic granuloma. A different behaviour in children than in adults. J Bone Joint Surg Br 84(6):870–872

# Chapter 12
# Chondroma (Enchondroma)

**Nicola Fabbri and Davide Donati**

**Definition:** Intramedullary neoplasm made of well-differentiated hyaline cartilage.
**Epidemiology:** Rather frequent (only preceded by histiocytic fibroma and osteochondroma). No predilection for either sex. It may be diagnosed at any age.

N. Fabbri, MD (✉)
Department of Surgery, Orthopaedic Service, Memorial Sloan-Kettering Cancer Center, 1275 York Avenue, New York, NY 10065, USA
e-mail: fabbrin@mskcc.org

D. Donati, MD (✉)
3rd Orthopaedic and Traumatologic Clinic Prevalently Oncologic, Istituto Ortopedico Rizzoli, Bologna, Italy
e-mail: davide.donati@ior.it

P. Picci et al. (eds.), *Atlas of Musculoskeletal Tumors and Tumorlike Lesions*,
DOI 10.1007/978-3-319-01748-8_12, © Springer International Publishing Switzerland 2014

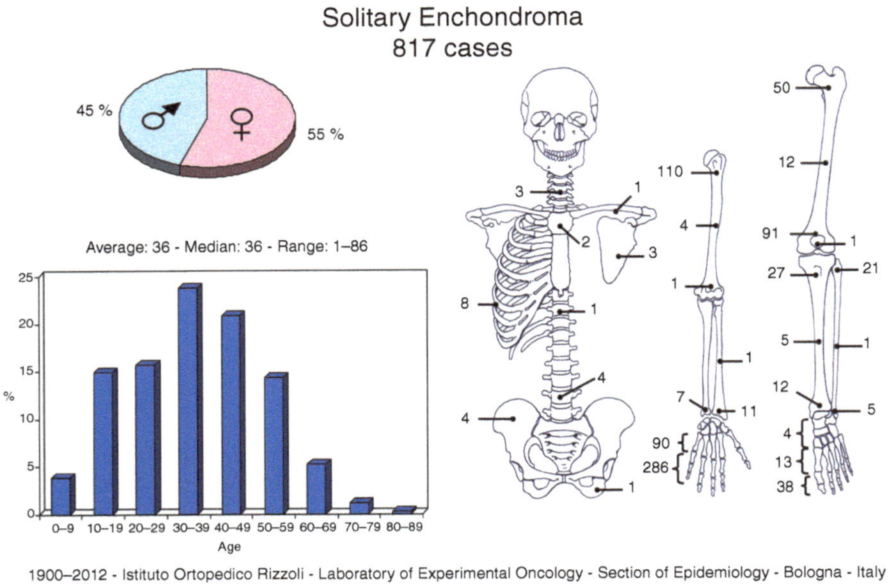


**Localization:** Chondroma occurs only in bones that are preformed in cartilage. Most cases are observed in tubular bones of the hand. Enchondroma is the most common bone tumor of the hands. The other chondromas are mainly distributed in the long bones, with preference for the femur. The real incidence of chondroma outside the hand is underestimated, as most of these lesions are asymptomatic.

**Clinical:** Because of its slow growth, minimal peritumoral reaction, and avascularity, chondroma is usually painless. In some active chondromas (mainly in young age), or after exercise, or because of pathologic fracture, immediate pain does occur. A mild expansion of the cortex can be seen or palpated in superficial bones, such as tubular bones of the hand or foot, ribs, and fibula where pathologic fractures are also common.

**Imaging:** The lesion is frequently central, sometimes eccentric or intracortical. It is an osteolysis, with rounded, lobulated, well-defined edges with a thin rind of reactive sclerosis. Usually the lesion contains granular, popcorn, ringlike opacities that represent calcification and ossification at the periphery of the lobules. Chondroma can reach considerable extension in the major long bones but rarely exceeds 6 cm; in small ones (hand, ribs, fibula), the cortex is thinned and the bone slightly to moderately expanded. CT scan demonstrates the lobular or multi-island nature of the lesion, its sharp limits, and its radiodensities, as well as the lack of permeative alterations of the cortex. MRI at best defines longitudinal extension; lesion is low signal in T1, high signal in T2; and calcifications are seen as black signal voids in both T1 and T2. Isotope scan is very hot in most lesions and sometimes reveals asymptomatic chondromas when performed for staging or follow-up of other conditions; it is useful in showing asymptomatic chondromas.

**Pathology:** Lesion consists of lobules of cartilage. The aspect is typical of hyaline cartilage. Calcified areas appear as white opaque granules. Reactive or enchondral ossifications manifest as white-yellow hard rings and streaks around and between the lobules. The limits of the lesion are often irregular, as lobules of cartilage push toward the cancellous bone and excavate little niches in the cortex but always well defined. The chondrocytes are sparse, with small, round, dense nuclei, of relatively uniform size. Occasionally, isogenous groups of cells can be seen. Double-nucleated cells are present, usually rare, but can be moderate. This does not really distinguish well benign enchondroma from chondrosarcoma. While diagnosis of cartilage tumor is usually easy even on clinico-imaging findings alone, the real problem is to differentiate chondroma from grade 1 chondrosarcoma. Histologically, this distinction is difficult:

(a) Grade 1 chondrosarcoma has a higher cellularity, more plump nuclei, and more than four to five double-nucleated cells per high-power field, but such features are subjective.
(b) Areas of chondroma and grade 1 chondrosarcoma may be found in the same tumor.
(c) The histological indicators of low-grade malignancy are meaningless if the lesion is in the hand, or in a child, or periosteal, or in chondromatosis. A useful differential element is represented by the relationship between the tumor and host bone. Chondroma can present with cartilage islands scattered in the bone, usually encased by a shell of mature lamellar bone. Chondrosarcoma, on the contrary, permeates marrow spaces and Haversian channels of the host in at least 90 % of cases. The permeative pattern is 99 % accurate, although cartilage permeation-like areas can be found in advanced osteoarthritis and in post-fracture epiphyses.

**Course and Staging:** Until skeletal maturity, chondroma grows slowly, and then it tends to stop. Thus, chondroma is stage 2 in children and stage 1 in adults. The incidence of malignant transformation of a solitary chondroma is unknown and controversial. Very rare in the hand, probably less rare in the trunk and limb girdles.

**Treatment and Prognosis:** Diagnosis can usually be made on clinico-radiographic features. The majority of chondromas do not require biopsy or surgical treatment. Enchondromas of the hand are sometimes treated with curettage and bone grafting because of pain, pathologic fracture, or cosmesis. Rarely, biopsy may be indicated if diagnosis of enchondroma is unclear. Follow-up with serial radiograms is helpful in the differential diagnosis vs grade 1 chondrosarcoma. After the epiphyseal plate is closed, enchondroma shows little to no growth; chondrosarcoma growth is continuous, grows 1–2 cm per year or more, and is associated with increasing pain, often waking up patient at night.

**Key Points**

| | |
|---|---|
| Clinical | Incidental findings |
| Radiological | Central, lobulated, granular, and ringlike calcifications |
| Histological | Lobules of benign cartilage ossified at the periphery without infiltration of the cancellous host bone |
| Differential diagnosis | Low-grade central chondrosarcoma |

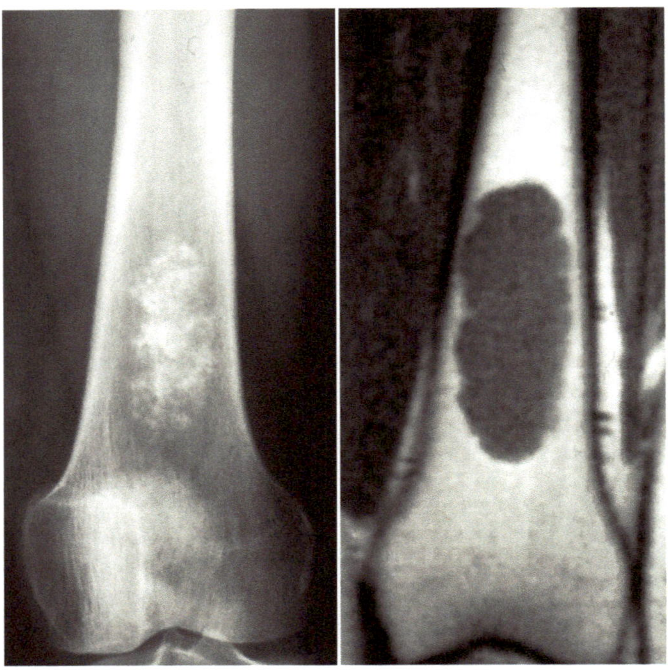

AP radiograph and coronal T1 MR of the distal femur. Well-limited lesion, centered in the medullary cavity, with typical cartilaginous calcifications (round with a clear center). The cortex and soft tissues are not involved

Radiograph of a finger. Well-limited lesion, containing cartilaginous calcifications

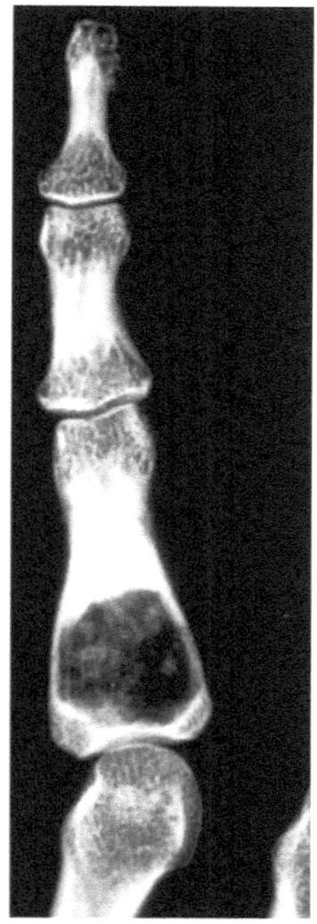

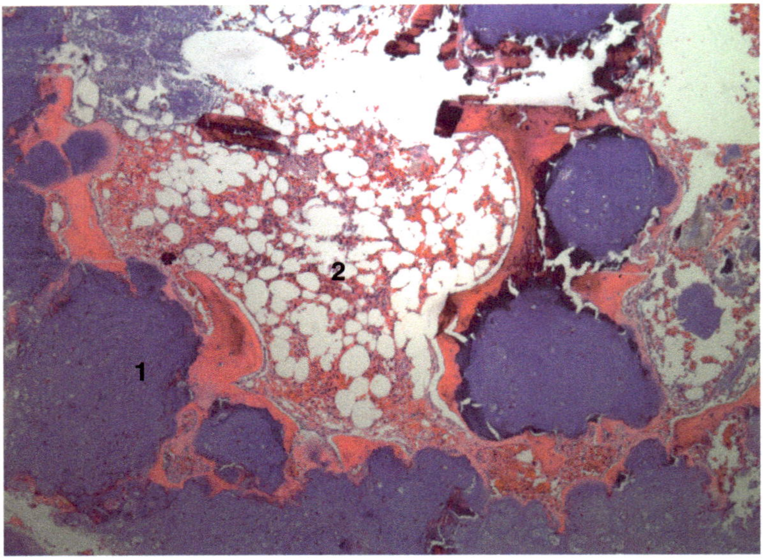


(1) Cartilaginous lobules with mature matrix. (2) Normal fatty marrow in between the cartilaginous lobules

## Selected Bibliography

Brien EW, Mirra JM, Kerr R (1997) Benign and malignant cartilage tumors of bone and joint: their anatomic and theoretical basis with an emphasis on radiology, pathology and clinical biology. I. The intramedullary cartilage tumors. Skeletal Radiol 26(6):325–353. Review

Gajewski DA, Burnette JB, Murphey MD, Temple HT (2006) Differentiating clinical and radiographic features of enchondroma and secondary chondrosarcoma in the foot. Foot Ankle Int 27(4):240–244

Kransdorf MJ, Peterson JJ, Bancroft LW (2007) MR imaging of the knee: incidental osseous lesions. Magn Reson Imaging Clin N Am 15(1):13–24. Review

Robbin MR, Murphey MD (2000) Benign chondroid neoplasms of bone. Semin Musculoskelet Radiol 4(1):45–58. Review

Romeo S, Hogendoorn PC, Dei Tos AP (2009) Benign cartilaginous tumors of bone: from morphology to somatic and germ-line genetics. Adv Anat Pathol 16(5):307–315. Review

# Chapter 13
# Periosteal Chondroma

**Nicola Fabbri and Davide Donati**

**Definition:** Benign cartilage neoplasm originating at the surface of the bone.

Rather rare, it is usually observed in children or young adults. It prefers the metaphyses of the long bones, particularly the proximal humerus. It is usually slightly to moderately painful because of nociception by the periosteum and commonly causes some swelling. Imaging shows a superficial erosion of the bone cortex, at times slightly scalloped, with regular borders. Such erosion is caused by a hemispherical periosteal cartilaginous mass, usually of small to moderate size. In the larger chondromas, tumor often contains granular or popcorn densities.

N. Fabbri, MD (✉)
Department of Surgery, Orthopaedic Service,
Memorial Sloan-Kettering Cancer Center,
1275 York Avenue, New York, NY 10065, USA
e-mail: fabbrin@mskcc.org

D. Donati, MD (✉)
3rd Orthopaedic and Traumatologic Clinic Prevalently Oncologic,
Istituto Ortopedico Rizzoli, Bologna, Italy
e-mail: davide.donati@ior.it

P. Picci et al. (eds.), *Atlas of Musculoskeletal Tumors and Tumorlike Lesions*,
DOI 10.1007/978-3-319-01748-8_13, 

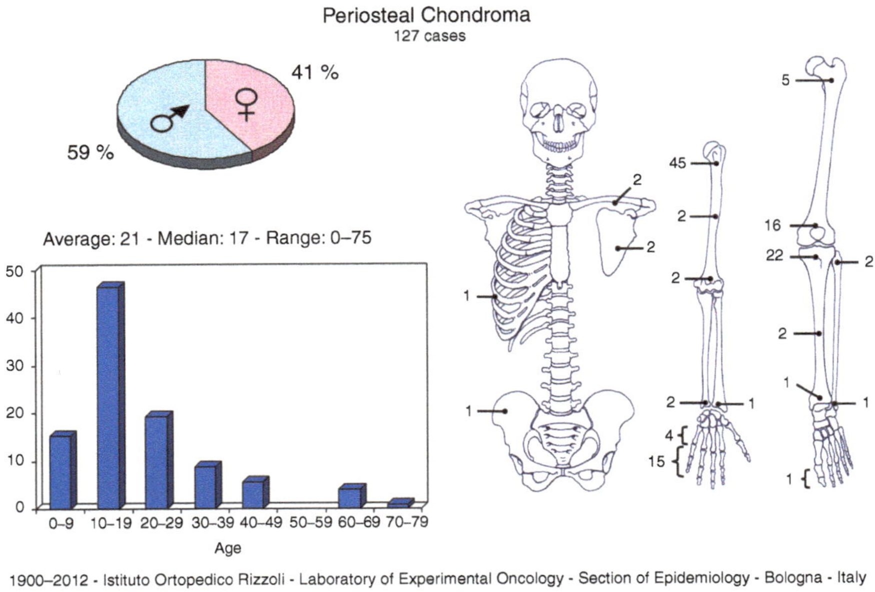


Histologically the tumor is very similar to enchondroma, but it more frequently displays features of cell proliferation (high cellularity, nuclear plumpness, and frequent double-nucleated cells). Being somewhat painful and causing some swelling in most instances, it usually requires surgical management consisting of either en bloc marginal excision or thorough curettage, equally effective.

**Key Points**

| | |
|---|---|
| Clinical | Some pain, young patients |
| Radiological | Subperiosteal, metaphyseal, with erosion of the cortex, granular calcifications |
| Histological | Lobules of benign cartilage. Possible hypercellularity |
| Differential diagnosis | Periosteal chondrosarcoma |

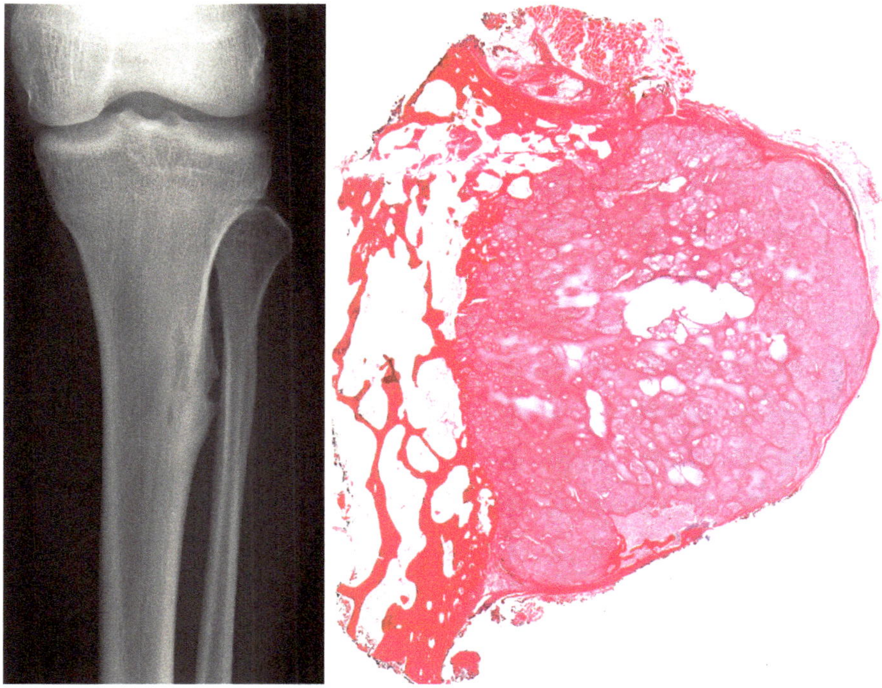

Male, 22 years old. Small (less than 3 cm) well-circumscribed lobulated lesion composed of hyaline cartilage. The lesion is beneath the periosteum with a sharp margin with the underlying cortex. The chondrocytes frequently are enlarged and hyperchromatic with increased cellularity and variability in nuclear size and shape. Out of context, these features suggest the diagnosis of chondrosarcoma

## Selected Bibliography

Boriani S, Bacchini P, Bertoni F, Campanacci M (1983) Periosteal chondroma. A review of twenty cases. J Bone Joint Surg Am 65(2):205–212

Nojima T, Unni KK, McLeod RA, Pritchard DJ (1985) Periosteal chondroma and periosteal chondrosarcoma. Am J Surg Pathol 9(9):666–677

Robinson P, White LM, Sundaram M, Kandel R, Wunder J, McDonald DJ, Janney C, Bell RS (2001) Periosteal chondroid tumors: radiologic evaluation with pathologic correlation. AJR Am J Roentgenol 177(5):1183–1188

Varma DG, Kumar R, Carrasco CH, Guo SQ, Richli WR (1991) MR imaging of periosteal chondroma. J Comput Assist Tomogr 15(6):1008–1010

Woertler K, Blasius S, Brinkschmidt C, Hillmann A, Link TM, Heindel W (2001) Periosteal chondroma: MR characteristics. J Comput Assist Tomogr 25(3):425–430

# Chapter 14
# Multiple Chondromas (Chondromatosis, Ollier's Disease) Associated Condition: Maffucci's Syndrome

**Nicola Fabbri and Davide Donati**

Multiple chondromas are infrequent, probably at least ten times less than solitary chondroma. The condition is not inherited and prevails in males. Their distribution and spectrum of presentation is variable. Only few chondromas are limited to the hands or one limb with minimal if any symptoms or, in other cases, diffusion is hemisomic, or extended to the entire body with hemisomic prevalence and overall features of diffuse chondrodysplasia, known as Ollier's disease. The most affected bones are the small tubular bones of the hand and foot, but chondromas may present anywhere in the skeleton. Knobby swelling, bowing deformities, and lower limb length discrepancy (even >10 cm) are the dominant symptoms. Relationship between the polyostotic limited form and the fully expressed diffused chondrodysplasia (Ollier's disease) remains unknown.

N. Fabbri, MD (✉)
Department of Surgery, Orthopaedic Service,
Memorial Sloan-Kettering Cancer Center,
1275 York Avenue, New York, NY 10065, USA
e-mail: fabbrin@mskcc.org

D. Donati, MD (✉)
3rd Orthopaedic and Traumatologic Clinic Prevalently Oncologic,
Istituto Ortopedico Rizzoli, Bologna, Italy
e-mail: davide.donati@ior.it

P. Picci et al. (eds.), *Atlas of Musculoskeletal Tumors and Tumorlike Lesions*,
DOI 10.1007/978-3-319-01748-8_14, © Springer International Publishing Switzerland 2014

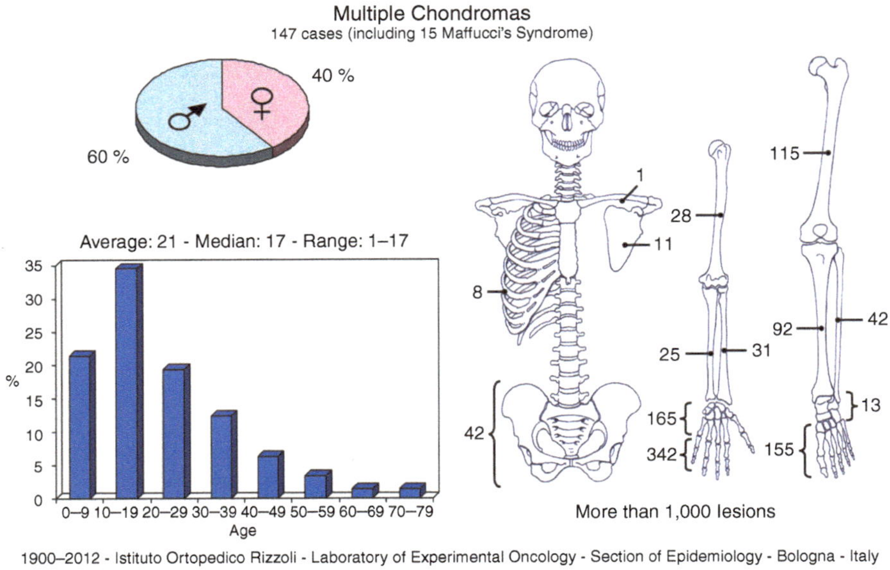


In Maffucci's syndrome (very rare), multiple, diffuse chondromas are associated with multiple hemangiomas, cutaneous, subcutaneous, or in the deep soft tissues (not in the bone). Basic imaging is the same as described for solitary chondroma. In the metaphyses, longitudinal columns of radiolucency extend toward the diaphysis. They are divided by longitudinal bony septae, giving a characteristic W-like pattern. Chondromas can be very extensive, sometimes bubbly or trabeculated, with expansion of the bone, very thin cortex, or lack of any cortex. In Maffucci's syndrome, phleboliths reveal on X-rays the angiomas. Histologically, compared to solitary enchondroma, lesions show features of more pronounced and persistent proliferative potential. Cartilage is more cellular and nuclei sometimes hyperchromatic, and the histology overlaps with low-grade chondrosarcoma by cytology alone. The stage is 2 in children, more frequently 1 in adults. Transformation to secondary sarcoma, most commonly chondrosarcoma, is frequent and probably ranges 20–30 % in Ollier and certainly higher in Maffucci (likely >50 %). Malignant transformation to sarcoma is usually seen in adults but may occur even before age 20, especially in Maffucci.

Both Ollier and Maffucci are conditions at increased risk to develop extra-skeletal malignancies, such as breast, liver, ovarian cancer and CNS tumors, suggesting an underlying genetic disorder predisposing to cancer in general. Surgical treatment is aimed to relieve symptoms rather than excise chondromas. Skeletal deformities and limb length discrepancy are addressed by osteotomies and/or lengthening procedures. Prognosis is burdened by the incidence of malignant change.

## 14.1 Concept of "Active" Chondromas and Differential Diagnosis with Chondrosarcoma

It is important to keep in mind that certain chondromas may show a histological pattern essentially similar to chondrosarcoma. In fact, multiple chondromas in Ollier and Maffucci, periosteal chondroma, enchondromas of the hands and feet, synovial chondromatosis, and soft tissue chondromas show histologic features consistent with grade 1 chondrosarcoma of the bone. In other words a grade 1 chondrosarcoma is cytologically indistinguishable from a benign lesion encountered in the abovementioned clinical settings. Therefore, diagnosis of malignant change is based on clinico-radiographic features and on the permeative growth pattern of tissue toward bone trabeculae. Secondarily, it is of paramount importance that the pathologist reviewing the slides has adequate clinical information, including site of biopsy and X-rays, and discusses the case with the orthopedic surgeon before making the diagnosis of chondrosarcoma. Quite important to the recognition of malignant change is to have a radiologic baseline taken in early adulthood (around age 20). Malignancy is characterized by a focal change in a bone of the baseline pattern often demonstrating a more "windblown" pattern often with a new soft tissue mass best seen on an MRI film. The patient also experiences chronic pain and may feel a new mass.

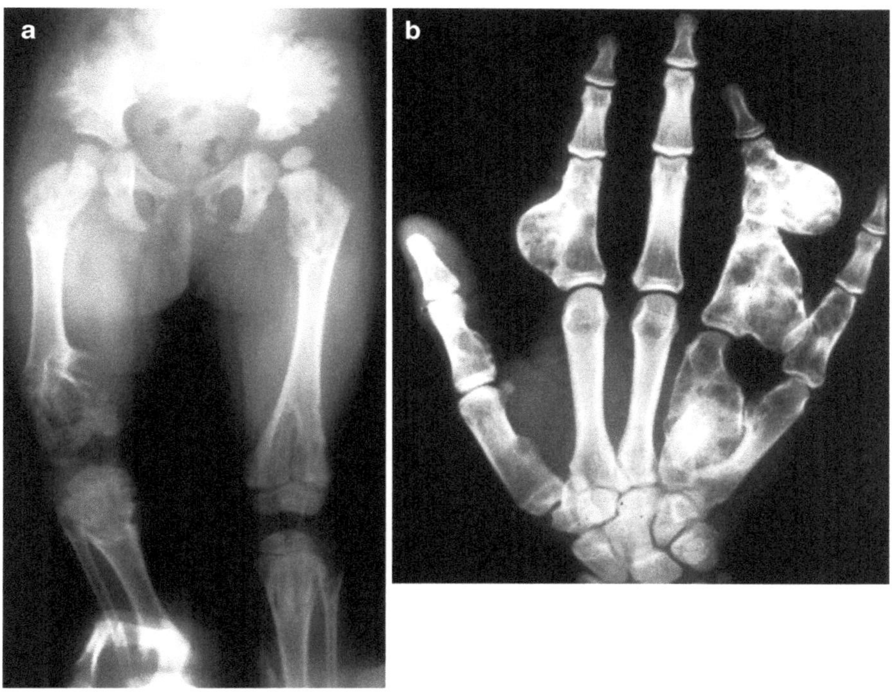


(**a**) Radiograph of the pelvis and proximal limbs. Multiple cartilaginous lesions, mainly metaphyseal, with bone deformations. (**b**) Radiograph of the hand: multiple chondromas

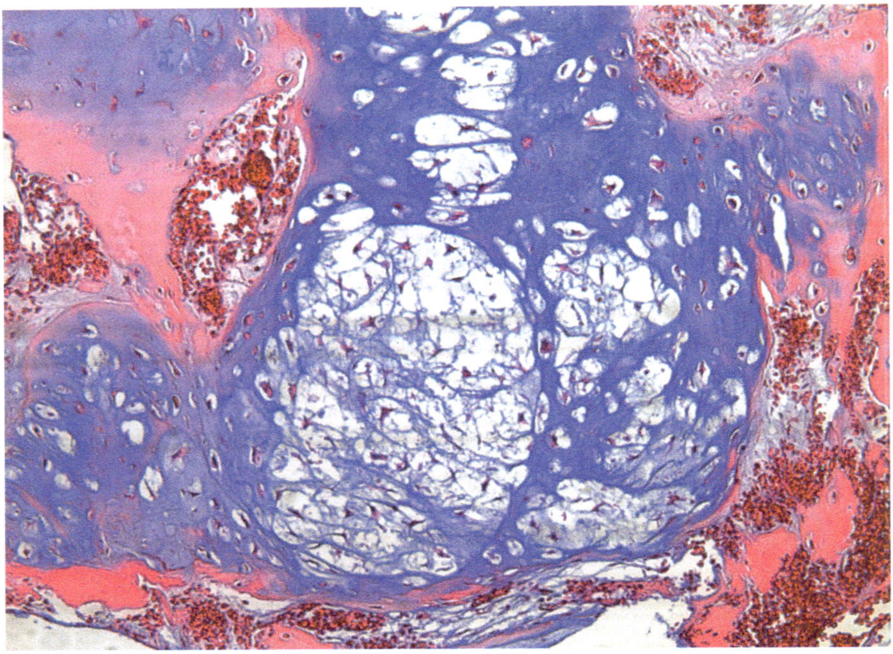

Cartilaginous lobules show evident atypia with a pseudo-malignant appearance of the lesion. This histology has to be considered benign if the lesion is in a child, or in Ollier/Maffucci disease, or in a periosteal site, in hands and feet, or in a joint (synovial chondromatosis)

## Selected Bibliography

Flemming DJ, Murphey MD (2000) Enchondroma and chondrosarcoma. Semin Musculoskelet Radiol 4(1):59–71. Review

Pansuriya TC, Kroon HM, Bovée JV (2010) Enchondromatosis insights on the different subtypes. Int J Clin Exp Pathol 3(6):557–569. Review

Pansuriya TC, Oosting J, Krenács T, Taminiau AH, Verdegaal SH, Sangiorgi L, Sciot R, Hogendoorn PC, Szuhai K, Bovée JV (2011) Genome-wide analysis of Ollier disease: is it all in the genes? Orphanet J Rare Dis 6:2

Schwartz HS, Zimmerman NB, Simon MA, Wroble RR, Millar EA, Bonfiglio M (1987) The malignant potential of enchondromatosis. J Bone Joint Surg Am 69(2):269–274

Silve C, Jüppner H (2006) Ollier disease. Orphanet J Rare Dis 1:37. Review

# Chapter 15
# Solitary Osteochondroma

**Nicola Fabbri and Davide Donati**

**Definition:** Benign cartilaginous neoplasm derived from an aberrant subperiosteal germ of the physeal cartilage, which grows and matures according to normal enchondral ossification.

It originates from a misplaced, subperiosteal island of physeal cartilage. It may also be seen in children secondary to radiation therapy. Presumably irradiation favors the exclusion of a cartilaginous island, by partially arresting and disorganizing proliferation of the physeal plate.

**Epidemiology:** Osteochondroma is very frequent. It prefers males by 1.5–2 to 1. Originating in early infancy, it is usually first noticed between 6 and 20 years of age.

N. Fabbri, MD (✉)
Department of Surgery, Orthopaedic Service, Memorial Sloan-Kettering Cancer Center, 1275 York Avenue, New York, NY 10065, USA
e-mail: fabbrin@mskcc.org

D. Donati, MD (✉)
3rd Orthopaedic and Traumatologic Clinic Prevalently Oncologic, Istituto Ortopedico Rizzoli, Bologna, Italy
e-mail: davide.donati@ior.it

P. Picci et al. (eds.), *Atlas of Musculoskeletal Tumors and Tumorlike Lesions*, DOI 10.1007/978-3-319-01748-8_15, 

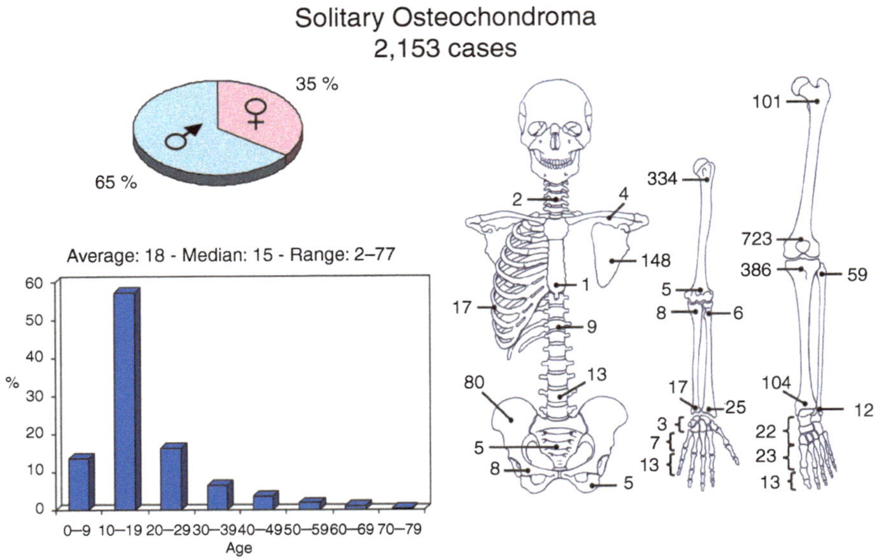


1900–2012 - Istituto Ortopedico Rizzoli - Laboratory of Experimental Oncology - Section of Epidemiology - Bologna - Italy

**Localization**: The most frequent localization is in the long bones: distal femur, proximal humerus, and proximal tibia. It originates from the metaphysis, but, with skeletal growth, it tends to move toward the diaphysis. In the trunk, more frequently involved are the scapula and ilium. Osteochondroma is exceptional in the hand and foot, and it does not occur in bones originating from membranous ossification (skull) nor in the epiphyses and carpal and tarsal bones (except the calcaneus).

**Clinical:** Swelling is the main symptom, slowly increasing during skeletal growth. Osteochondroma is usually painless. More rarely, pain may be due to bursitis or activity-related discomfort, especially in large lesions. Occasionally, a bursa forms on the osteochondroma, due to chronic friction; when fluid collection occurs in this bursa, an apparent and rapid increase of the volume, accompanied by pain, can arise the suspicion of a malignant change. Exceptionally, osteochondroma compresses a peripheral nerve or the dural sac, thus causing neurological symptoms; or it rubs against a large artery thus producing a false aneurysm (femoropopliteal artery). Also exceptionally, as a result of traumatic fracture of its stalk, osteochondroma becomes painful and mobile, clinically simulating a muscular ossification or a loose articular body.

**Imaging:** Osteochondroma is a bony excrescence with well-defined limits, having a thin outer cortex and an internal cancellous structure. The pathognomonic radiographic feature is that the cortex of the host bone flares into the cortex of the osteochondroma, and the cancellous bone of the osteochondroma blends with the cancellous bone of the metaphysis. In large lesions, areas or rarefied bone may alternate with irregular blotches of intense radio density, due to remnants of calcified cartilage, focal thickening of bone trabeculae, and bone necrosis. Some are pedunculated with a globose, cauliflower-like summit or with a sharp hornlike extremity. Others have a broad sessile base. Pedunculated osteochondromas are usually inclined toward the diaphysis. Rarely osteochondroma reaches huge sizes (even 15–20 cm), which are not proof of malignancy. By chronic compression, osteochondroma can cause scalloping and bowing of an adjacent bone. CT, MRI, sometimes angiography, and ultrasound may be useful (a) to confirm diagnosis in atypical cases, (b) for preoperative planning (relationship with the vascular bundle), and (c) in cases of suspect malignant change (thickness of the cartilage cap, fluid collection in reactive bursitis). Isotope scan is hot in active osteochondromas during childhood and adolescence and remains weakly positive or becomes negative after skeletal maturity, becoming again positive in malignant changes and in some bursitis.

**Pathology:** In children, a cartilage cap covers osteochondroma with a thickness ranging from a few mm to 1 cm or more, and it appears as a light blue cartilage similar to that of the physeal plate. In the adult, this cap decreases in thickness and in some areas it disappears; residual cartilage is white and similar to articular cartilage. Limits of the cartilage with the underlying bone are well-defined. The inner part of osteochondroma is irregularly cancellous, with fatty or occasionally hemopoietic marrow. When osteochondroma is covered by a bursa, this may contain serous or hematic effusion, rarely osteocartilaginous loose bodies. In its active stage, the cartilage cap presents, although irregular, the same features of the normal growth plate. Some cellularities, plumpness of the nuclei, and hypertrophy of the cells are to be expected in children and adolescents. The bony trabeculae of osteochondroma are originated by enchondral ossification of cartilage. Cancellous bone may include remnants of calcified cartilage and/or areas of necrotic bone. Some parosteal reactive ossifications (so-called exostosis of the great toe, bizarre osteocartilaginous pseudotumor of the hand) may be composed by proliferating cartilage undergoing enchondral ossification and thus may be somewhat similar to osteochondroma; these lesions, however, lack the overall above-mentioned clinico-radiographic features characteristic of osteochondroma.

**Course and Staging:** Growth occurs during childhood and adolescence. After skeletal maturity, osteochondroma stops growing. Thus, it is a benign stage 2 lesion in children and adolescents, becoming stage 1 in the adult. The change of a solitary exostosis into peripheral chondrosarcoma is rare (probably <1 %) and does not occur before puberty. The risk of transformation depends on the site. It is less rare in the trunk and limb girdles, rare around the knee, and almost exceptional in the more distal extremities.

**Treatment and Prognosis:** In asymptomatic lesions, there is no indication for surgery. An exception can be made for osteochondromas in the trunk or limb girdles, when an adult patient may elect to have osteochondroma prophylactically removed to avoid the risk of secondary transformation to chondrosarcoma. When a child presents with a large osteochondroma located close to neurovascular bundles, surgery may be indicated to avoid further enlargement possibly encasing the neurovascular structures, making late surgery more complex.

**Key Points**

| | |
|---|---|
| Clinical | Increasing swelling during growth age |
| Radiological | Bony excrescence without interruption of the cancellous bone: “finger glove sign” |
| Histological | Cartilage cap (max 1.5–2 cm) covering the normal cancellous bone |
| Differential diagnosis | Low-grade peripheral chondrosarcoma |

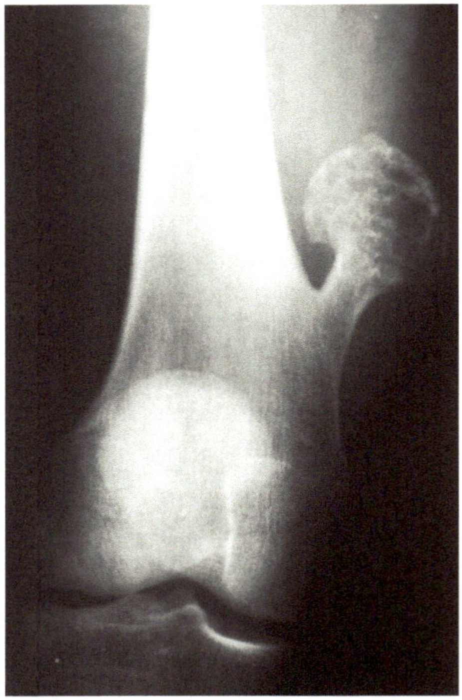

Pedunculated osteochondroma. Radiographs of the distal femur. The cortex of the osteochondroma is in continuity with that of normal bone

AP radiograph. Sessile osteochondroma of the proximal humerus

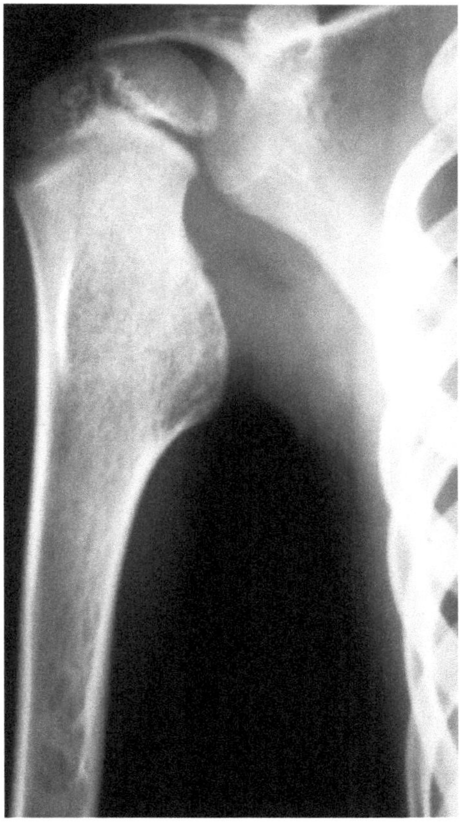

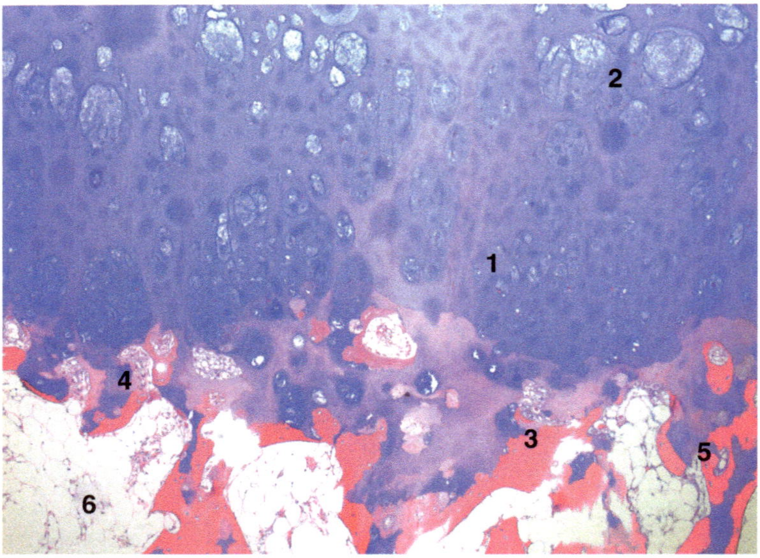


It commonly arises, in bones formed by enchondral ossification, at the region of the edge of the epiphyseal plate. During the growth phase, the cartilage cap presents the same aspects of normal growth cartilage, although less regular. Progressive transformation of proliferating cartilage in the underlying bone also mimics the epiphyseal growth mechanism, but it is much less orderly. (1) Irregular columns of swollen chondrocytes. (2) Degeneration of chondrocytes. (3) Calcified cartilage containing empty lacunae left after chondrocyte autolysis. (4) Vascular mesenchymal cells invading the calcified cartilage and maturating into osteoblasts. (5) Seams of osteoid formed by the osteoblasts are laid on the framework of the calcified cartilage. (6) Normal marrow

## Selected Bibliography

Brien EW, Mirra JM, Luck JV Jr (1999) Benign and malignant cartilage tumors of bone and joint: their anatomic and theoretical basis with an emphasis on radiology, pathology and clinical biology. II. Juxtacortical cartilage tumors. Skeletal Radiol 28(1):1–20. Review

Canella P, Gardini F, Boriani S (1981) Exostosis: development, evolution and relationship to malignant degeneration. Ital J Orthop Traumatol 7(3):293–298

Florez B, Mönckeberg J, Castillo G, Beguiristain J (2008) Solitary osteochondroma long-term follow-up. J Pediatr Orthop B 17(2):91–94

Saglik Y, Altay M, Unal VS, Basarir K, Yildiz Y (2006) Manifestations and management of osteochondromas: a retrospective analysis of 382 patients. Acta Orthop Belg 72(6):748–755

Valdivielso-Ortiz A, Barber I, Soldado F, Aguirre-Canyadell M, Enriquez G (2010) Solitary osteochondroma: spontaneous regression. Pediatr Radiol 40(10):1699–1701

# Chapter 16
# Multiple Exostoses

**Nicola Fabbri and Davide Donati**

Multiple exostoses are infrequent (incidence 1–2/100,000). Male sex is preferred, by 2 to 1. Exostoses usually manifest before 10 years of age, earlier as compared to solitary osteochondroma. Heredity is present in 2/3 of cases. Transmission is autosomal dominant. Basic research has identified several genetic abnormalities determining the disease. Most common mutations involve gene EXT1 on chromosome 8 and gene EXT2 on chromosome 11; a third gene named EXT3 has been identified on chromosome 19. Recent studies suggest a role for other EXT genes. Multiple skeletal lesions are usually diffused and relatively symmetrical. Typically, exostoses involve the bone circumferentially, mostly surrounding the metaphyseal regions, causing swelling, and sometimes limiting joint motion. In severe forms, limb shortening and deformity are associated. There is a large spectrum of presentation, from no deformity to severe impairment of upper and lower extremities. The relationship between type of genetic abnormality, severity of the disease, and risk of malignant transformation is under investigation in several centers. Lesions present the same pathologic features as solitary osteochondroma.

Treatment principles are prevention and correction of deformity and shortening, by removal variably combined with stapling, osteotomies, and lengthening procedures. Most patients do surprisingly well and have satisfactory function without surgery.

N. Fabbri, MD (✉)
Department of Surgery, Orthopaedic Service, Memorial Sloan-Kettering Cancer Center, 1275 York Avenue, New York, NY 10065, USA
e-mail: fabbrin@mskcc.org

D. Donati, MD (✉)
3rd Orthopaedic and Traumatologic Clinic Prevalently Oncologic, Istituto Ortopedico Rizzoli, Bologna, Italy
e-mail: davide.donati@ior.it

P. Picci et al. (eds.), *Atlas of Musculoskeletal Tumors and Tumorlike Lesions*,
DOI 10.1007/978-3-319-01748-8_16, 

The incidence of sarcomatous change in adult patients is low, ranging about 0.5–2 %. Preferred sites for sarcoma are the trunk, limb girdles, and knees. As for solitary osteochondroma, prognosis is dependent on the risk of malignant transformation. Patients should be followed lifetime, monitoring the deeper exostoses (pelvis, spine) with serial radiograms every 2–3 years.

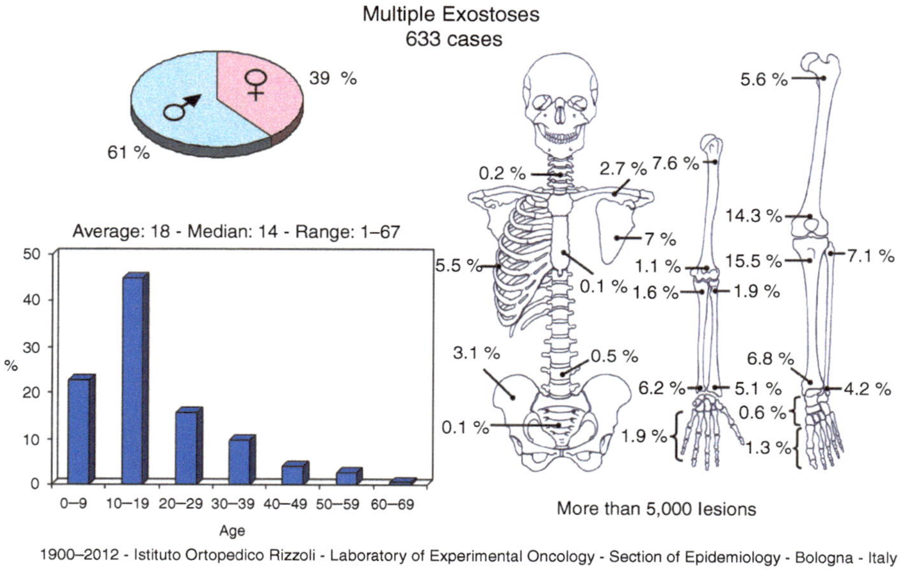

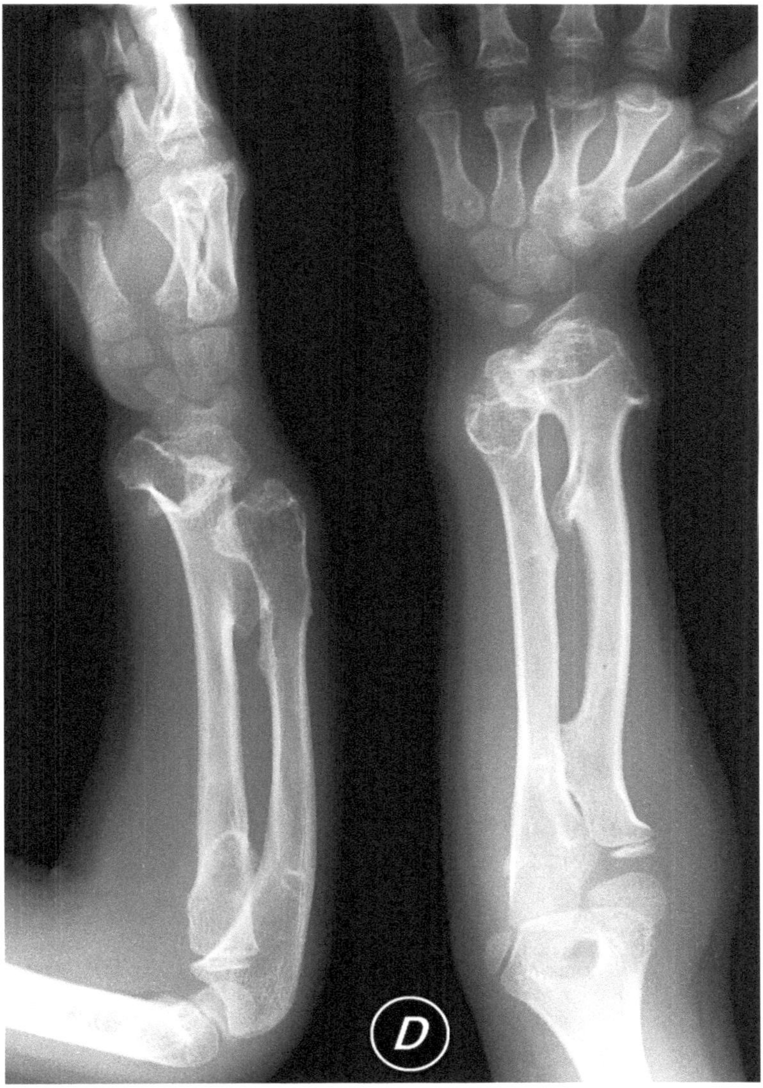


Radiographs of the arm and hand. Multiple exostoses. Bone deformation and shortening

## Selected Bibliography


Bovée JV (2008) Multiple osteochondromas. Orphanet J Rare Dis 3:3. Review

Hameetman L, Bovée JV, Taminiau AH, Kroon HM, Hogendoorn PC (2004) Multiple osteochondromas: clinicopathological and genetic spectrum and suggestions for clinical management. Hered Cancer Clin Pract 2(4):161–173

Kitsoulis P, Galani V, Stefanaki K, Paraskevas G, Karatzias G, Agnantis NJ, Bai M (2008) Osteochondromas: review of the clinical, radiological and pathological features. In Vivo 22(5):633–646. Review

Schmale GA, Wuyts W, Chansky HA, Raskind WH (1993–2000) Hereditary multiple osteochondromas. In: Pagon RA, Bird TD, Dolan CR, Stephens K (eds) GeneReviews [Internet]. University of Washington, Seattle. Aug 3 [updated 2008 Sep 5]

Stieber JR, Dormans JP (2005) Manifestations of hereditary multiple exostoses. J Am Acad Orthop Surg 13(2):110–120. Review

# Chapter 17
# Osteoid Osteoma

**Laura Campanacci**

**Definition:** Small benign tumor, usually very painful, made of osteoid and woven bone, surrounded by a halo of reactive bone.
**Epidemiology:** Relatively frequent in young patients (3 % of primary bone tumors). It has predilection for males (ratio of 2:1). Rarely observed before 5 and after 30 years of age.

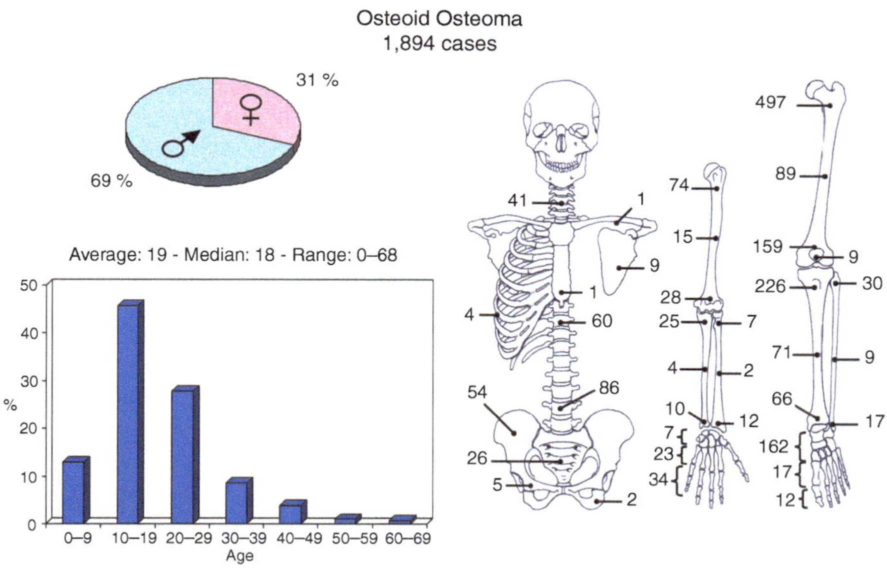


1900–2012 - Istituto Ortopedico Rizzoli - Laboratory of Experimental Oncology - Section of Epidemiology - Bologna - Italy

L. Campanacci, MD
3rd Orthopaedic and Traumatologic Clinic Prevalently Oncologic,
Istituto Ortopedico Rizzoli, Bologna, Italy
e-mail: laura.campanacci@ior.it

P. Picci et al. (eds.), *Atlas of Musculoskeletal Tumors and Tumorlike Lesions*,
DOI 10.1007/978-3-319-01748-8_17, © Springer International Publishing Switzerland 2014

**Localization:** It prevalently occurs in the appendicular skeleton, while it is rare in the trunk, with the exception of the spine (mostly localized in the posterior arch). In the limbs, it prevails in the proximal femur, but it is also common in the other long bones where it is more often diaphyseal or metadiaphyseal. It may occur also in the short bones, like talus. When occurring in long shafts, it involves the cortex, but also in cancellous areas, it is localized at the surface rather than in the center of the bone.
**Clinical:** The almost constant and often only symptom is pain, with a typical tendency to increase during the night, relieved by nonsteroidal anti-inflammatory drugs. When localized close to a joint, limited motion and chronic synovitis can be observed. In the spine, it may cause muscular spasm with stiff scoliosis.
**Imaging:** The basic radiographic element is a small (1–2 cm) rounded area of osteolysis ("nidus"), surrounded by a halo of bone sclerosis. The lesion is located at the center of this sclerosis that can be so thick that the small nidus cannot be seen except by CT. In cancellous bone areas, the halo of sclerosis may be less prominent. The isotope bone scan is constantly positive, showing a small, rounded area of intense uptake centered in a more diffused halo, thus resembling a "headlight in the fog." CT always demonstrates the nidus and its localization, thus permitting an adequate surgical plan. Angiography may demonstrate the high vascularity of the nidus. MRI, instead, is of little value because it is inferior to CT in depicting the nidus and showing inflammation of the surrounding tissue; it may also confuse the diagnosis.
**Histopathology:** Small roundish and hyperemic tumor, softer than the surrounding bone. Microscopic view shows a packed mesh of thin, contorted woven bone trabeculae, with osteoblastic rimming, osteoclasts, and numerous dilated capillaries. The osteoid may tend to mature in the center of the nidus. This calcified center corresponds to the radio dense nucleus of the nidus. The surrounding host bone appears as a reactive bone, more or less mature and sclerotic. The soft tissues around the lesion show chronic inflammation.
**Course and Staging:** Left untreated, it increases very slowly. Actually, there are cases, managed by nonsteroidal anti-inflammatory drugs, where pain subsided after some years of treatment and eventually even the nidus regressed. The tumor can be staged 1 or 2 in all cases.
**Treatment and Prognosis:** Surgery has been historically the mainstay of treatment; accurate intralesional curettage has been the treatment of choice in most centers until the late 1990s. Nowadays CT-guided percutaneous radiofrequency or laser ablation is considered the treatment of choice. The success rate of this approach is usually more than 90 % based on pain relief. Surgery remains an option in cases refractory to percutaneous ablation.

**Key Points**

| | |
|---|---|
| Clinical | Pain increasing during the night, relieved by nonsteroidal anti-inflammatory drugs |
| Radiological | The "nidus" surrounded by bone sclerosis |
| Histological | Woven bone formed by active osteoblasts and rich vascularization |
| Differential diagnosis | Intracortical osteosarcoma (very rare), osteoblastoma |

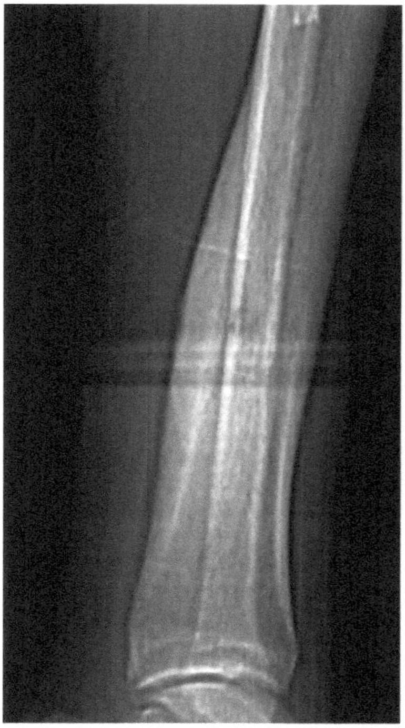

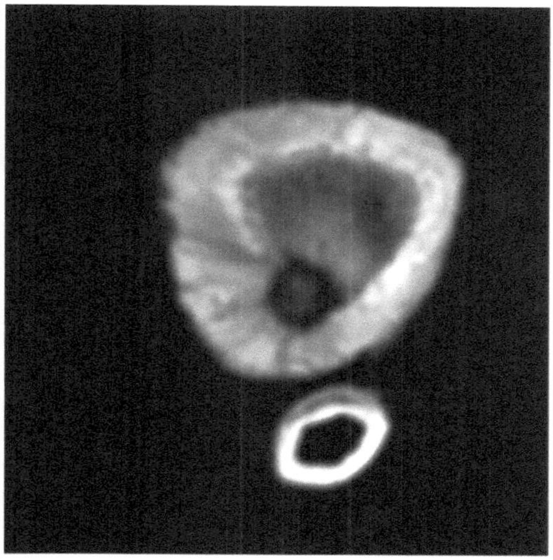

Scout view and CT. On the scout, thick periosteal bone formation is detected. The nidus is easy to visualize on CT. It is surrounded by reactive sclerosis and periosteal bone formations

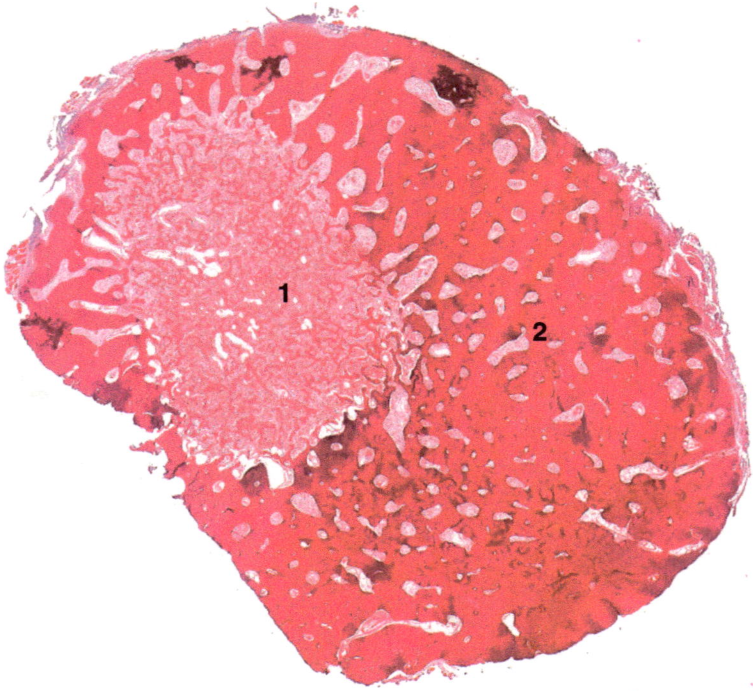


Macrosection of osteoid osteoma, with the nidus (*1*) and the sclerotic reactive bone (*2*)

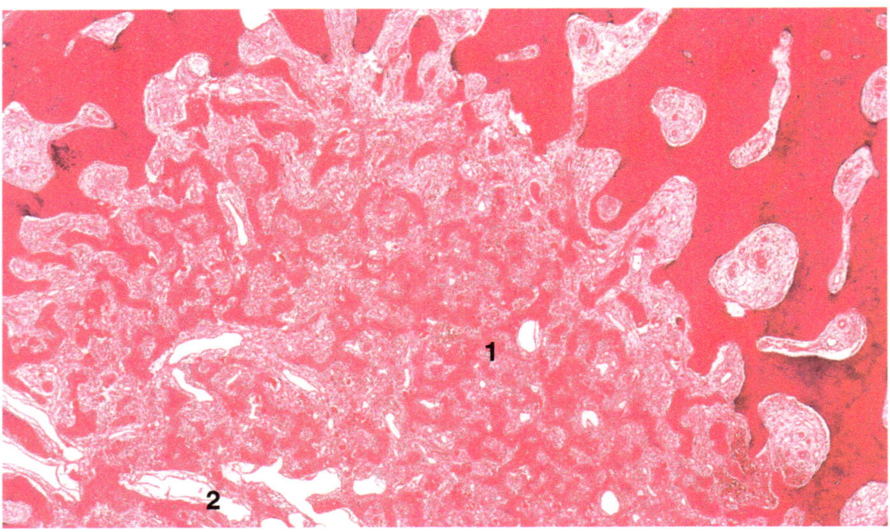


(*1*) Randomly arranged thick bars of osteoid with variable amounts of mineralization. (*2*) Capillaries and short fibroblasts and edema comprise an inter-bone trabecular granulation tissue

## Selected Bibliography

Allen SD, Saifuddin A (2003) Imaging of intra-articular osteoid osteoma. Clin Radiol 58(11): 845–852. Review

Becce F, Theumann N, Rochette A, Larousserie F, Campagna R, Cherix S, Guillou L, Mouhsine E, Anract P, Drapé JL, Feydy A (2010) Osteoid osteoma and osteoid osteoma-mimicking lesions: biopsy findings, distinctive MDCT features and treatment by radiofrequency ablation. Eur Radiol 20(10):2439–2446

Ghanem I (2006) The management of osteoid osteoma: updates and controversies. Curr Opin Pediatr 18(1):36–41. Review

Kitsoulis P, Mantellos G, Vlychou M (2006) Osteoid osteoma. Acta Orthop Belg 72(2):119–125. Review

Lee EH, Shafi M, Hui JH (2006) Osteoid osteoma: a current review. J Pediatr Orthop 26(5): 695–700. Review

# Chapter 18
# Osteoblastoma

**Laura Campanacci**

**Definition:** Benign tumor made of osteoblasts producing an osteoid and woven bone.
**Epidemiology:** It is rare (20 % of osteoid osteomas, 10 % of osteosarcomas). Again it predilects males (2–3:1). Rarely observed prior to 8 and after 40 years of age.

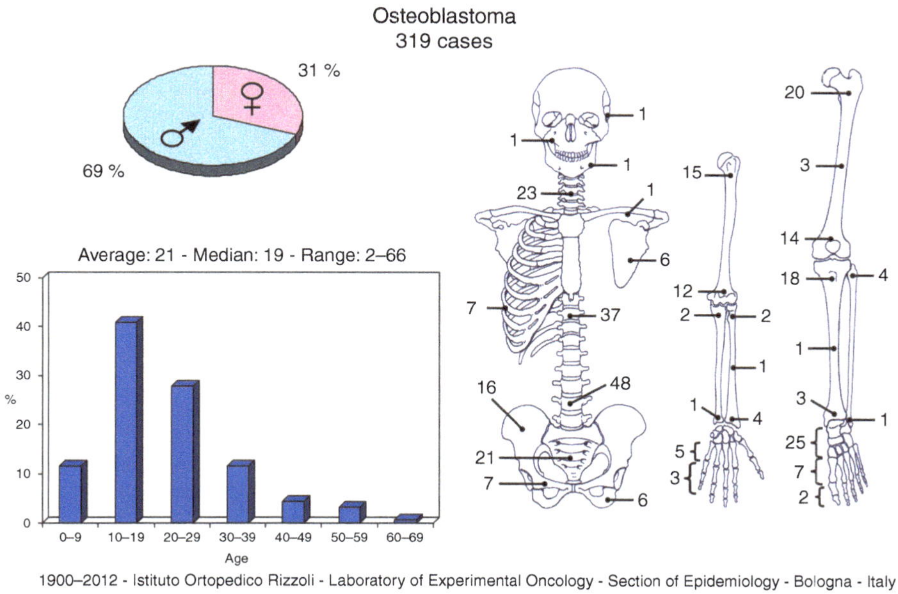


L. Campanacci, MD
3rd Orthopaedic and Traumatologic Clinic Prevalently Oncologic,
Istituto Ortopedico Rizzoli, Bologna, Italy
e-mail: laura.campanacci@ior.it

P. Picci et al. (eds.), *Atlas of Musculoskeletal Tumors and Tumorlike Lesions*,
DOI 10.1007/978-3-319-01748-8_18, © Springer International Publishing Switzerland 2014

**Localization:** Osteoblastoma shows evident predilection for the vertebral column (posterior arch) and the sacrum, but it may occur in any skeletal site.
**Clinical:** In the spine, it presents symptoms similar to osteoid osteoma (pain, scoliosis) with frequent signs of root compression. Usually it grows slowly but aggressive lesions manifest a rapid growth with severe symptoms due to the peritumoral inflammation.
**Imaging:** It is an osteolytic tumor containing a variable extent of osseous-type mineralization. Its size varies from 2 to 10 cm, the majority being between 3 and 5 cm. The tumor may be central, eccentric, and rarely periosteal. It tends to be roundish, with margins often demarcated by a rind of bone sclerosis, not as dense as in osteoid osteoma. The cortex may be destroyed with intense periosteal reaction. In aggressive lesions, the limits may appear blurred. Rarely the tumor blows the bone out or contains cystic spaces, similar to an ABC. A regional osteoporosis may be associated. The isotope bone scan is very hot. CT at best depicts intratumoral radio densities. MRI may show extensive peritumoral inflammatory reaction. Angiography reveals tumor vascularity.
**Histopathology:** The tissue is compact, reddish brown, and of soft-to-gritty consistency. Occasionally wide cavities typical of ABC are observed. The cortex is thinned, expanded, and sometimes absent, with a pseudo-capsule covering the tumor.

Microscopically, the tumor consists of large osteoblasts producing an osteoid and woven bone. Trabeculae are usually thin, with a regular "organoid" pattern. Osteoblasts rim the trabeculae. Cytologic features of activity (large cytoplasm, plump dark nuclei, and evident nucleolus) may be present. Mitotic figures are rare and typical. Large cells with bizarre hyperchromatic nuclei may be seen; they are never in mitosis and are interpreted as regressive cells (the so-called pseudomalignant osteoblastoma). Intertrabecular tissue contains a loose fibrovascular stroma, with abundant capillaries. The interface between the tumor and surrounding bone is sharp with no permeative pattern (d. d. vs. osteosarcoma).
**Course and Staging:** Most osteoblastoma are actively growing but well contained (stage 2). Occasionally, they are more invasive, bulging into the soft tissues (stage 3). Rarely the tumor appears almost quiescent and heavily mineralized so that it can be approximated to a stage 1 lesion. The vast majority of "osteoblastomas" that end up metastasizing to the lungs, leading to patient demise, were probably osteoblastoma-like osteosarcomas from the beginning. It is very important to assess the matrix of the "osteoblastomatous" lesion with the host bone. If it permeates the marrow spaces and traps the host lamellar bone, the lesion is an osteosarcoma, osteoblastoma-like.
**Treatment:** In stage 1 (latent) or stage 2 (active) osteoblastoma, intralesional curettage with local adjuvants is used. In stage 3 lesions, (aggressive) marginal or wide resection is indicated. In vertebral localizations, aggressive curettage can be used completed by radiation therapy. Selective preoperative arterial embolization may be useful to reduce hemorrhage during surgery.

**Key Points**

| | |
|---|---|
| Clinical | Pain and swelling, depending on the site. Frequent in the spine (posterior aspect) |
| Radiological | Mixed lesion (lytic/mineralized) |
| Histological | Osteoblasts producing an osteoid and a woven bone in a regular organoid pattern |
| Differential diagnosis | Low-grade central osteosarcoma |

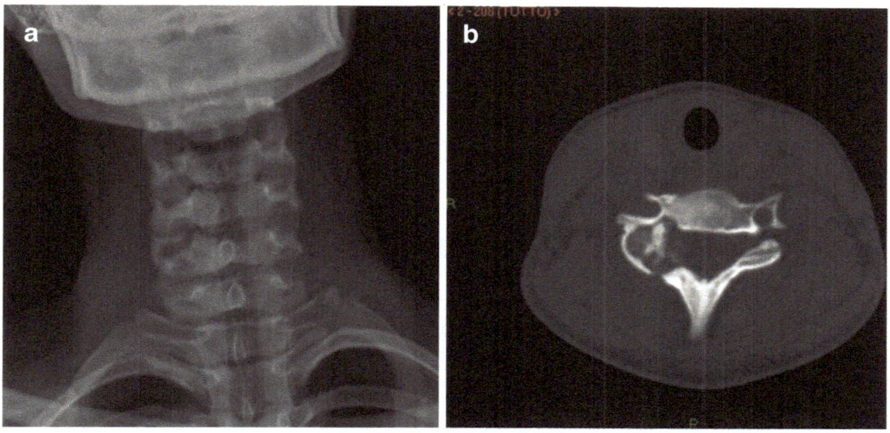


Radiograph and CT of the cervical spine. The lesion is well limited, contains ossifications, and is surrounded by reactive sclerosis

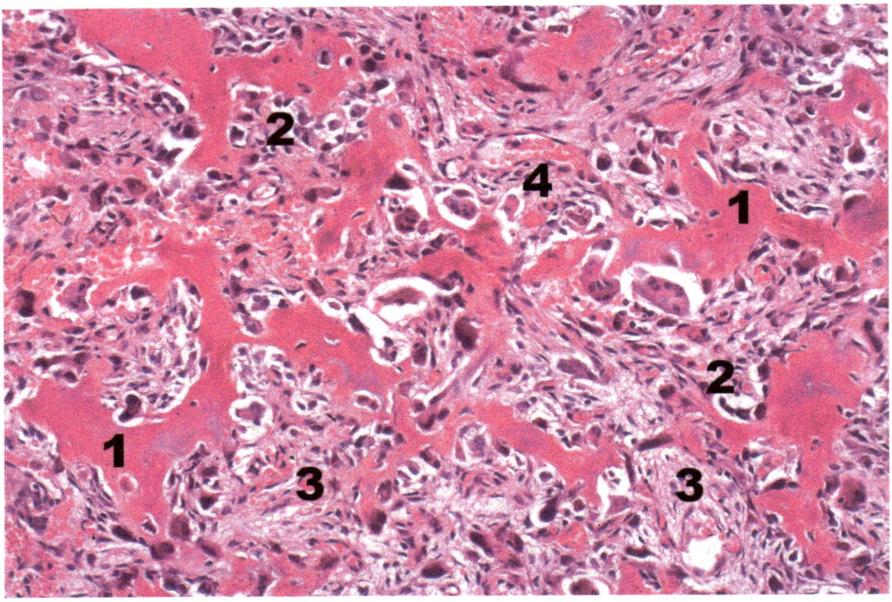


(1) Irregular bars of neoplastic osteoid. (2) Plump, deeply stained and slightly pleomorphic osteoblasts encircling woven bone trabeculae. (3) Proliferation of mesenchymal cells that tend toward osteoblastic differentiation. (4) Rich capillary vascularization

# Selected Bibliography

Harrop JS, Schmidt MH, Boriani S, Shaffrey CI (2009) Aggressive "benign" primary spine neoplasms: osteoblastoma, aneurysmal bone cyst, and giant cell tumor. Spine (Phila Pa 1976) 34(22 Suppl):S39–S47. Review

Mirra JM, Kendrick RA, Kendrick RE (1976) Pseudomalignant osteoblastoma versus arrested osteosarcoma: a case report. Cancer 37(4):2005–2014

Papagelopoulos PJ, Galanis EC, Sim FH, Unni KK (1999) Clinicopathologic features, diagnosis, and treatment of osteoblastoma. Orthopedics 22(2):244–247

Ruggieri P, McLeod RA, Unni KK, Sim FH (1996) Osteoblastoma. Orthopedics 19(7):621–624. Review

Schmidt MH (2008) Osteoid osteoma and osteoblastoma of the spine. Neurosurg Clin N Am 19(1):65–70. Review

White LM, Kandel R (2000) Osteoid-producing tumors of bone. Semin Musculoskelet Radiol 4(1):25–43. Review

# Chapter 19
# Aneurysmal Bone Cyst (ABC)

**Laura Campanacci**

**Definition:** Benign tumoral condition characterized by cystic cavities lined by spindle cells, inflammatory cells, giant cells, and histiocytes with hemosiderin deposition.

Production of variable amounts of reactive bone is frequently encountered. The cavities are filled with flowing blood. This lesion, when primary, is now considered to be tumoral, because in about 70 % of cases it has been demonstrated that the presence of typical translocation t(16; 17) (q22; p13) and t(17; 17) (q22; p13) that produce the chimeric proteins CDH11-USP6 and COL1A1-USP6, respectively. When this lesion is described side by side with another bone tumor (e.g., giant cell tumor, chondroblastoma, or fibrous dysplasia), it is considered reactive and does not display any kind of genetic abnormality.

**Importance:** It is often mistaken for a malignant tumor because of its radiological and pathological aggressiveness.

**Epidemiology:** Primary lesions represent about 2 % of primary bone tumors. No differences in gender. They are found mostly in the second decade of life (85 % under 20 years of age; rare above 50).

L. Campanacci, MD
3rd Orthopaedic and Traumatologic Clinic Prevalently Oncologic,
Istituto Ortopedico Rizzoli, Bologna, Italy
e-mail: laura.campanacci@ior.it

P. Picci et al. (eds.), *Atlas of Musculoskeletal Tumors and Tumorlike Lesions*,
DOI 10.1007/978-3-319-01748-8_19, © Springer International Publishing Switzerland 2014

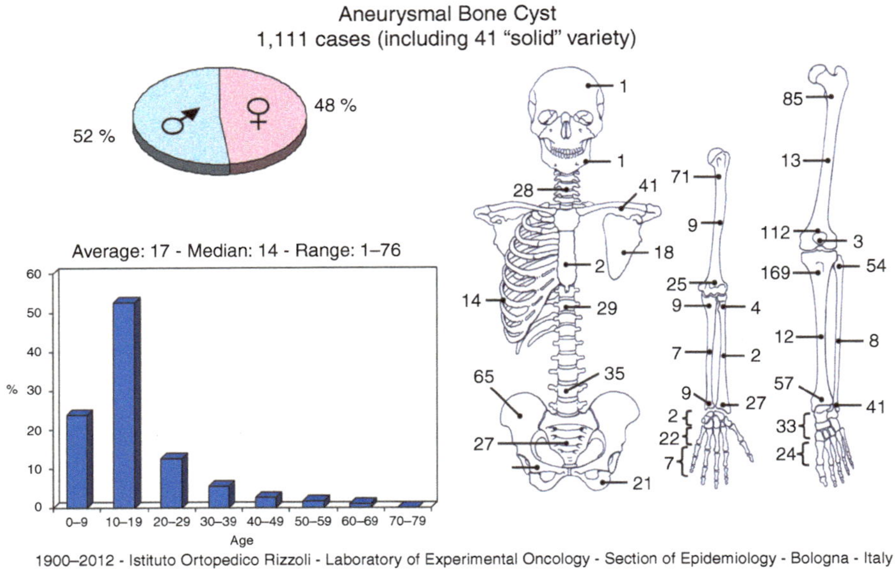


**Localization:** Any osseous segment may be affected, but the most common sites are the metadiaphyses of the long bones and the spine.
**Clinical:** Pain and swelling generally of less than 3 months duration. Frequent reports of a previous trauma in the same skeletal site. Onset or progression of symptoms has been observed in some cases during pregnancy.
**Imaging:** The characteristic radiographic appearance of ABC is a subperiosteal, poorly defined osteolysis, elevating and inflating the periosteum and progressively eroding the cortex. CT and MRI are often helpful in showing fluid levels within the cyst. Angiograms show intense and persistent accumulation of contrast media. Isotope scan commonly reveals an increased uptake with a cold central area.
**Histopathology:** Grossly it appears as a blood-filled sponge limited by fibrous septa. The wall of the cystic cavities is not formed by endothelial cells but by mesenchymal tissue rich in fibroblasts, histiocytes, thin capillaries, and scattered multinucleated giant cells. Immature trabeculae of reactive bone may be seen.
**Course and Staging:** The lesion may progress very rapidly, but in some cases, it may heal spontaneously, after a fracture or a biopsy. Recent cytogenetic data have demonstrated the presence of fusion oncogenes in a subgroup of ABCs. Nevertheless, its origin is controversial; ABC is commonly staged as a benign tumor.

**Treatment and Prognosis:** After intralesional surgery, local recurrence occurs in 20 % of cases. Radiation therapy (30–40 Gy) has been proven effective in inducing cyst ossification, but it is charged by the risk of sarcoma induction or growth plate damage in children. Selective arterial embolization of the nutrient vessels represents an alternative method in difficult locations (spine or pelvis).

**Key Points**

| | |
|---|---|
| Clinical | Pain and swelling in young people |
| Radiological | Eccentric, lytic lesion; fluid-fluid levels |
| Histological | Cystic cavities and septa not lined by endothelial cells |
| Differential diagnosis | Telangiectatic osteosarcoma |

**Chromosomal Translocations**

| | | |
|---|---|---|
| t(16;17) (q22;q13) | CDH11-USP6 | 40–50 % |
| t(17;17) (q12;q13) | COL1A1-USP6 | Rare |

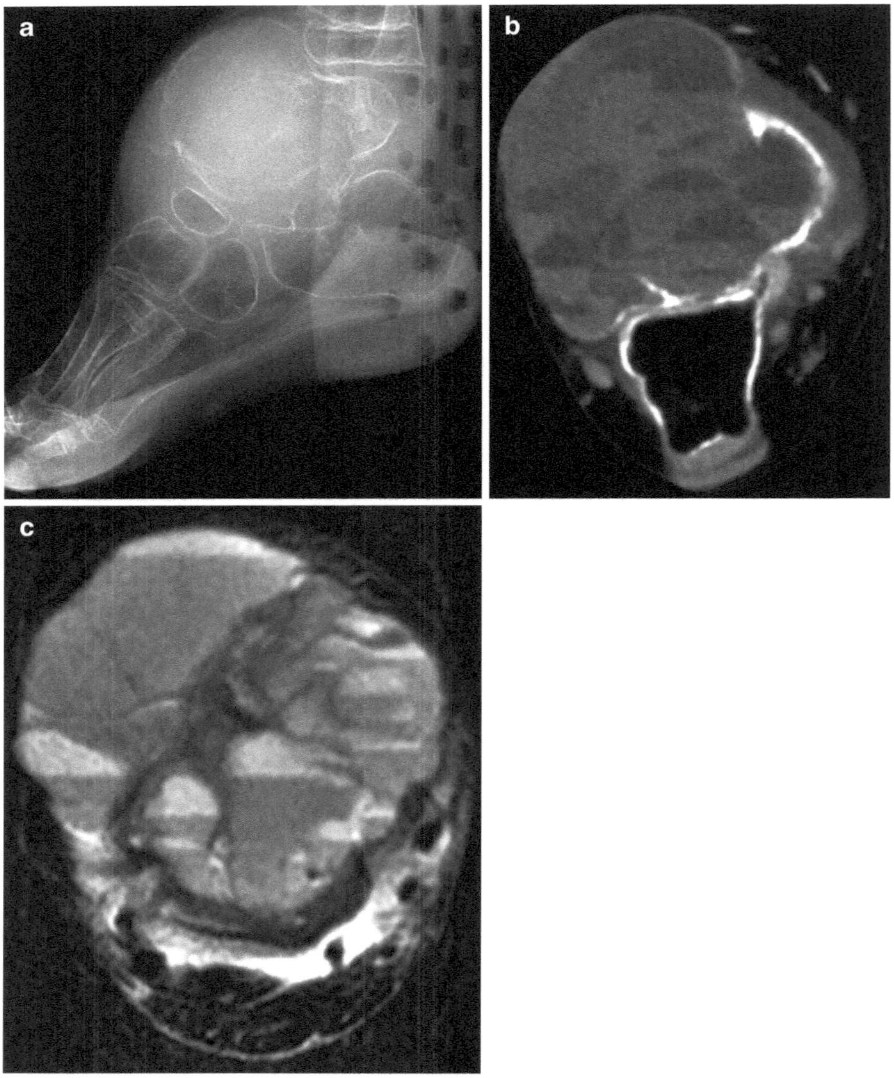


Radiograph, CT, and MR T2 axial image. The lesion is limited by a thin calcified periosteal reaction, indicating, despite a huge mass, slow growth. Fluid-fluid levels are detected on CT but better seen on MR

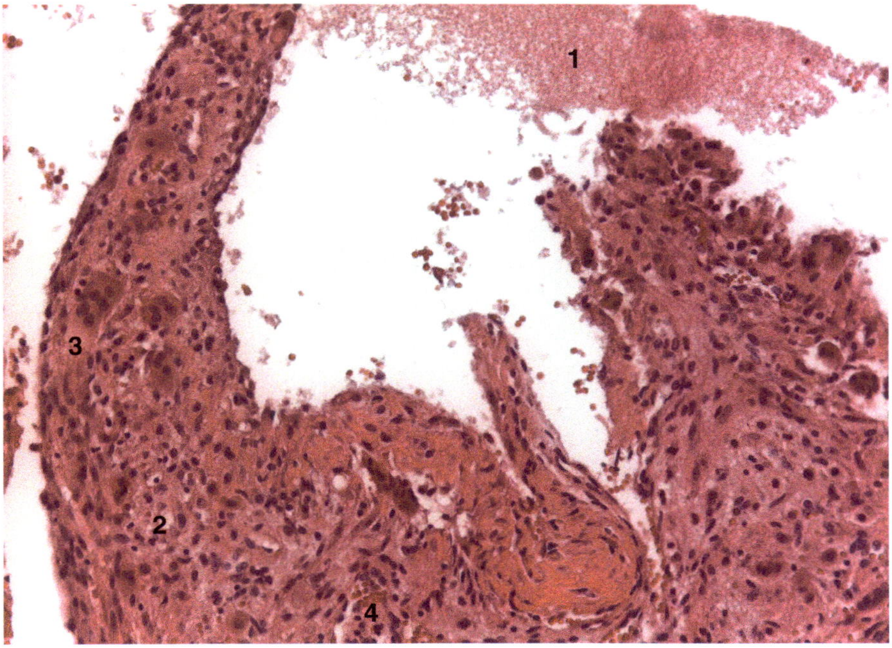


The typical spongy structure is observed: multiloculated cystic spaces of varying size filled with blood are interspaced with solid areas of benign-appearing fibrous tissue. (1) Blood-filled lacunae. (2) The walls of the hematic cavities are formed by histio-fibroblastic tissue. (3) Multinucleated giant cells. (4) Thin blood capillaries

## Selected Bibliography

Kransdorf MJ, Sweet DE (1995) Aneurysmal bone cyst: concept, controversy, clinical presentation, and imaging. AJR Am J Roentgenol 164(3):573–580. Review

Mankin HJ, Hornicek FJ, Ortiz-Cruz E, Villafuerte J, Gebhardt MC (2005) Aneurysmal bone cyst: a review of 150 patients. J Clin Oncol 23(27):6756–6762. Review

Mendenhall WM, Zlotecki RA, Gibbs CP, Reith JD, Scarborough MT, Mendenhall NP (2006) Aneurysmal bone cyst. Am J Clin Oncol 29(3):311–315. Review

Oliveira AM, Hsi BL, Weremowicz S, Rosenberg AE, Dal Cin P, Joseph N, Bridge JA, Perez-Atayde AR, Fletcher JA (2004) USP6 (Tre2) fusion oncogenes in aneurysmal bone cyst. Cancer Res 64(6):1920–1923

Panagopoulos I, Mertens F, Löfvenberg R, Mandahl N (2008) Fusion of the COL1A1 and USP6 genes in a benign bone tumor. Cancer Genet Cytogenet 180(1):70–73

# Chapter 20
# Giant Cell Tumor

**Marco Manfrini**

**Definition:** Intramedullary tumor composed of mononuclear cells and osteoclast-like multinucleated giant cells, with a variable and unpredictable potential for growth.
**Epidemiology:** Roughly, 5 % of all bone tumors and 20 % of all benign tumors. Slight preference for females. Usually young adults (about 65 %, ages 20–40), uncommon after 50 years of age, and rare in the immature skeleton (about 2 % of the cases).

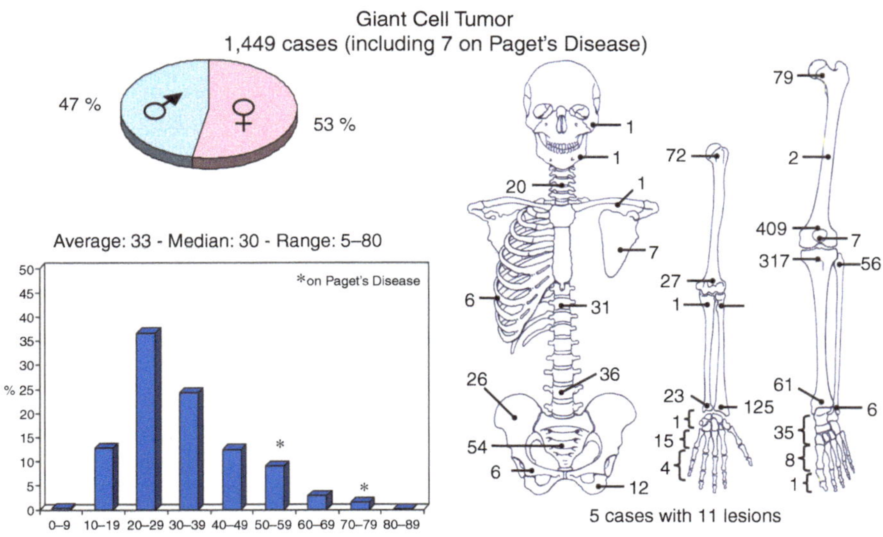


1900–2012 - Istituto Ortopedico Rizzoli - Laboratory of Experimental Oncology - Section of Epidemiology - Bologna - Italy

M. Manfrini, MD
Department of Surgery, 3rd Orthopaedic and Traumatologic
Clinic Prevalently Oncologic, Istituto Ortopedico Rizzoli, Bologna, Italy
e-mail: marco.manfrini@ior.it

P. Picci et al. (eds.), *Atlas of Musculoskeletal Tumors and Tumorlike Lesions*,
DOI 10.1007/978-3-319-01748-8_20, © Springer International Publishing Switzerland 2014

**Location:** Long bones, metaepiphyseal, and eccentrical: distal femur, proximal tibia, and distal radius (65 % of the cases). Rare in the sacrum and pelvis. When occurring in long bones before closure of growth plate, the tumor is metaphyseal. Exceptionally multicentric (less than 0.5 %, rule out hyperparathyroidism).
**Clinical:** Pain dominates the scenario, but swelling and muscular hypotrophy is also frequently encountered; severe stiffness and pathologic fracture are less common.
**Imaging:** On x-ray, purely lytic with relatively well-defined margins ("puddle on the sand") but not so sharply defined as other benign tumors and somewhat variable depending upon the tumor stage. The cortex is usually thinned, often expanded, and sometimes violated with little if any periosteal reaction. Soft tissue recurrence usually has an ossified rim. In an isotope bone scan, the uptake usually corresponds with the radiographic limits of the lesion, often it may show a cooler center with hotter rim; in aggressive forms, the uptake is beyond the radiographic limits. The CT images show a usually isodense lesion that occupies the bone with small outpockets in the borderline spongiosa. In some cases, intralesional cystic areas with fluid levels may be evident. MRI images may reveal a nonhomogenous signal, low in T1 and high in T2.
**Histopathology:** Gross: meaty reddish-purple tissue merged with soft yellow areas and often cystic areas (ABC-like areas). Histology: mononuclear cells and giant cells are uniformly distributed. Nuclei of mononuclear cells and giant cells are identical. There is mitotic activity but no atypical mitoses or cytological atypia. Necrotic areas and collection of foam cells are common, sometimes unusually rich of spindle cell areas. The stromal cells represent the neoplastic and proliferative component of the lesion. They secrete several chemokines which attract blood monocytes and stimulate their migration into tumor tissues. These monocytes fuse to form the osteoclast-like, multinucleated giant cells.
**Course and Staging:** Course unpredictable if untreated: sometimes indolent, sometimes rather aggressive. Histological grading has no value. Local recurrence rate depends upon stage and treatment. About 2–3 % is metastatic, usually to the lung. Malignant transformation: spontaneous 1 %, radiation induced 5–15 %. Presentation: about 10 % as stage 1, 60 % as stage 2, and 30 % as stage 3.
**Treatment:** Curettage represents the preferred surgery in stage 1, stage 2, and some stage 3 lesions. Recurrence rate ranges between 10 and 15 % with local adjuvants (cement, phenol, liquid nitrogen) and 15 and 30 % without local adjuvants. En bloc resection is recommended in stage 3 lesions with massive bone destruction, displaced pathologic fracture, and expendable bones. Radiation therapy adjuvant following intralesional surgery has been occasionally used in difficult locations (spine, pelvis, and sacrum). Recent studies indicate a possible role for medical treatment using anti-RANK-L antibodies (see Chap. 21).

**Key Points**

| | |
|---|---|
| Clinical | Pain and swelling in adult age |
| Radiological | Eccentric lytic metaepiphyseal lesion with small periosteal reaction |
| Histological | Mononuclear and giant cells with similar nuclei |
| Differential diagnosis | ABC in youngers, lytic bone tumors in adults (but usually not in the epiphysis, i.e., giant cell-rich OS) |

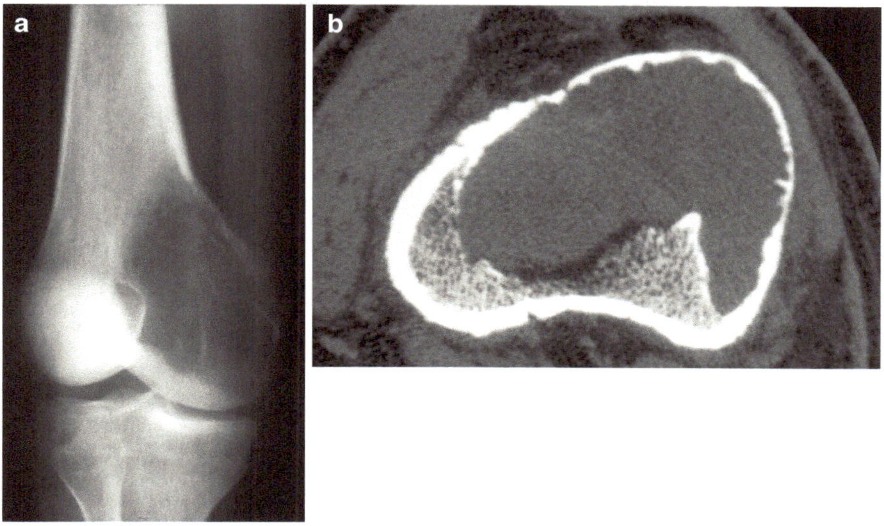


Radiograph and CT of the knee. Purely lytic, well limited but without sclerosis, epi-metaphyseal tumor. The cortex is thin, but not destroyed

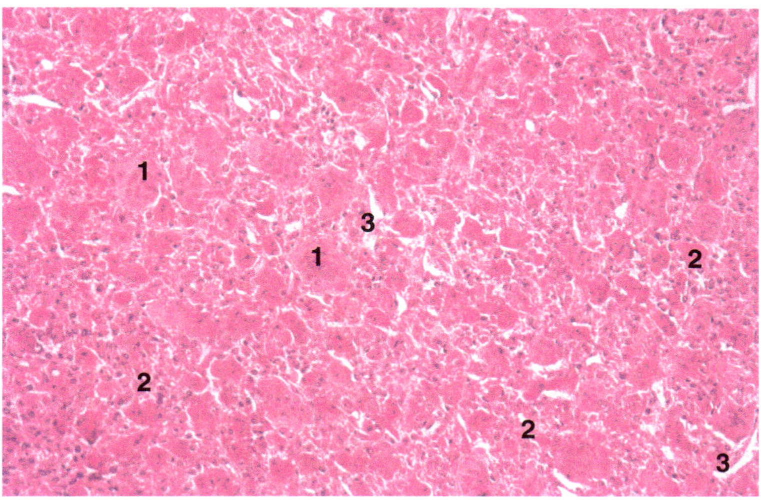


The lesion is composed of very cellular tissue with a dense population of cells of average size and with a single nucleus, constellated nearly everywhere by a large number of multinucleated giant cells. (1) The multinucleated giant cells stud the tissue. Giant cells have a large number of nuclei (even 50–100 in the section plane) gathered at the center. (2) Basic cells. They have the same nuclei as those of the giant cells. (3) Blood vessels

## Selected Bibliography

Errani C, Ruggieri P, Asenzio MA, Toscano A, Colangeli S, Rimondi E, Rossi G, Longhi A, Mercuri M (2010) Giant cell tumor of the extremity: a review of 349 cases from a single institution. Cancer Treat Rev 36(1):1–7

Kwon JW, Chung HW, Cho EY, Hong SH, Choi SH, Yoon YC, Yi SK (2007) MRI findings of giant cell tumors of the spine. AJR Am J Roentgenol 189(1):246–250. Review

Mendenhall WM, Zlotecki RA, Scarborough MT, Gibbs CP, Mendenhall NP (2006) Giant cell tumor of bone. Am J Clin Oncol 29(1):96–99. Review

Turcotte RE (2006) Giant cell tumor of bone. Orthop Clin North Am 37(1):35–51. Review

Yasko AW (2002) Giant cell tumor of bone. Curr Oncol Rep 4(6):520–526. Review

# Chapter 21
# Biology of Giant Cell Tumor

**Maria Serena Benassi**

Giant cell tumor (GCT) of bone is a well-vascularized tissue made up of a network of spindle-shaped mononuclear stromal cells that represent the proliferating pattern, round mononuclear monocytes, and multinucleated giant cells.

Mononuclear stromal cells are osteoblast-like cells, originating from bone marrow mesenchymal stem cells, positive to collagen 1, bone sialoproteins, osteonectin, osteocalcin, and alkaline phosphatase.

Giant cells are osteoclast-like cells originating from hematopoietic stem cells and expressing tartrate-resistant acid phosphatase (TRAP), cathepsin K, vitronectin receptor, calcitonin receptor, and CD68 and CD14 macrophage markers (Thomas and Skubitz 2009).

Besides tumor-derived factors (growth factors, BMP, etc.), mononuclear proliferating cells produce a component of tumor-necrosis factor (TNF) family, the membrane and soluble receptor activator for nuclear factor kB (RANKL), that interacts with RANK expressed on monocyte lineage and osteoclast precursors and, in presence of cytokines and macrophage colony-stimulating factor (M-CFS), induces osteoclast differentiation, thus shifting bone remodeling toward osteolysis (Wittrant et al. 2004).

In osteolytic tumors solubilization and digestion of inorganic and organic components by acid protease activity releases from bone matrix a very complex network of soluble factors (i.e. TGFb) and hormones that activate mononuclear cell proliferation creating a vicious circle where suitable targets for therapy can be identified (Broadhead et al. 2011). Previous results (Gamberi et al. 2004; Fujisaki et al. 2006) showed increased expression of urokinase-type plasminogen activation system, MMP9, and carbonate dehydratase 2 (CAII) activity in aggressive GCT when compared to latent forms (Enneking's surgical system), and a recent study demonstrated that targeting osteoclast lineage by blocking RANKL binding to RANK, cell adhesion and migration, or enzymatic activities is a current therapeutic approach in bone

M.S. Benassi, PhD
Laboratory of Experimental Oncology, Istituto Ortopedico Rizzoli, Bologna, Italy
e-mail: mariaserena.benassi@ior.it

P. Picci et al. (eds.), *Atlas of Musculoskeletal Tumors and Tumorlike Lesions*,
DOI 10.1007/978-3-319-01748-8_21, 

tumor treatment. Interestingly, the authors also pointed out the additional direct role of RANKL in tumor cell proliferation through an independent osteoclast way (Heymann 2012).

Cytogenetically, the most common chromosome aberrations in GCTs are telomeric associations involving multiple chromosome ends that may be responsible for structural chromosomal aberrations, an important event in GCT recurrences. Genomic instability and aberrations almost exclusively in aggressive GCTs have been reported as well as translocations involving chromosomes 1, 2, 11p, 12, 18p, 13p, 15p, and 19 (Moskovszky et al. 2009; Gorunova 2009). Advances in molecular biology have allowed a better knowledge of the mechanisms involved in malignant progression and the detection of key molecules or pathways involved in tumor-host interaction, useful to identify a subset of high-risk GCT patients prone to receive a different clinical management in terms of a closer follow-up and medical adjuvant therapy.

By integrated mRNA, miRNA, and proteomic data, we provided information on the most activated pathways and specific endpoints related to clinical course of disease also showing the “cross talk” between proteins and intracellular pathways.

There is evidence that the strong relationship between tumor cells and bone microenvironment facilitates tumor development and/or the metastatic process. Genes involved in matrix organization and adhesion processes such as tenascin C (TNC), osteocalcin (BGLAP), chondroadherin (CHAD), integrin (ITAGBL1), fibronectin protein 2 (FLRT2), and fibromodulin were found deregulated in aggressive/metastatic when compared to latent/localized GCTs and, when the clinical course of patients was matched on the bases of mRNA and protein levels, a strong correlation between high TNC expression and poor prognosis in terms of malignant transformation and relapse was seen (Pazzaglia et al. 2010).

Even more interestingly, expression levels of TNC, as well as NFIB and FLRT2 correlated inversely with miRNA data and CPBD database provided a list of activated endpoints by integrating pathways and other functional interaction resources (Mosakhani 2013).

On the other hand, proteomic studies carried out on GCT samples revealed an aggressive-related signature including molecules that control cellular red-ox state and survival as well as growth factor signaling pathways. In particular, Kaplan-Meier analysis revealed that the probability of disease-free survival significantly decreased for patients with glutathione-peroxidase 1 overexpression, while EGFR signaling in neoplastic stromal cells may contribute to disease progression through promoting stromal cell proliferation and osteoclastogenesis (Conti et al. 2011; Balla et al. 2011).

## Bibliography

Balla P, Moskovszky L, Sapi Z, Forsyth R, Knowles H, Athanasou NA, Szendroi M, Kopper L, Rajnai H, Pinter F, Petak I, Benassi MS, Picci P, Conti A, Krenacs T (2011) Epidermal growth factor receptor signalling contributes to osteoblastic stromal cell proliferation, osteoclastogenesis and disease progression in giant cell tumour of bone. Histopathology 59(3):376–389

Broadhead ML, Clark JC, Dass CR, Choong PF, Myers DE (2011) Therapeutic targeting of osteoclast function and pathways. Expert Opin Ther Targets 15(2):169–181
Conti A, Rodriguez GC, Chiechi A, Blazquez RM, Barbado V, Krènacs T, Novello C, Pazzaglia L, Quattrini I, Zanella L, Picci P, De Alava E, Benassi MS (2011) Identification of potential biomarkers for giant cell tumor of bone using comparative proteomics analysis. Am J Pathol 178(1):88–97
Fujisaki K, Tanabe N, Suzuki N, Mitsui N, Oka H, Ito K, Maeno M (2006) The effect of IL-1alpha on the expression of matrix metalloproteinases, plasminogen activators, and their inhibitors in osteoblastic ROS 17/2.8 cells. Life Sci 78(17):1975–1982
Gamberi G, Benassi MS, Ragazzini P, Pazzaglia L, Ponticelli F, Ferrari C, Balladelli A, Mercuri M, Gigli M, Bertoni F, Picci P (2004) Proteases and interleukin-6 gene analysis in 92 giant cell tumors of bone. Ann Oncol 15(3):498–503
Gorunova L, Vult von Steyern F, Storlazzi CT, Bjerkehagen B, Follerås G, Heim S, Mandahl N, Mertens F (2009) Cytogenetic analysis of 101 giant cell tumors of bone: nonrandom patterns of telomeric associations and other structural aberrations. Genes Chromosomes Cancer 48(7):583–602
Heymann D (2012) Anti-RANKL therapy for bone tumours. Basic, pre-clinical and clinical evidence. J Bone Oncol 1:2–11
Mosakhani N, Pazzaglia L, Benassi MS, Borze I, Quattrini I, Picci P, Knuutila S (2013) MicroRNA expression profiles in metastatic and non-metastatic giant cell tumor of bone. Histol Histopathol 28(5):671–8
Moskovszky L, Szuhai K, Krenács T, Hogendoorn PC, Szendroi M, Benassi MS, Kopper L, Füle T, Sápi Z (2009) Genomic instability in giant cell tumor of bone. A study of 52 cases using DNA ploidy, relocalization FISH, and array-CGH analysis. Genes Chromosomes Cancer 48(6):468–479
Pazzaglia L, Conti A, Chiechi A, Novello C, Magagnoli G, Astolfi A, Pession A, Krenacs T, Alberghini M, Picci P, Benassi MS (2010) Differential gene expression in classic giant cell tumours of bone: Tenascin C as biological risk factor for local relapses and metastases. Histopathology 57(1):59–72
Thomas DM, Skubitz KM (2009) Giant cell tumour of bone. Curr Opin Oncol 21(4):338–344
Wittrant Y, Théoleyre S, Chipoy C, Padrines M, Blanchard F, Heymann D, Rédini F (2004) RANKL/RANK/OPG: new therapeutic targets in bone tumours and associated osteolysis. Biochim Biophys Acta 1704(2):49–57

# Chapter 22
# Chondroblastoma

**Andrea Ferraro**

**Definition:** Benign tumor of childhood, usually epiphyseal, made of chondroblasts.
**Epidemiology:** Infrequent (less than 1 % of all bone neoplasms, 4–5 % of all benign lesions); it prefers the male sex by 2–3 to 1. Most cases (around 60 %) are in the second decade of life: rare after 35 years of age and exceptional before age 10.

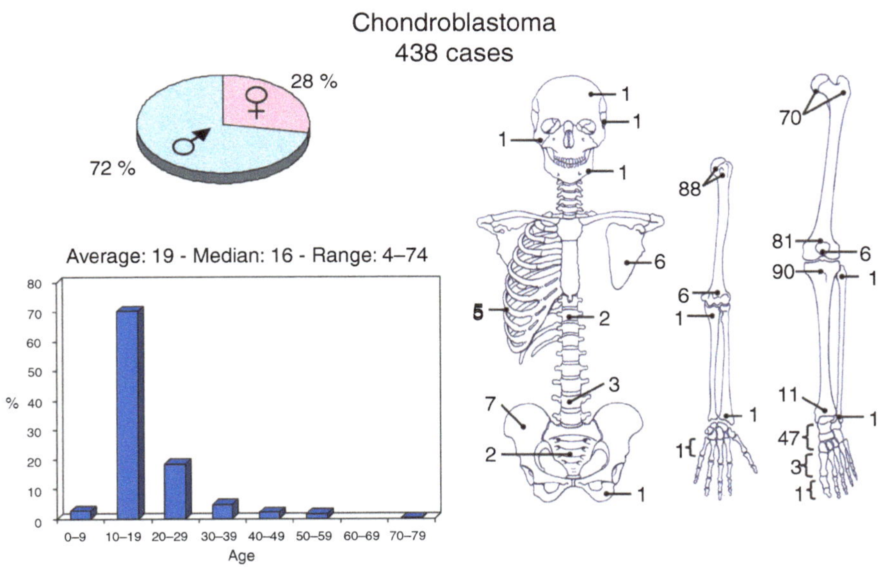


1900–2012 - Istituto Ortopedico Rizzoli - Laboratory of Experimental Oncology - Section of Epidemiology - Bologna - Italy

A. Ferraro, MD
3rd Orthopaedic and Traumatologic Clinic Prevalently Oncologic,
Istituto Ortopedico Rizzoli, Bologna, Italy
e-mail: andrea.ferraro@ior.it

P. Picci et al. (eds.), *Atlas of Musculoskeletal Tumors and Tumorlike Lesions*,
DOI 10.1007/978-3-319-01748-8_22, © Springer International Publishing Switzerland 2014

**Location:** Usually at the end of long bones (femur, tibia, humerus), involving the epiphysis, both the epiphysis and the metaphysis, or, less frequently, an apophysis. Exceptionally, it can be purely metaphyseal. Less common locations are tarsal bones and pelvis, quite rare elsewhere.

**Clinical:** Moderate to severe pain nearly always present, relatively common also joint effusion, muscular hypotrophy and stiffness.

**Imaging:** Round or oval radiolucent lesion, small to moderate in size (1–7 cm) within the epiphysis or an apophysis or even extending across the plate; sharp margins because of a sclerotic rim. The cortex may be expanded but preserved in most cases. Usually no periosteal reaction. Calcification inside the defect in 30–40 %. A soft tissue recurrence tends to be well circumscribed by a shell of ossification. Isotope scan: coinciding with radiographic limits and merging into uptake of the physeal plate. CT: fine punctate calcification; presence of cystic areas. MRI: homogenous intratumoral signal, transphyseal extent.

**Histopathology:** Gross: pink gray to tan brown, overall "wet sawdust" aspect; areas of chondroid matrix, calcification, and hemorrhage may be easily recognized. Histology: combination of mononuclear cells and giant cell. Mononuclear cells have oval nuclei, sometimes with a coffee bean appearance. Chondroid matrix present in most. Chicken-wire calcifications present in a third of cases (but pathognomonic). Mitotic figures common, never atypical mitoses. Secondary ABC-like areas in about 35 % of cases.

**Course and Staging:** Slow growth, normally stage 2 at presentation; stage 1 rare, stage 3 infrequent (recurrences, pelvic lesions).

**Treatment:** Curettage: stage 1, stage 2, and some stage 3 lesions. Avoid opening of the joint whenever possible to prevent articular contamination. En bloc resection: stage 3 lesions with massive bone destruction and/or tissue mass, large recurrences. Recurrence rate: about 10 % after curettage, minimal following en bloc resection.

**Key Points**

| | |
|---|---|
| Clinical | Pain in young, possible important inflammatory symptoms |
| Radiological | Epiphyseal or metaepiphyseal with calcifications |
| Histological | Chondroid matrix with "chicken-wire" calcifications |
| Differential diagnosis | Clear cell chondrosarcoma |

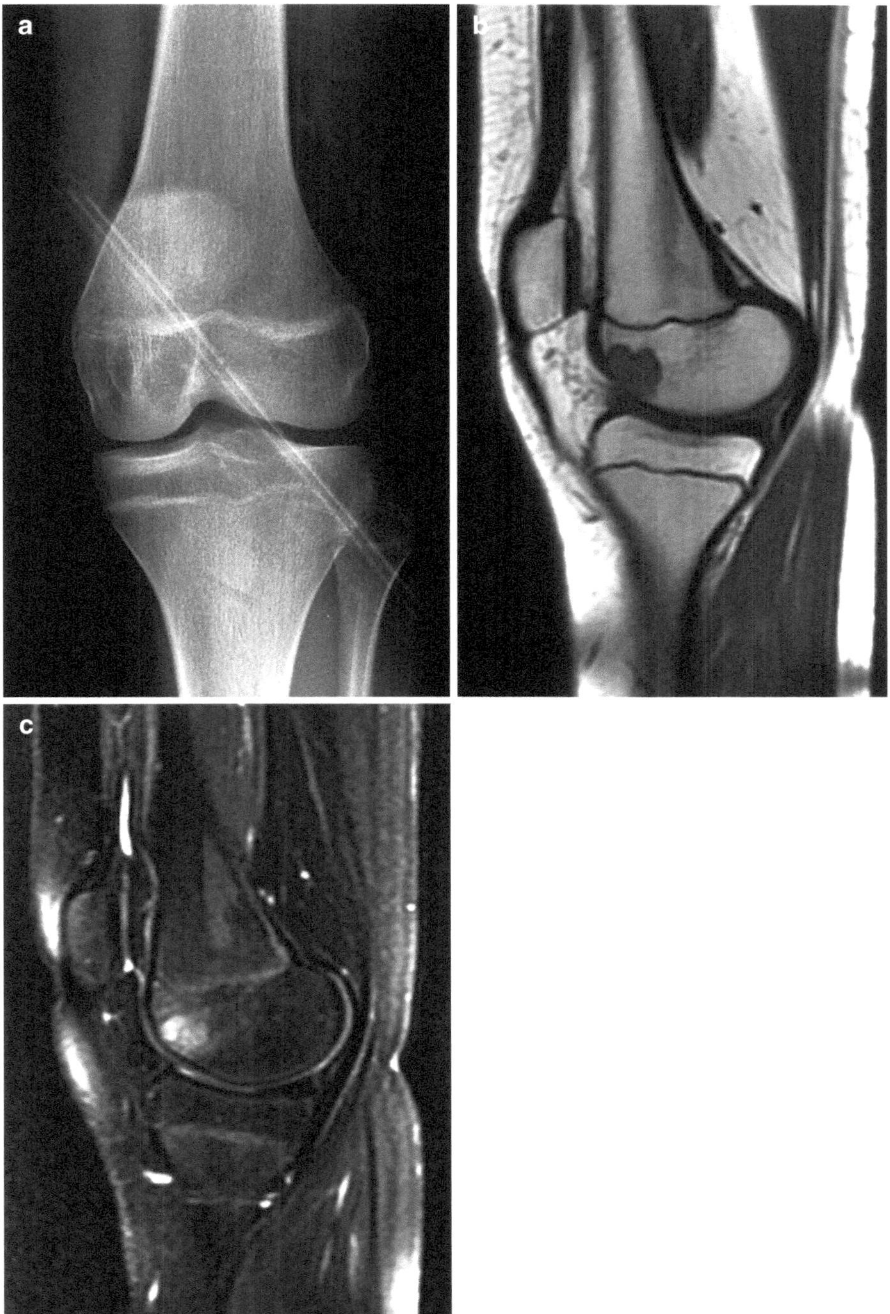


Radiograph and sagittal T1 and T2FS MR images. Well-limited epiphyseal lesion. An inflammatory reaction is well detected on T2 image

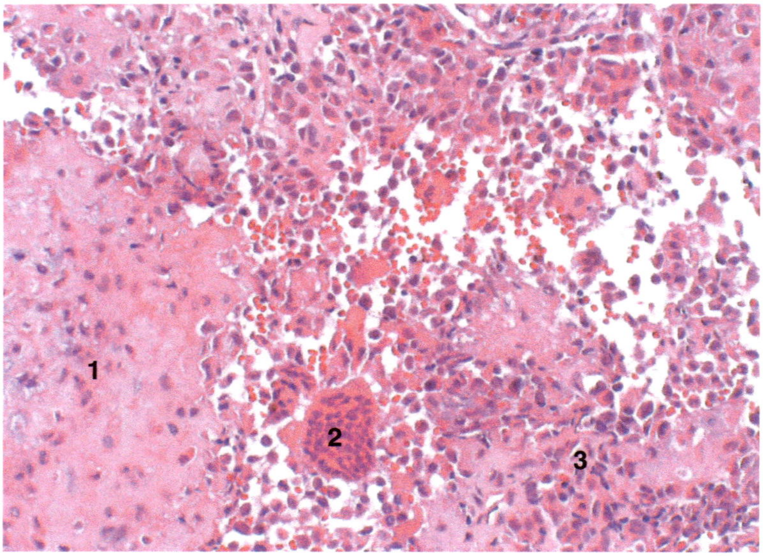


(*1*) Immature and irregularly shaped chondroblasts. They have well-defined cytoplasmatic boundaries and rounded well-stained nuclei. (*2*) Multinucleated giant cells. (*3*) Focal area of intercellular calcification. This aspect is of considerable diagnostic value

## Selected Bibliography

Bacchini P, Inwards C, Biscaglia R, Picci P, Bertoni F (1999) Chondroblastoma-like osteosarcoma. Orthopedics 22(3):337–339

Kaim AH, Hügli R, Bonél HM, Jundt G (2002) Chondroblastoma and clear cell chondrosarcoma: radiological and MRI characteristics with histopathological correlation. Skeletal Radiol 31(2):88–95

Lin PP, Thenappan A, Deavers MT, Lewis VO, Yasko AW (2005) Treatment and prognosis of chondroblastoma. Clin Orthop Relat Res 438:103–109

Sailhan F, Chotel F, Parot R (2009) SOFOP. Chondroblastoma of bone in a pediatric population. J Bone Joint Surg Am 91(9):2159–2168

Springfield DS, Capanna R, Gherlinzoni F, Picci P, Campanacci M (1985) Chondroblastoma. A review of seventy cases. J Bone Joint Surg Am 67(5):748–755

# Chapter 23
# Chondromyxoid Fibroma

**Andrea Ferraro**

**Definition:** Benign cartilaginous tumor made of lobulated, fibromyxoid, and chondroid tissue.
**Epidemiology:** It is definitely rare (0.5 % of all bone tumors, about 2 % of all benign neoplasms) and prefers the male sex by 1.5–1. Generally seen between 5 and 30 years of age, it has a predilection for the second and third decades of life.

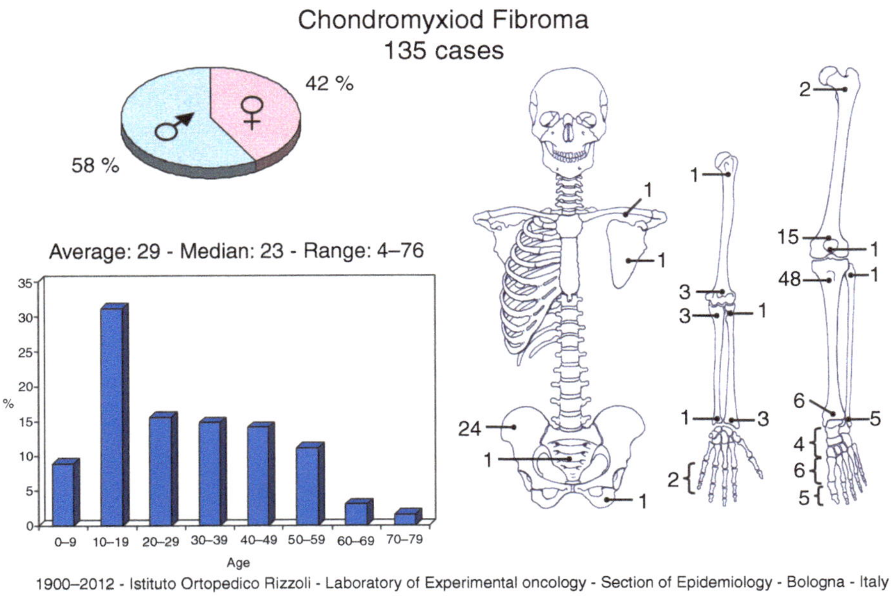


A. Ferraro, MD
3rd Orthopaedic and Traumatologic Clinic Prevalently Oncologic,
Istituto Ortopedico Rizzoli, Bologna, Italy
e-mail: andrea.ferraro@ior.it

P. Picci et al. (eds.), *Atlas of Musculoskeletal Tumors and Tumorlike Lesions*,
DOI 10.1007/978-3-319-01748-8_23, © Springer International Publishing Switzerland 2014

**Location:** Typically located in the long bone metaphysis, it can invade the epiphysis, especially in adults. Preferred sites proximal tibia (30 % of cases), small bones of the foot, and pelvis. Rare in the upper limb and in the trunk besides pelvis.
**Clinical:** Long-standing mild to moderate pain sometimes associated with bland local swelling is the typical scenario, because of the rather slow growth. Occasionally asymptomatic and discovered as incidental finding on roentgenograms.
**Imaging:** Small (generally <5 cm), metaphyseal, and eccentric radiolucent defect, usually with the long axis parallel to the bone of origin and occupying one half to two thirds of the diameter of the shaft; sometimes, especially in small bones, fusiform expansion of its entire contour. Sharply marginated by a lobulated shell of endosteal reactive bone because of a sclerotic rim. The cortex is usually canceled, with the tumor "bubbling out" into soft tissue. Little and peripheral periosteal reaction, chronic in nature and similar to that of periosteal chondroma. Intratumoral calcification is unusual. Isotope scan: moderately hot and corresponding to radiographic extent. CT: metaphyseal, eccentric, and subperiosteal, heavily marginated radiolucency, usually without any mineralization. MRI: homogenous intratumoral signal.
**Histopathology:** Gross: the tumor is rubbery soft, distinctly lobular, and clearly separated from the surrounding bone. Tissue is whitish or bluish and semitranslucent in chondromyxoid areas, tannish red in the undifferentiated and vascularized zones. Histology**:** lobular pattern better appreciated at low power, a light center, and a dark periphery. Basic elements stellate cells (blue) and myxoid background; center of lobule is hypocellular, periphery hypercellular. While the smaller lobules are exclusively or prevalently dark, the larger lobules are predominantly light, the dark part being limited to a narrow layer at the periphery of the lobules and in between them. These dark bands are absent where the lobules fuse together. Mitotic figures are not common. Cellular atypia may be present (15–20 % of cases). Occasionally, chondroblastoma-like cells at periphery.
**Course and Stage:** The tumor grows slowly and is generally small, at presentation usually stage 2 and occasionally stage 3. Few reports of malignant transformation.
**Treatment:** Aggressive curettage is usually indicated and frequently curative for either stage 2 and stage 3 tumors; recurrence rate probably lower than GCT and chondroblastoma. Consider en bloc resection for recurrent lesions and expendable bones.

**Key Points**

| | |
|---|---|
| Clinical | Mild symptoms |
| Radiological | Lobulated, subperiosteal lytic lesion |
| Histological | Lobules of myxoid cartilage surrounded by more cellular bundles with vessels |
| Differential diagnosis | Histiocytic fibroma (at imaging) |

TC: the lesion is well limited by a sclerotic line, metaphyseal, and eccentric, and contains small calcifications (usually non-detected on radiographs)

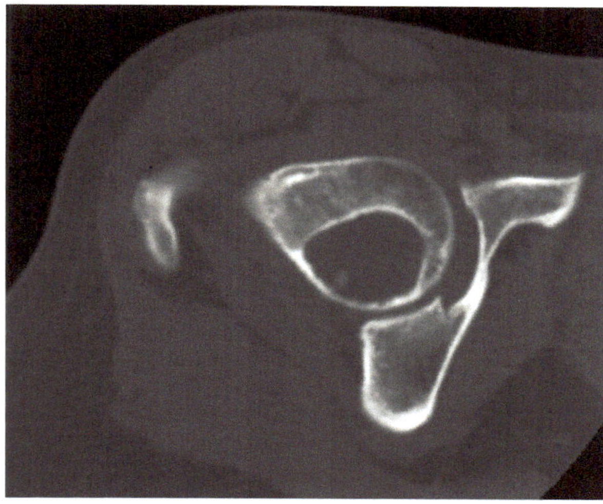

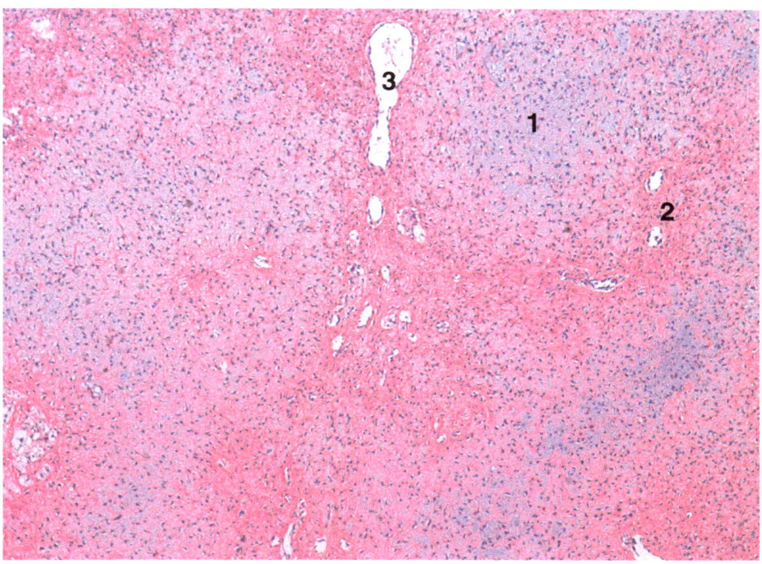


Histologic characteristics are particularly evocative at low power view. This shows the typical lobular architecture of the tumor. (*1*) Clear center of the lobuli. They are composed of spindle-stellate cells interspersed in an abundant and fluid ground substance. These areas represent the more mature phase of the tumor and are avascular. (*2*) Dark periphery of the lobuli. There are more densely cellular areas composed of well stained and usually not very pleomorphic cells with plump nuclei. Numerous and ectatic blood vessels run through these peri- and interlobular bands. These areas represent the younger more proliferative and undifferentiated phase of the tumor. (*3*) Blood vessels

## Selected Bibliography

Gherlinzoni F, Rock M, Picci P (1983) Chondromyxoid fibroma. The experience at the Istituto Ortopedico Rizzoli. J Bone Joint Surg Am 65(2):198–204

Lersundi A, Mankin HJ, Mourikis A, Hornicek FJ (2005) Chondromyxoid fibroma: a rarely encountered and puzzling tumor. Clin Orthop Relat Res 439:171–175

Romeo S, Duim RA, Bridge JA, Mertens F, de Jong D, Dal Cin P, Wijers-Koster PM, Debiec-Rychter M, Sciot R, Rosenberg AE, Szuhai K, Hogendoorn PC (2010) Heterogeneous and complex rearrangements of chromosome arm 6q in chondromyxoid fibroma: delineation of breakpoints and analysis of candidate target genes. Am J Pathol 177(3):1365–1376

Wu CT, Inwards CY, O'Laughlin S, Rock MG, Beabout JW, Unni KK (1998) Chondromyxoid fibroma of bone: a clinicopathologic review of 278 cases. Hum Pathol 29(5):438–446

Zustin J, Akpalo H, Gambarotti M, Priemel M, Rueger JM, Luebke AM, Reske D, Lange C, Pueschel K, Lohmann C, Rüther W, Amling M, Alberghini M (2010) Phenotypic diversity in chondromyxoid fibroma reveals differentiation pattern of tumor mimicking fetal cartilage canals development: an immunohistochemical study. Am J Pathol 177(3):1072–1078

# Chapter 24
# Desmoid Fibroma

**Laura Campanacci**

## 24.1 Desmoplastic Fibroma of Bone

**Definition:** It is a benign neoplasm composed of spindle cells and collagen, similar to desmoid tumor of soft tissues.
**Importance:** Slow-growing benign tumor with a high tendency to local recurrence.
**Epidemiology:** Tumor occurs rarely, more frequently in males. The majority of the cases are aged less than 30 years at diagnosis.

L. Campanacci, MD
3rd Orthopaedic and Traumatologic Clinic Prevalently Oncologic,
Istituto Ortopedico Rizzoli, Bologna, Italy
e-mail: laura.campanacci@ior.it

P. Picci et al. (eds.), *Atlas of Musculoskeletal Tumors and Tumorlike Lesions*,
DOI 10.1007/978-3-319-01748-8_24, © Springer International Publishing Switzerland 2014

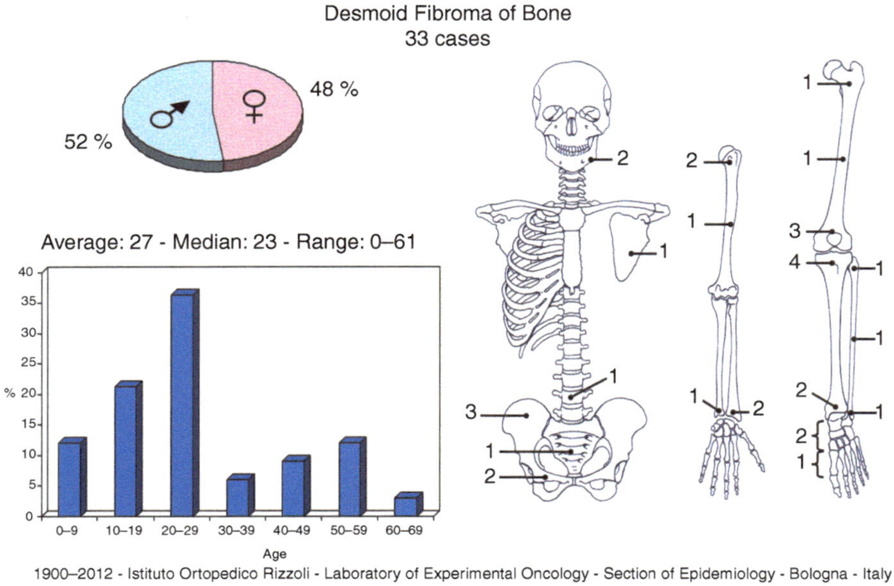


**Localization:** Desmoid fibroma is more frequent in the mandible, long bones, and pelvis. In the long bones, it is more often observed in the meta-diaphysis, but it may be diaphyseal. During adult age, it usually extends to the epiphysis.

**Clinical:** Symptoms are mild and late (the lesion grows slowly and is not very vascularized) and consist of moderate pain, possible pathologic fractures.

**Imaging:** The radiographic picture shows a large osteolysis, usually with expansion of the bone. The original cortex may be replaced by a thin shell of newly formed bone. Occasionally and focally, the tumor breaches the cortex with no clear demarcation toward the surrounding tissues. Rather typically, osteolysis has a finely trabeculated, reticulated, or bubbly appearance. Reactive chronic hyperostosis can produce a rind of sclerosis at the edges of the tumor. On angiography and CT, the tumor only moderately takes the contrast, and it appears relatively "cold" on isotope scan. MRI gives a low signal in T1 and T2 suggestive of dense collagenous tissue.

**Histopathology:** Hypocellular spindle cell tumor associated with large amounts of collagen. Nuclei have a bland appearance with no atypia and few mitotic figures. Areas of relative hypercellularity can be present, but are not common. β-Catenin pathway does not seem to have the same role in the tumorigenesis of desmoid tumor, as it has in the desmoid type-fibromatosis.

**Course and Staging:** Growth of the tumor is very slow; it may even take several years prior to causing symptoms. The stage is 2 or 3.

**Treatment and Prognosis:** Although historical data indicated wide en-bloc resection as the treatment of choice, recent experiences show that aggressive curettage is about equally effective providing better functional results. Desmoid fibroma does not metastasize.

## Key Points

| | |
|---|---|
| Clinical | Mild symptoms |
| Radiological | Slow growing lytic lesion with possible expansion of the cortex |
| Histological | Bland spindle cells resembling desmoid tumor of soft tissue |
| Differential diagnosis | Low-grade fibrosarcoma |

## Immunohistochemical Panel

| | |
|---|---|
| VIM | + |
| MS Act | ± |
| Smooth M act | ± |
| Beta-catenin | + (nuclear staining, occasionally) |

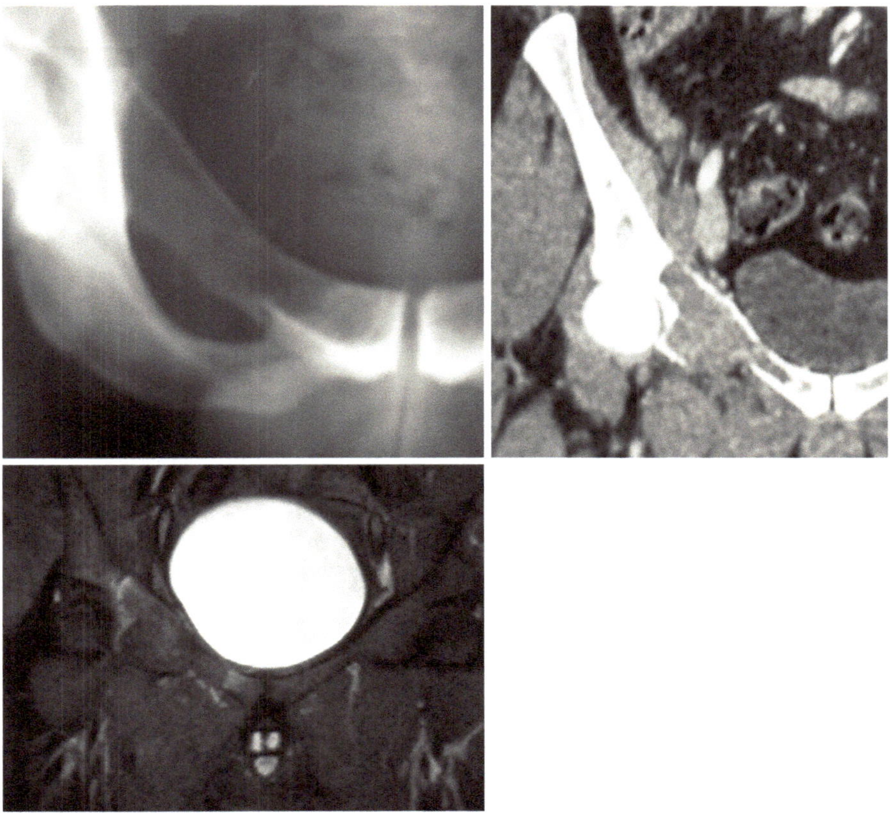

Lytic well-limited lesion of the iliopubic ramus. The cortex is partially missing. The lesion has a low signal on T2 image, indicating fibrous tissue

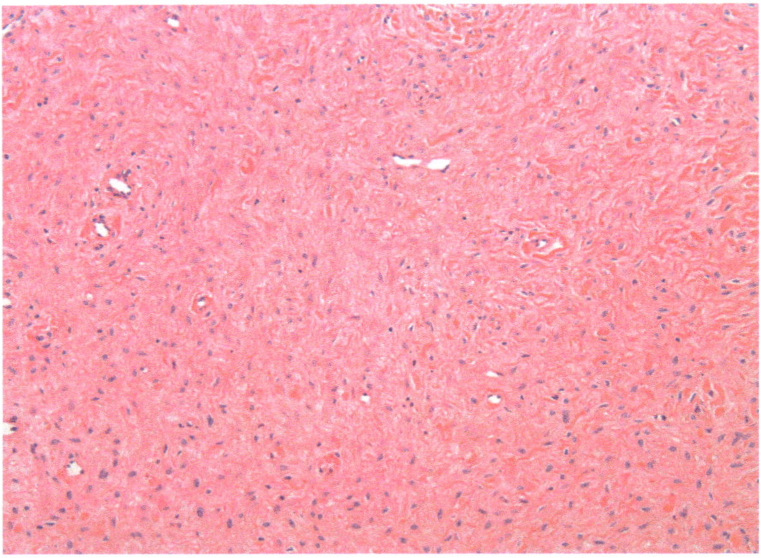

Hypocellular proliferation of spindle cells with slender nuclei associated with large amounts of fibrillar, hyalinized, or keloid-like collagen

## Selected Bibliography

Frick MA, Sundaram M, Unni KK, Inwards CY, Fabbri N, Trentani F, Bacchini P, Bertoni F (2005) Imaging findings in desmoplastic fibroma of bone: distinctive T2 characteristics. AJR Am J Roentgenol 184(6):1762–1767

Gould CF, Ly JQ, Lattin GE Jr, Beall DP, Sutcliffe JB III (2007) Bone tumor mimics: avoiding misdiagnosis. Curr Probl Diagn Radiol 36(3):124–141. Review

Hauben EI, Jundt G, Cleton-Jansen AM, Yavas A, Kroon HM, Van Marck E, Hogendoorn PC (2005) Desmoplastic fibroma of bone: an immunohistochemical study including beta-catenin expression and mutational analysis for beta-catenin. Hum Pathol 36(9):1025–1030

Taconis WK, Schütte HE, van der Heul RO (1994) Desmoplastic fibroma of bone: a report of 18 cases. Skeletal Radiol 23(4):283–288

Vaz G, Richard A, Guyen O, Bejui-Hugues J, Carret JP (2005) Desmoplastic fibroma or bone desmoid tumor: two cases. Rev Chir Orthop Reparatrice Appar Mot 91(8):782–787

# Chapter 25
# Chondrosarcomas (CHS)

**Nicola Fabbri and Davide Donati**

**Definition:** Malignant cartilaginous matrix-producing tumors.

N. Fabbri, MD (✉)
Department of Surgery, Orthopaedic Service, Memorial Sloan-Kettering Cancer Center,
1275 York Avenue, New York, NY 10065, USA
e-mail: fabbrin@mskcc.org

D. Donati, MD (✉)
3rd Orthopaedic and Traumatologic Clinic Prevalently Oncologic,
Istituto Ortopedico Rizzoli, Bologna, Italy
e-mail: davide.donati@ior.it

P. Picci et al. (eds.), *Atlas of Musculoskeletal Tumors and Tumorlike Lesions*,
DOI 10.1007/978-3-319-01748-8_25, © Springer International Publishing Switzerland 2014

CHS varieties and their prevalence on 100 cases

| Chondrosarcomas varieties | % |
|---|---|
| Central | 55–60 |
| Peripheral | 15–20 |
| Dedifferentiated central | 15 |
| Clear cells | 3 |
| Periosteal | 2–3 |
| Mesenchymal | 1–2 |
| Dedifferentiated peripheral | 1 |

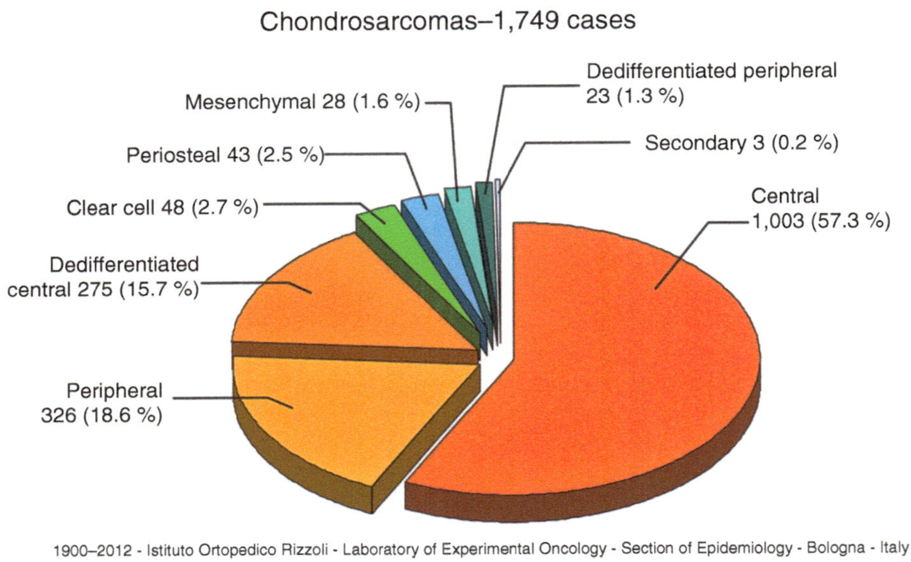

# Chapter 26
# Central Chondrosarcoma

**Nicola Fabbri and Davide Donati**

**Epidemiology:** Males. Adult age. Almost absent in children.

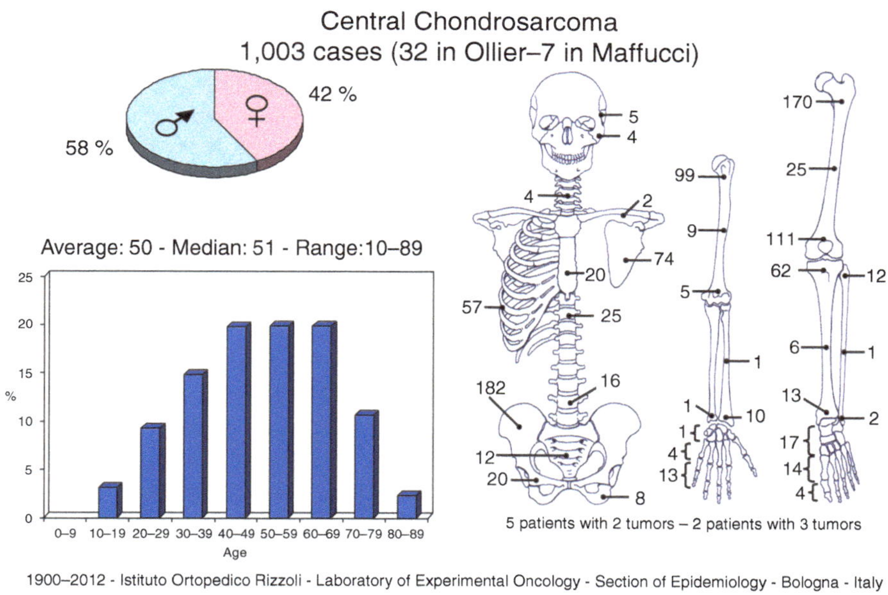


N. Fabbri, MD (✉)
Department of Surgery, Orthopaedic Service, Memorial Sloan-Kettering Cancer Center,
1275 York Avenue, New York, NY 10065, USA
e-mail: fabbrin@mskcc.org

D. Donati, MD (✉)
3rd Orthopaedic and Traumatologic Clinic Prevalently Oncologic,
Istituto Ortopedico Rizzoli, Bologna, Italy
e-mail: davide.donati@ior.it

P. Picci et al. (eds.), *Atlas of Musculoskeletal Tumors and Tumorlike Lesions*,
DOI 10.1007/978-3-319-01748-8_26, © Springer International Publishing Switzerland 2014

**Localization:** Proximal femur, pelvis, proximal humerus, scapula, and proximal tibia. In long bones, it starts metaphyseal extending as far as half or more of the entire bone.
**Clinical:** Deep, discontinuous, mild pain with small swelling. In the pelvis and at more advanced phases, large soft tissue mass is present. Pathologic fractures rare.
**Imaging:** On x-ray – intraosseous, osteolytic, geographic lesion with diffused, irregular, granules, nodules, and radiopaque rings of calcifications. Sometimes, a bubbling or "honeycomb" feature, metallic, or compact opacity. Cortex is scalloped, thinned, and destroyed. Often, it is thickened because the cortical bone reacts to the slow neoplastic infiltration with hyperostosis. Periosteal reaction is faint, with short perpendicular spiculae as a velvet or "beard" appearance. Margins may be sharp, with partial sclerotic rim or ill-defined, faded. It grows where there is less resistance (along the medullary canal). Soft tissue mass are not calcified. On bone scan – increased uptake more than radiographic extension. On angiography – avascular, but there may be a peripheral vascularization around the mass. On CT – typical calcified radiolucency, endosteal scalloping, buttressing, no enhancement, and often well contained. May show remnants of prior enchondroma. On MRI – gray homogeneous signal contrasts clearly with white marrow signal on T1, typical bright lesion on T2, and unexpected medullary extension on coronal projection.
**Histopathology:** Lobular, faceted mass of grayer, softer, juicer, more transparent than normal cartilage. Frequent gelatinous, grayish white, mucoid, hemorrhagic, and necrotic areas. Spots or calcified rings to the periphery of the lobules, hard and gritty, with a chalky yellowish aspect. Grade 1–20 %, well-differentiated in cartilage. Slightly larger nuclei, rounded nuclei of different size, mitotic figure absent, frequent cells with double nuclei, more frequent calcifications, and infiltrating surrounding bony trabeculae. Grade 2–60 %, hypercellular lesion with hyperchromatic nuclei, frequent binucleate cells, myxoid areas with spindle-stellate cells gathered in short cords or little groups spread in abundant, semi-liquid, ground substance, extensive infiltration of the medullary spaces, thin pseudocapsule with neoplastic satellites. Grade 3–20 %, high cellularity with pleomorphic, hyperchromatic, gigantic, bizarre, with three or more nuclei. Mitotic figures. Diffusely infiltrating the bone marrow spaces with destruction of bony trabeculae and the soft tissue. Solid intravenous plugs.
**Course and Staging:** Very slow growth. Typical progression in malignancy: transformation from a low to high grade or to another malignant tumor. Grade 1: very rare metastases, recurrence even after 10 years. Grade 2: frequent early or late metastases, recurrence within 5 years. Grade 3: higher rate of early metastases, recurrence often within 1 year, usually stage IIB.
**Treatment and Prognosis:** Wide or radical resections are curative. High risk of recurrence with inadequate margins and when incisional biopsy is not performed carefully because tumoral cells may be implanted in the soft tissue. Lung metastases must be excised. Radio- and chemotherapy are not used because they are poorly effective. Death is rare in grade 1 lesions, but it occurs in 30 % and 60 %, respectively, of grade 2 and 3 lesions.

**Key Points**

| | |
|---|---|
| Clinical | Pain, adults, no children |
| Radiological | Lytic lesion with granular calcifications. Expanded and invaded cortex |
| Histological | Lobules of cartilage infiltrating host trabeculae |
| Differential diagnosis | Chondroma, dedifferentiated chondrosarcoma |

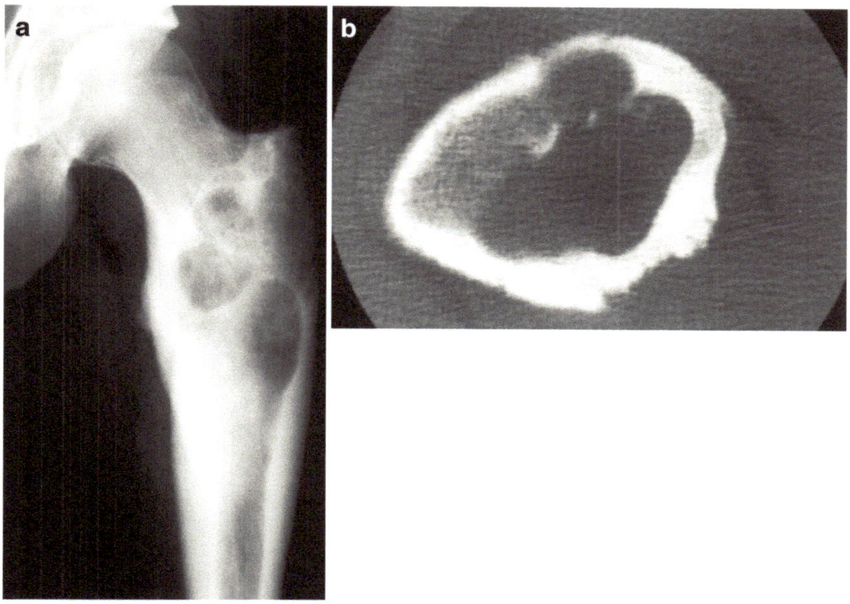


Radiograph and CT. Central well-limited lesion, containing rare cartilaginous calcifications, thinning the cortex, with an irregular periosteal bone formation

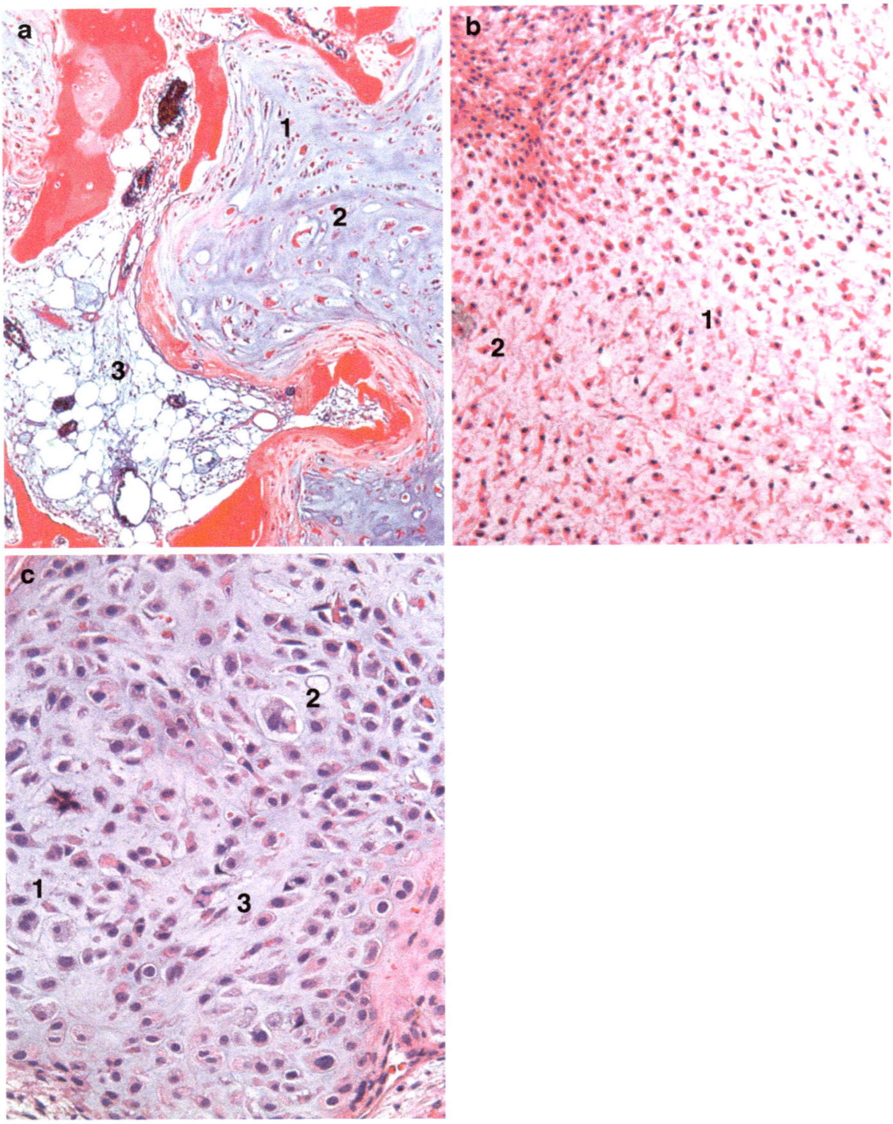


(a) Central chondrosarcoma – Grade 1 Nearly always well differentiated in cartilage. (*1*) Cells are more numerous in relation to most chondromas. (*2*) Slightly larger and pleomorphic nuclei, generally maintaining their rounded shape. (*3*) Infiltration of the bony trabeculae. (b) Central chondrosarcoma – Grade 2 This is the most frequent variety. Cartilage tissue shows aspects of frank atypia, with hyperchromatic nuclei. Binucleated cells are very frequently observed. In nearly half of the cases the tumor is partially or totally myxoid: (*1*) Cells have generally a spindle-stellate shape, dispersed, or in small groups or in short cords in single file. Cytoplasm is ossiphile and clearly visible. The nuclei are fairly plump and hyperchromatic. (*2*) An abundant, semi-liquid, and tenuously basophilic ground substance is present. (c) Central chondrosarcoma – Grade 3 Increased cellularity and severe atypia and pleomorphism are present. (*1*) Cartilaginous cells are very atypical and very numerous. (*2*) Severe pleomorphism and intense hyperchromia of the nuclei are present. (*3*) Mitotic figures can be found

## Selected Bibliography

Gelderblom H, Hogendoorn PC, Dijkstra SD, van Rijswijk CS, Krol AD, Taminiau AH, Bovée JV (2008) The clinical approach towards chondrosarcoma. Oncologist 13(3):320–329. Review. Erratum in Oncologist 2008 May;13(5):618

Gitelis S, Bertoni F, Picci P, Campanacci M (1981) Chondrosarcoma of bone. The experience at the Istituto Ortopedico Rizzoli. J Bone Joint Surg Am 63(8):1248–1257

Murphey MD, Walker EA, Wilson AJ, Kransdorf MJ, Temple HT, Gannon FH (2003) From the archives of the AFIP: imaging of primary chondrosarcoma: radiologic-pathologic correlation. Radiographics 23(5):1245–1278. Review

Riedel RF, Larrier N, Dodd L, Kirsch D, Martinez S, Brigman BE (2009) The clinical management of chondrosarcoma. Curr Treat Options Oncol 10(1–2):94–106. Review

Varma DG, Ayala AG, Carrasco CH, Guo SQ, Kumar R, Edeiken J (1992) Chondrosarcoma: MR imaging with pathologic correlation. Radiographics 12(4):687–704

# Chapter 27
# Peripheral (Secondary) Chondrosarcoma

**Nicola Fabbri and Davide Donati**

**Definition:** It originates from an osteochondroma on the bone surface.
**Epidemiology:** Males. Less frequent than central but at a younger age. Adults.
**Clinical:** Slow growing, hard, painful swelling, and adherent to bone. At times, asymptomatic and no history of previous osteochondroma. Invading vertebral canal radicular pain and paraplegia may be present.
**Location:** Pelvis (iliac wing), proximal femur (metaphyseal region), vertebral column (posterior arch), proximal humerus, and ribs.

N. Fabbri, MD (✉)
Department of Surgery, Orthopaedic Service, Memorial Sloan-Kettering Cancer Center,
1275 York Avenue, New York, NY 10065, USA
e-mail: fabbrin@mskcc.org

D. Donati, MD (✉)
3rd Orthopaedic and Traumatologic Clinic Prevalently Oncologic,
Istituto Ortopedico Rizzoli, Bologna, Italy
e-mail: davide.donati@ior.it

P. Picci et al. (eds.), *Atlas of Musculoskeletal Tumors and Tumorlike Lesions*,
DOI 10.1007/978-3-319-01748-8_27, © Springer International Publishing Switzerland 2014 

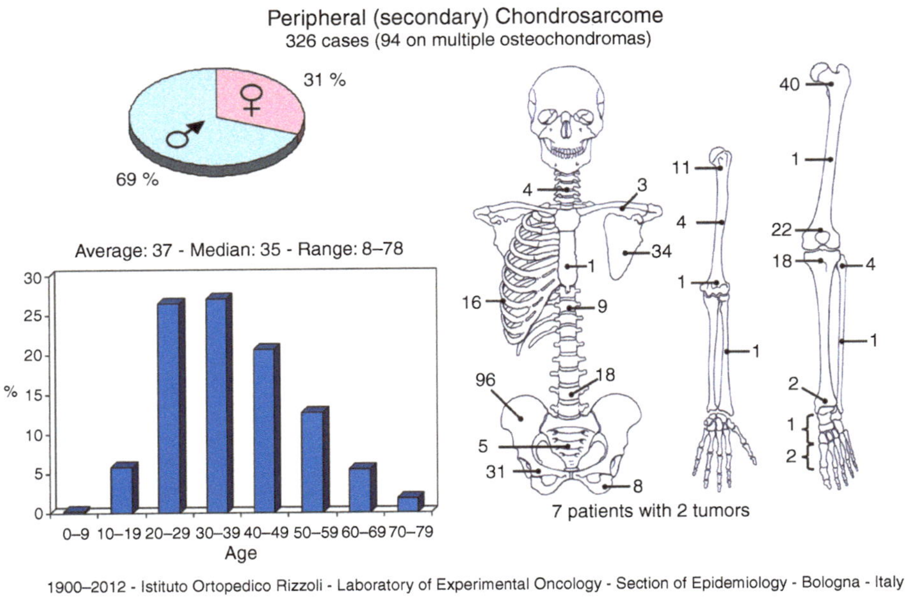


**Imaging:** On x-ray – typical aspect of an osteochondroma with more abundant calcifications or ossifications, intense, and diffused radiopacities, with thicker superficial uncalcified layer and with fuzzy margins towards the soft tissue. At times, the implant base of the osteochondroma is still visible or the tumor invades the medullary canal. On bone scan – intense uptake. On CT – many, large, noncalcified tumoral lobules, ringlike, popcorn-like radiopacities, and higher thickness of the cap of cartilaginous tissue. On MRI – lobulated, ill-defined, inhomogeneous muscular signal intensity on T1, extremely heterogeneous signal intensity on T2 with large foci of signal void due to the calcified areas, and with a thick peripheral layer of white signal due to the cap of the lesion.
**Histopathology:** Large, bumpy, cauliflower-like, with a thin pseudocapsule and thick cap of cartilage. Cartilage involves the spongy bone of osteochondroma; it is softer, juicy, grayish, and translucent. Calcifications are very intense in low grade C.: granules, rings, spots of yellowish-white aspect, chalky and gritty, with hard consistency. Lobules of well-differentiated cartilage: many large cells in groups with abundant matrix, with double hyperchromatic nuclei. Formation of nodules, mitosis, and cystic to myxoid changes may be seen. Nests of tumoral cells are observed in the pseudocapsule. Cartilage infiltrates the marrow spaces of cancellous bone of the osteochondroma.
**Course and Staging:** It grows more slowly than central C, recurrence is observed from a few months to 10 years, <20 % of cases have lung and late metastases. Dedifferentiated peripheral C is rare (4 %). Usually, stage IB.
**Treatment:** Wide resection. Inadequate surgery may cause scattered neoplastic nodules in scarring tissue. Radio- and chemotherapy are not effective. Amputation is necessary when it is too large and otherwise inoperable. Peripheral C is less malignant than central because grade 1 forms are frequent and grade 3 rare.

**Key Points**

| | |
|---|---|
| Clinical | Slow increasing swelling |
| Radiological | Osteochondroma with lytic areas |
| Histological | Thicker cartilaginous cup with infiltration of the marrow spaces of the osteochondroma |
| Differential diagnosis | Osteochondroma, dedifferentiated peripheral chondrosarcoma |

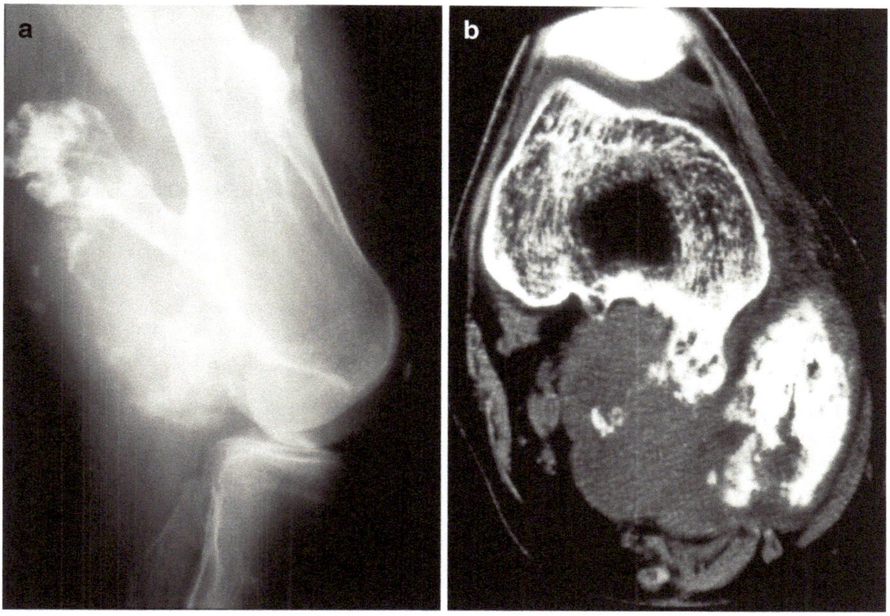


Radiograph and CT: the cortex of the osteochondroma is in continuity with the one of the normal bone. There is a large noncalcified malignant mass

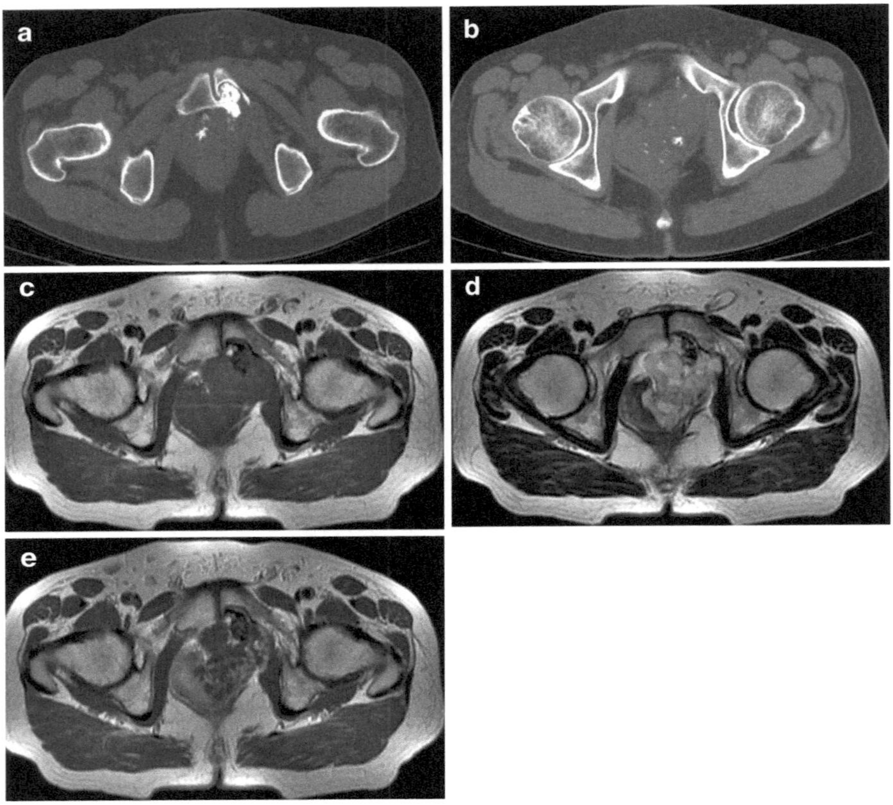


CT (**a, b**) and axial R images (**c**: T1, **d**: T2, and **e**: T1 after contrast medium injection). The cortex continuity is well visible on CT (**a**) and MR. The mass contains typical arciform cartilaginous calcifications (**b**), made of nodules, with a high signal on T2 MR, and no uptake after injection

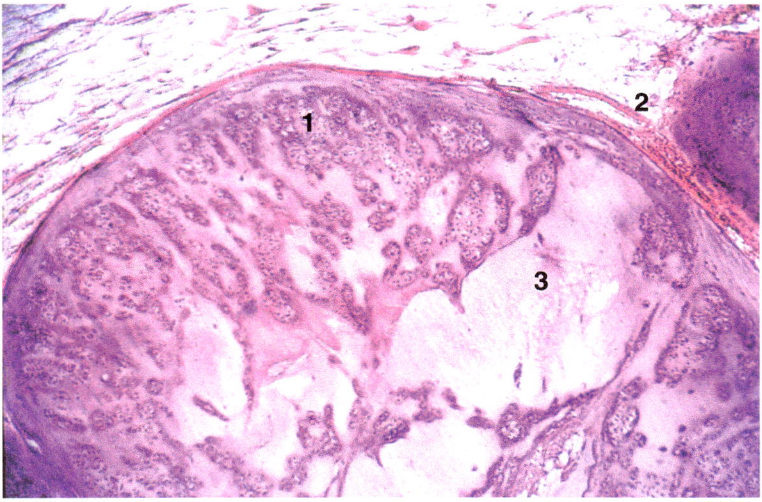


(*1*) Clustered chondrocytes at the periphery of the lobules (*2*) Lobulated architecture (*3*) Myxoid change and thickness of the cartilaginous cap exceeding 2 cm

## Selected Bibliography

Ahmed AR, Tan TS, Unni KK, Collins MS, Wenger DE, Sim FH (2003) Secondary chondrosarcoma in osteochondroma: report of 107 patients. Clin Orthop Relat Res 411:193–206

Bernard SA, Murphey MD, Flemming DJ, Kransdorf MJ (2010) Improved differentiation of benign osteochondromas from secondary chondrosarcomas with standardized measurement of cartilage cap at CT and MR imaging. Radiology 255(3):857–865

Gitelis S, Bertoni F, Picci P, Campanacci M (1981) Chondrosarcoma of bone. The experience at the Istituto Ortopedico Rizzoli. J Bone Joint Surg Am 63(8):1248–1257

Lin PP, Moussallem CD, Deavers MT (2010) Secondary chondrosarcoma. J Am Acad Orthop Surg 18(10):608–615. Review

Norman A, Sissons HA (1984) Radiographic hallmarks of peripheral chondrosarcoma. Radiology 151(3):589–596

# Chapter 28
# Dedifferentiated Chondrosarcoma

**Nicola Fabbri and Davide Donati**

**Definition:** Cartilaginous malignant tumor in which a high-grade non-cartilaginous sarcoma occurs.
**Epidemiology:** 15 % >50 years old. Males.

**Dedifferentiated Central Chondrosarcoma**
**275 cases (2 in Ollier - 2 in Maffucci)**

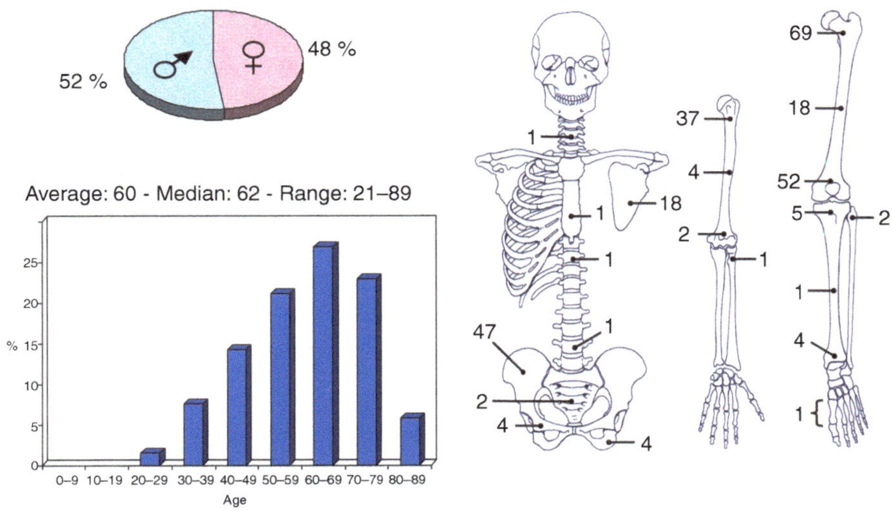


1900–2012 - Istituto Ortopedico Rizzoli - Laboratory of Experimental Oncology - Section of Epidemiology - Bologna - Italy

N. Fabbri, MD (✉)
Department of Surgery, Orthopaedic Service, Memorial Sloan-Kettering Cancer Center,
1275 York Avenue, New York, NY 10065, USA
e-mail: fabbrin@mskcc.org

D. Donati, MD (✉)
3rd Orthopaedic and Traumatologic Clinic Prevalently Oncologic,
Istituto Ortopedico Rizzoli, Bologna, Italy
e-mail: davide.donati@ior.it

P. Picci et al. (eds.), *Atlas of Musculoskeletal Tumors and Tumorlike Lesions*,
DOI 10.1007/978-3-319-01748-8_28, © Springer International Publishing Switzerland 2014

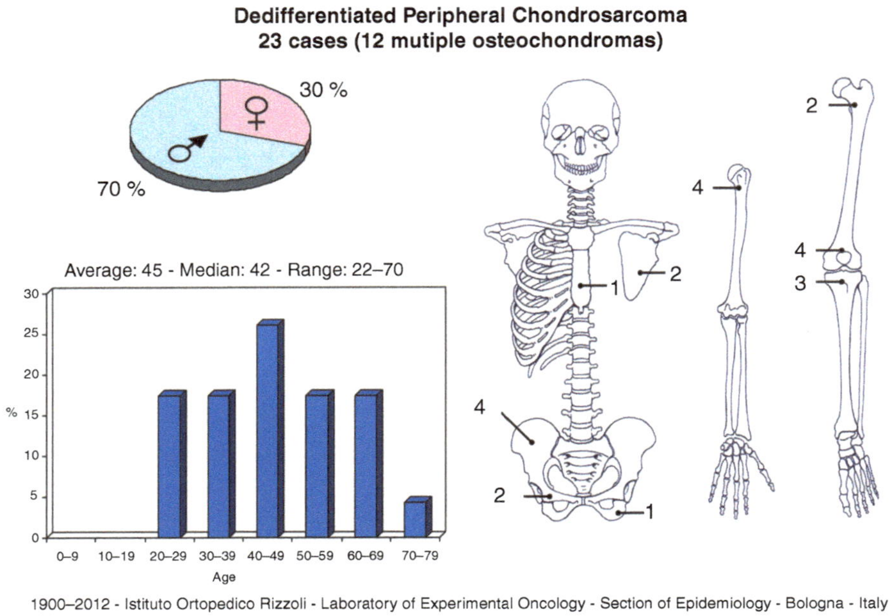


**Location:** Proximal femur, pelvis, and proximal humerus.

**Clinical:** On a history of long duration with moderate symptoms, a rapid progression of pain and swelling occurs. Pathologic fractures frequent.

**Imaging:** On x-ray: (1) old cartilaginous lesion: moderate expansion of the bone with thickened, scalloped cortex, with many calcifications; (2) new lesion: destroying the cortex, involving the soft tissues, and dissolving the calcifications, with ill-defined margins. Three different features are as follows: (a) pathologic fracture in a new aggressive lytic lesion canceling preexisting calcifications of an intensely radiopaque old cartilaginous lesion, (b) usual chondrosarcoma aspect with a small high-grade lytic lesion, (c) typical high-grade sarcoma feature with small remnants of chondrosarcoma. On CT, two lesions of different density, matrix, and enhancement. On MRI, two different signal intensities on T1 and T2.

**Histopathology:** Two types of tissues: (1) low-grade cartilaginous tumor and (2) high-grade malignancy like UPS, OS, FS. The transition is sharp.

**Course and Staging:** Fast growth, high risk of recurrence with inadequate surgery, and high rate of metastases, often observed at diagnosis. Usually, stage IIB or III.

**Treatment:** Wide or radical resection. Chemotherapy protocols of OS have been used with effectiveness. Often, an amputation is necessary to obtain adequate margins. Poor prognosis.

**Key Points**

| | |
|---|---|
| Clinical | Symptoms rapidly increasing. Pathologic fracture possible |
| Radiological | Very aggressive aspects on a low-grade cartilaginous lesion |
| Histological | Two different aspects: low-grade cartilage and high-grade sarcoma |
| Differential diagnosis | Central or peripheral high-grade chondrosarcoma |

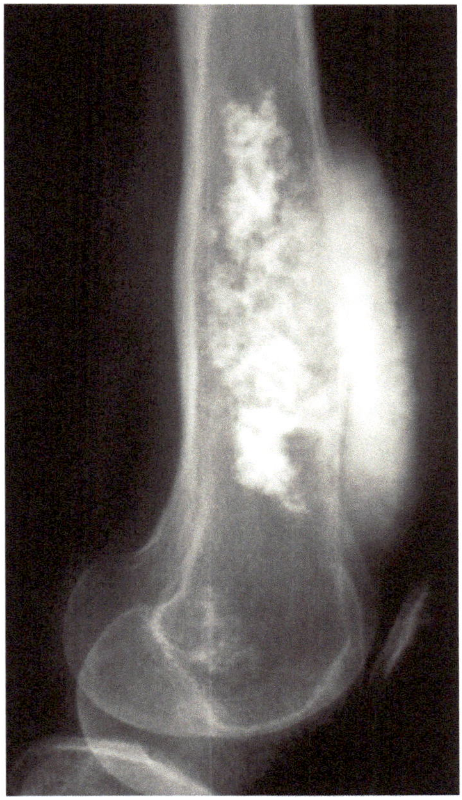

Radiograph: lateral view of the femur. The preexisting low-grade cartilaginous tumor contains arciform calcifications. The dedifferentiated part of the tumor forms bone, develops in the soft tissues, and corresponds to an osteosarcoma

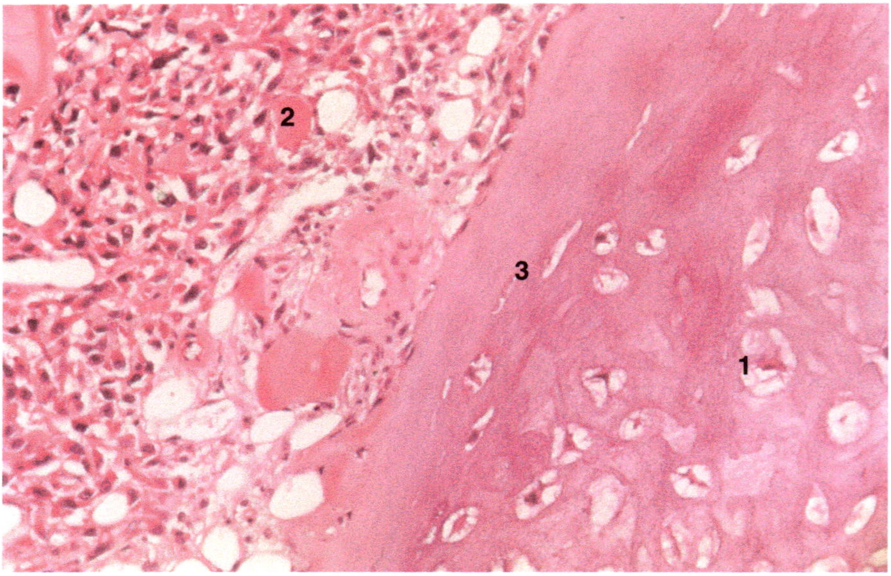


There are two clearly distinct tumor tissues. One is a well-differentiated cartilaginous tumor, the other has a different histiotype (usually an osteosarcoma, a UPS or a fibrosarcoma), characterized by high-grade malignancy. The transition from one type of tissue to the other is abrupt. (*1*) Chondrosarcomatous tissue. (*2*) High-grade malignant tumor, which in this case is an osteosarcoma. (*3*) Between the two zones the transition is abrupt

# Selected Bibliography

Capanna R, Bertoni F, Bettelli G, Picci P, Bacchini P, Present D, Giunti A, Campanacci M (1988) Dedifferentiated chondrosarcoma. J Bone Joint Surg Am 70(1):60–69

Littrell LA, Wenger DE, Wold LE, Bertoni F, Unni KK, White LM, Kandel R, Sundaram M (2004) Radiographic, CT, and MR imaging features of dedifferentiated chondrosarcomas: a retrospective review of 174 de novo cases. Radiographics 24(5):1397–1409. Review

Mercuri M, Picci P, Campanacci L, Rulli E (1995) Dedifferentiated chondrosarcoma. Skeletal Radiol 24(6):409–416

Staals EL, Bacchini P, Bertoni F (2006) Dedifferentiated central chondrosarcoma. Cancer 106(12):2682–2691

Staals EL, Bacchini P, Mercuri M, Bertoni F (2007) Dedifferentiated chondrosarcomas arising in preexisting osteochondromas. J Bone Joint Surg Am 89(5):987–993

# Chapter 29
# Periosteal Chondrosarcoma

**Nicola Fabbri and Davide Donati**

**Definition:** Chondrosarcoma originating on the bone surface.
**Epidemiology:** Rare, males, and adults.

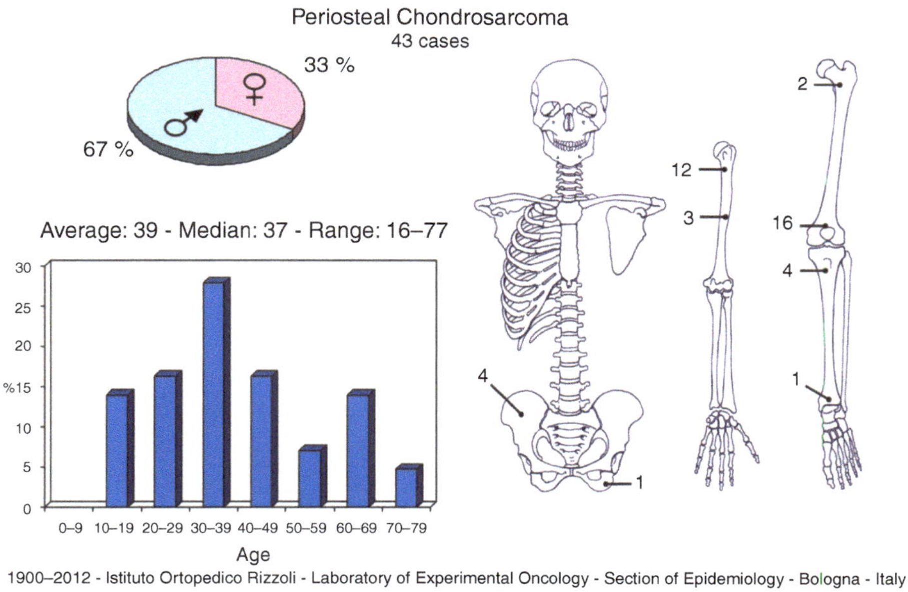


N. Fabbri, MD (✉)
Department of Surgery, Orthopaedic Service, Memorial Sloan-Kettering Cancer Center,
1275 York Avenue, New York, NY 10065, USA
e-mail: fabbrin@mskcc.org

D. Donati, MD (✉)
3rd Orthopaedic and Traumatologic Clinic Prevalently Oncologic,
Istituto Ortopedico Rizzoli, Bologna, Italy
e-mail: davide.donati@ior.it

P. Picci et al. (eds.), *Atlas of Musculoskeletal Tumors and Tumorlike Lesions*,
DOI 10.1007/978-3-319-01748-8_29, © Springer International Publishing Switzerland 2014

**Location:** Mainly in the limbs: distal femur, proximal tibia, and proximal humerus.
**Clinical:** Swelling with little or no pain.
**Imaging:** On x-ray – globose mass lying on the outer surface of the cortex, very radiolucent or with granules, rings, spots of cartilage, and rarely bunches of faded ossifications. Outer cortical erosion, saucerlike cortex, thickness of the adjacent cortex, buttress of periosteal reaction, internal sclerotic rim, and sharp margins. On CT – confirms periosteal site of the tumor without involvement of the medullary canal.
**Histopathology:** Like peripheral C, it is well differentiated. Rarely myxoid. Grade 1 is considered a periosteal chondroma, grade 2 is frequent, and grade 3 is rare.
**Course and Staging:** Rare recurrence if surgery is adequate; late, very rare metastasis. Usually stage IA.
**Treatment:** Wide resection. Prognosis is good.

### Key Points

| | |
|---|---|
| Clinical | Adults, swelling |
| Radiological | Subperiosteal, metaphyseal, with erosion of the cortex, granular calcifications, periosteal reaction |
| Histological | Lobules of low-grade cartilage |
| Differential diagnosis | Periosteal chondroma, periosteal osteosarcoma |

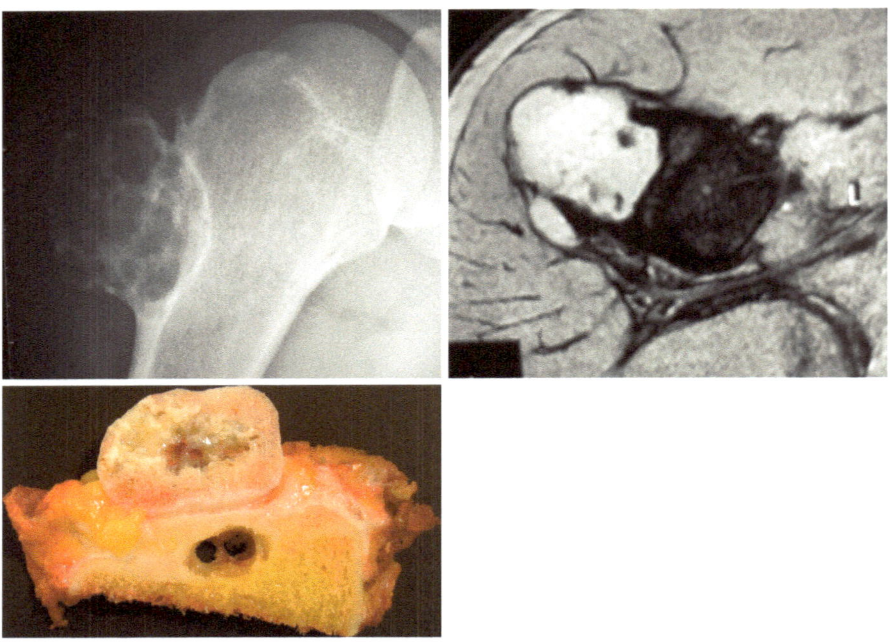

Radiograph, axial GE MR image, and specimen. The lesion contains calcifications and is centered on the cortex, which is eroded in a well-limited way

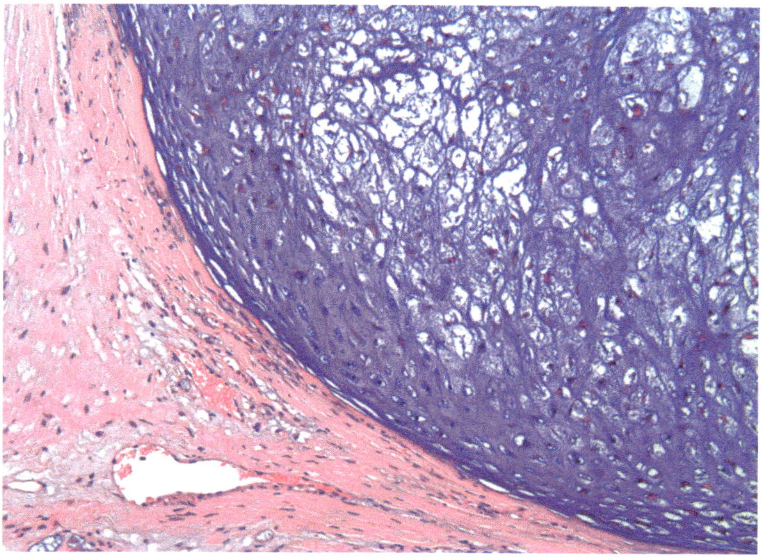

Large (usually more than 5 cm) cartilaginous tumor composed of lobules of cartilage, with possible myxoid change in the matrix. Permeation into the surrounding soft tissue is unequivocal evidence of malignancy

## Selected Bibliography

Chaabane S, Bouaziz MC, Drissi C, Abid L, Ladeb MF (2009) Periosteal chondrosarcoma. AJR Am J Roentgenol 192(1):W1–W6. Review

Mitchell A, Rudan JR, Fenton PV (1996) Juxtacortical dedifferentiated chondrosarcoma from a primary periosteal chondrosarcoma. Mod Pathol 9(3):279–283

Papagelopoulos PJ, Galanis EC, Mavrogenis AF, Savvidou OD, Bond JR, Unni KK, Sim FH (2006) Survivorship analysis in patients with periosteal chondrosarcoma. Clin Orthop Relat Res 448:199–207

Robinson P, White LM, Sundaram M, Kandel R, Wunder J, McDonald DJ, Janney C, Bell RS (2001) Periosteal chondroid tumors: radiologic evaluation with pathologic correlation. AJR Am J Roentgenol 177(5):1183–1188

Vanel D, De Paolis M, Monti C, Mercuri M, Picci P (2001) Radiological features of 24 periosteal chondrosarcomas. Skeletal Radiol 30(4):208–212

# Chapter 30
# Clear Cell Chondrosarcoma

**Nicola Fabbri and Davide Donati**

**Definition:** Chondrosarcoma with large amounts of clear cells.
**Epidemiology:** Rare, males, and adults.

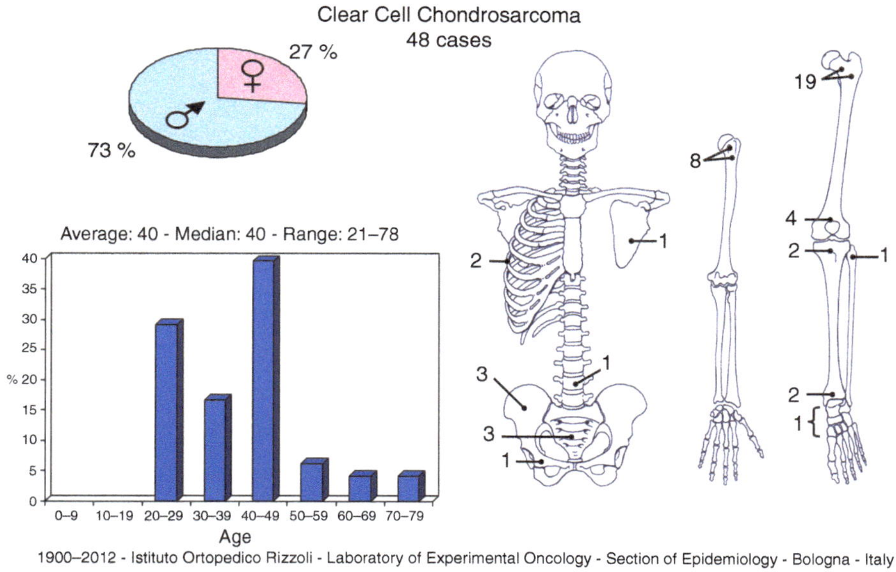


N. Fabbri, MD (✉)
Department of Surgery, Orthopaedic Service, Memorial Sloan-Kettering Cancer Center,
1275 York Avenue, New York, NY 10065, USA
e-mail: fabbrin@mskcc.org

D. Donati, MD (✉)
3rd Orthopaedic and Traumatologic Clinic Prevalently Oncologic,
Istituto Ortopedico Rizzoli, Bologna, Italy
e-mail: davide.donati@ior.it

P. Picci et al. (eds.), *Atlas of Musculoskeletal Tumors and Tumorlike Lesions*,
DOI 10.1007/978-3-319-01748-8_30, © Springer International Publishing Switzerland 2014

**Location:** Epiphysis or apophysis of long bones (proximal femur and humerus), flat, and short bones.
**Clinical:** Modest pain.
**Imaging:** On x-ray, like a chondroblastoma. Osteolytic lesion with small calcifications, sharp margins, irregular, and ill-defined sclerotic rim. On CT, spots or small granules typical of the cartilaginous tumors.
**Histopathology:** Lobular tissue with clear cells: central nucleus, extremely vacuolated cytoplasm, strongly PAS positive. Rare mitotic figures. Peripheral reactive giant cells, possible areas of well-differentiated chondrosarcoma, and intercellular calcification like chondroblastoma, cystic spaces, and reactive osteoid.
**Course and staging:** Slow growth, recurrence is possible when intralesional margins are obtained, metastases are exceptional. Usually, stage IA.
**Treatment:** Wide resection. Prognosis is good.

**Key Points**

| | |
|---|---|
| Clinical | Epiphysis and apophysis in adults |
| Radiological | Osteolytic, like a chondroblastoma |
| Histological | Lobules of clear cells, possible areas of hyaline cartilage, giant cells, calcifications, reactive osteoid |
| Differential diagnosis | Chondroblastoma |

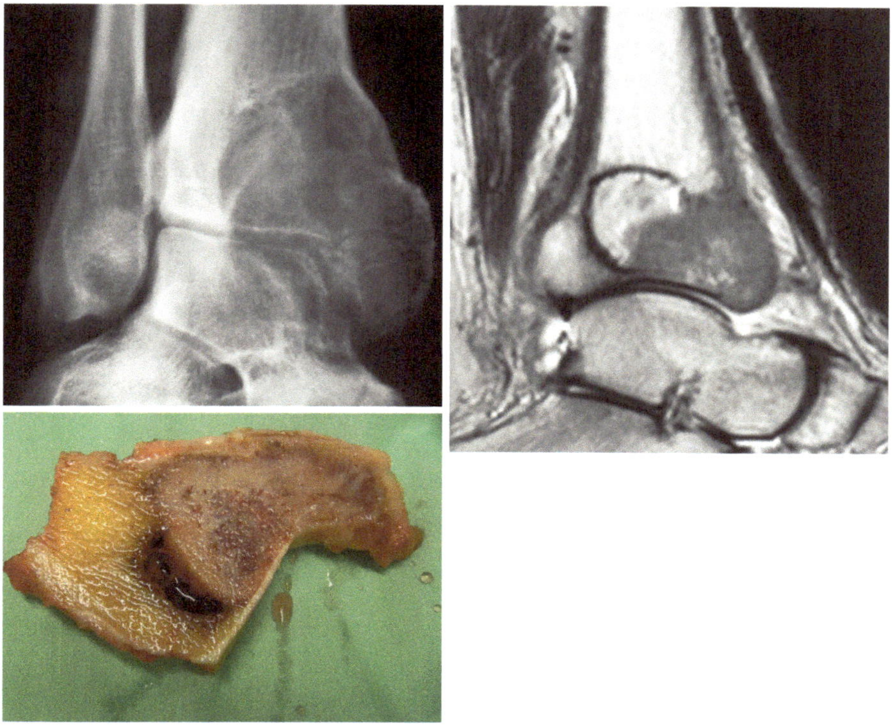

Radiograph, sagittal T1 MR image and specimen. Epiphyseal well-limited lesion. The high signal component well visible on MR corresponds to an associated hematoma on the specimen

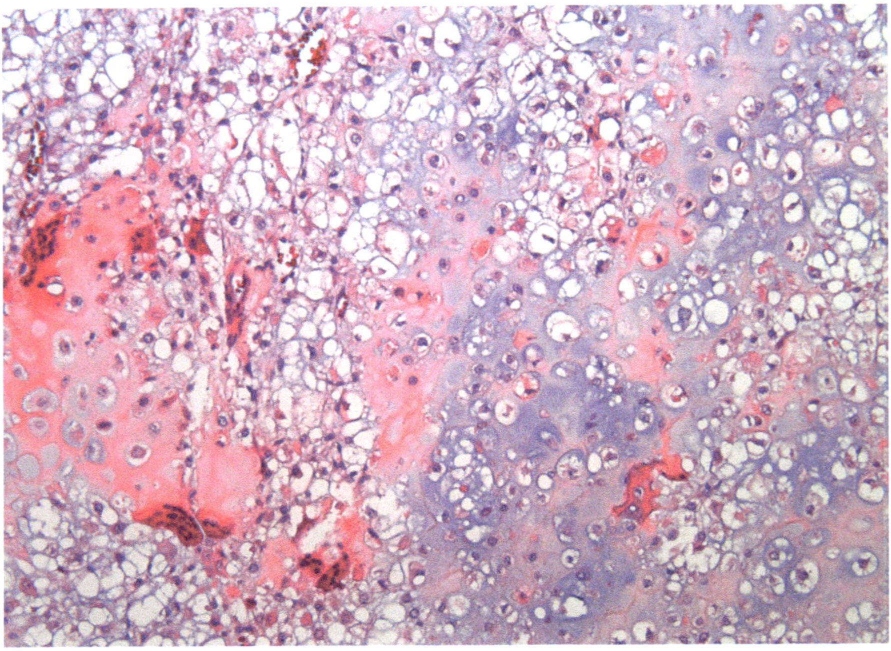

Proliferation of cells with abundant optically empty or eosinophylic cytoplasm and centrally located vesicular nuclei. Production of immature bone by tumoral cells is a peculiar feature. Scattered giant cells are present throughout

## Selected Bibliography

Aigner T, Dertinger S, Belke J, Kirchner T (1996) Chondrocytic cell differentiation in clear cell chondrosarcoma. Hum Pathol 27(12):1301–1305

Bosse A, Ueda Y, Wuisman P, Jones DB, Vollmer E, Roessner A (1991) Histogenesis of clear cell chondrosarcoma. An immunohistochemical study with osteonectin, a non-collagenous structure protein. J Cancer Res Clin Oncol 117(1):43–49

Corradi D, Bacchini P, Campanini N, Bertoni F (2006) Aggressive clear cell chondrosarcomas: do distinctive characteristics exist?: a report of 4 cases. Arch Pathol Lab Med 130(11): 1673–1679

Present D, Bacchini P, Pignatti G, Picci P, Bertoni F, Campanacci M (1991) Clear cell chondrosarcoma of bone. A report of 8 cases. Skeletal Radiol 20(3):187–191

Swanson PE (1997) Clear cell tumors of bone. Semin Diagn Pathol 14(4):281–291. Review

# Chapter 31
# Mesenchymal Chondrosarcoma

**Nicola Fabbri and Davide Donati**

**Epidemiology:** Very rare, no sex predilection, young adults, and elderly patients.

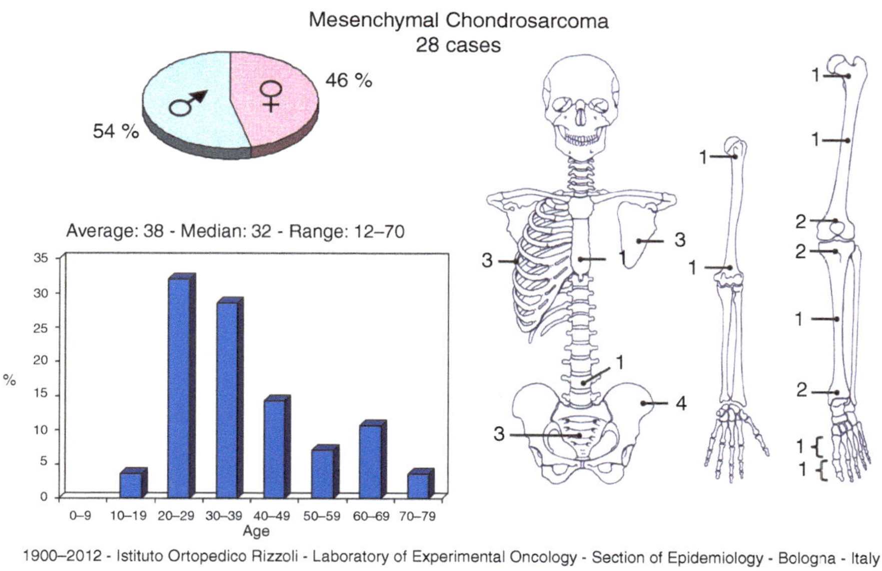


N. Fabbri, MD (✉)
Department of Surgery, Orthopaedic Service, Memorial Sloan-Kettering Cancer Center,
1275 York Avenue, New York, NY 10065, USA
e-mail: fabbrin@mskcc.org

D. Donati, MD (✉)
3rd Orthopaedic and Traumatologic Clinic Prevalently Oncologic,
Istituto Ortopedico Rizzoli, Bologna, Italy
e-mail: davide.donati@ior.it

P. Picci et al. (eds.), *Atlas of Musculoskeletal Tumors and Tumorlike Lesions*, 137
DOI 10.1007/978-3-319-01748-8_31, © Springer International Publishing Switzerland 2014

**Location:** Trunk and craniofacial bones. Rare in the limbs.
**Clinical:** Pain and swelling.
**Imaging:** On x-ray – osteolytic lesion with permeative destruction of bone, ill-defined margins, breaking the cortex, with soft tissue mass, and with faded typical calcifications.
**Histopathology:** Dense round cells proliferation with hemangiopericytoma-like architecture. Within the tissue there are small foci or larger islands of well-differentiated hyaline cartilage.
**Course and Staging:** Fast growth, high rate of recurrence with inadequate surgery, and frequent metastases. Usually, IIB.
**Treatment:** Wide or radical resection. Chemotherapy is used but its real effectiveness is not yet known. Prognosis is poor.

**Key Points**

| | |
|---|---|
| Clinical | Very rare in bone, adults, pain, and swelling |
| Radiological | Aggressive osteolysis, with possible calcification |
| Histological | Mixed population with isles of cartilagine and undifferentiated round cells |
| Differential diagnosis | None |

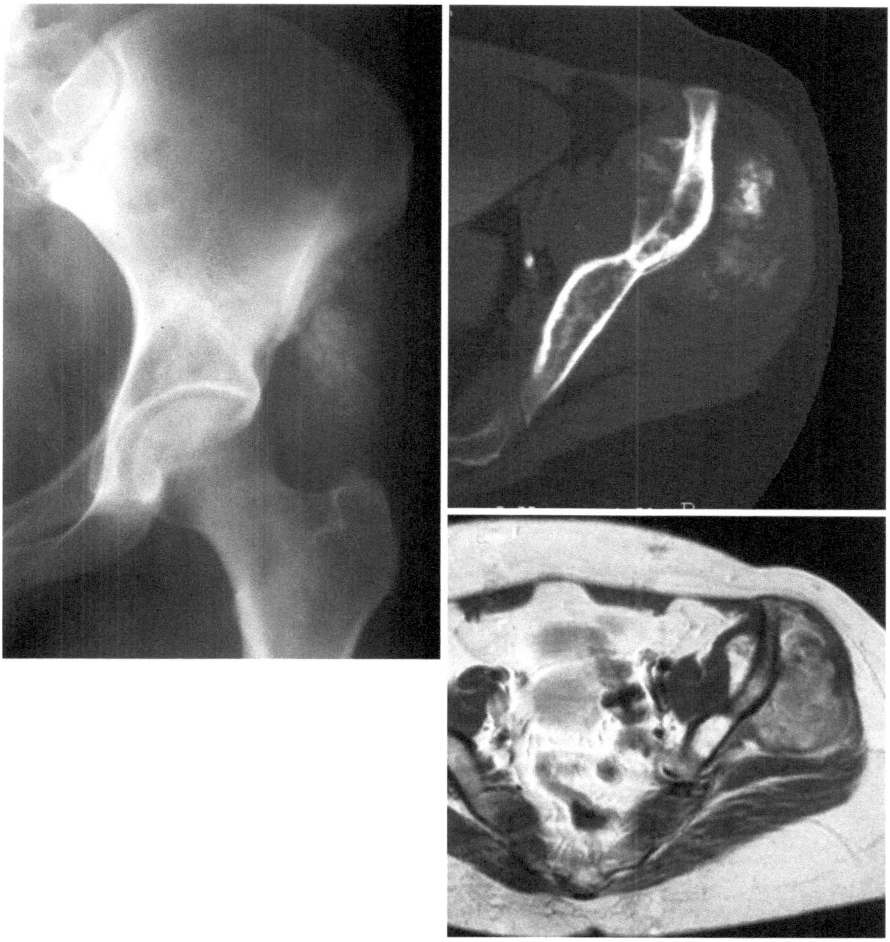

Radiograph, CT, and axial T1 injected MR image. The tumor contains cartilaginous calcifications and takes up contrast medium

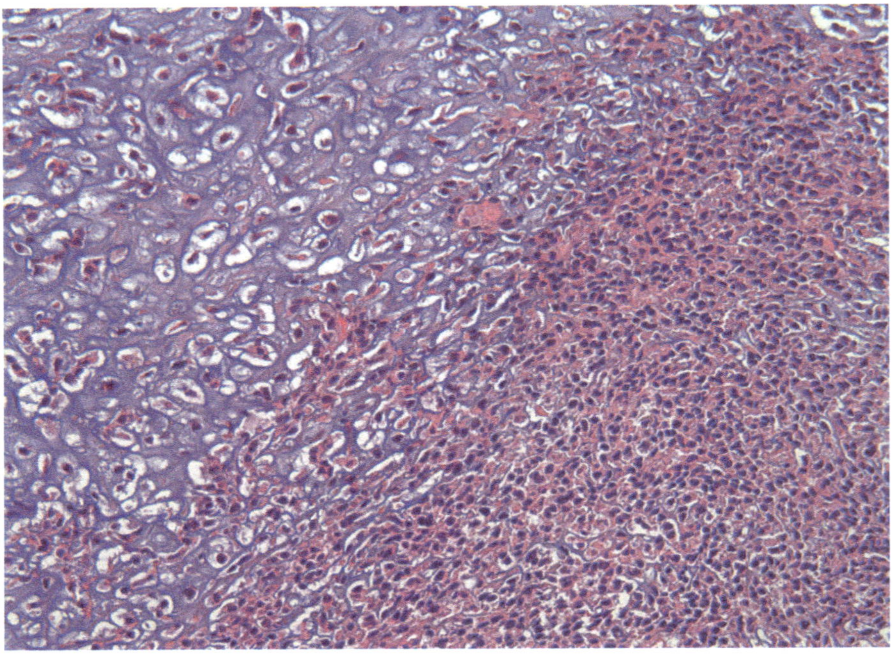

Islands of well-differentiated cartilage are surrounded by a proliferation of small round to ovoid undifferentiated cells

## Selected Bibliography

Bertoni F, Picci P, Bacchini P, Capanna R, Innao V, Bacci G, Campanacci M (1983) Mesenchymal chondrosarcoma of bone and soft tissues. Cancer 52(3):533–541

Cesari M, Bertoni F, Bacchini P, Mercuri M, Palmerini E, Ferrari S (2007) Mesenchymal chondrosarcoma. An analysis of patients treated at a single institution. Tumori 93(5):423–427

Fanburg-Smith JC, Auerbach A, Marwaha JS, Wang Z, Rushing EJ (2010) Reappraisal of mesenchymal chondrosarcoma: novel morphologic observations of the hyaline cartilage and endochondral ossification and beta-catenin, Sox9, and osteocalcin immunostaining of 22 cases. Hum Pathol 41(5):653–662

Nussbeck W, Neureiter D, Söder S, Inwards C, Aigner T (2004) Mesenchymal chondrosarcoma: an immunohistochemical study of 10 cases examining prognostic significance of proliferative activity and cellular differentiation. Pathology 36(3):230–233

Swanson PE, Lillemoe TJ, Manivel JC, Wick MR (1990) Mesenchymal chondrosarcoma. An immunohistochemical study. Arch Pathol Lab Med 114(9):943–948

# Chapter 32
# Biology of Central and Peripheral Chondrosarcoma

**Maria Serena Benassi**

Central and peripheral chondrosarcomas (CHS) present complex karyotypes and a genetic instability that results from early specific gene mutations followed by secondary genetic changes such as TP53 mutation and CDKN2 deletions.

Central CHS may develop from enchondroma where Indian hedgehog (IHH) and parathyroid hormone-like hormone (PTHLH) signaling pathways, involved in longitudinal growth of long bones and chondrocyte differentiation, are deregulated. Moreover, the frequent mutations in isocitrate dehydrogenase genes IDH1 and IDH2 found in enchondroma and central CHS suggest that these may be considered as early events in tumorigenesis, although further studies are needed to reveal the exact mechanism by which these mitochondrial deficiencies lead to tumor development (Szuhai et al. 2012).

Additional genetic or epigenetic alterations occur during progression toward central CHS including activation of signaling pathways IHH/PTHLH/Bcl-2 (Tiet et al. 2006); Src, Akt, and PDGFR (Schrage et al. 2009; Grignani et al. 2011); IGF; as well as hypoxic and glycolytic pathways (Bovée et al. 2005, 2010). An important role for overexpression of Bcl-2 family members in chemoresistance of CHS was recently shown (van Oosterwijk et al. 2012).

Cytogenetic studies highlighted the role of chromosome 9p21 and 12q13–15 aberrations in progression from low- to high-grade central CHS, resulting in a loss of CDKN2A tumor suppressor gene activity, amplification of CDK4, and pRB pathway deregulation. TP53 mutation and overexpression of the transcription factor Jun-B were also associated with malignant transformation, while cyclooxygenase (COX2), matrix metalloproteinase (MMP), and cathepsin endpoints may be considered as candidate targets for adjuvant treatment (Bovée et al. 2010).

Multiple and sporadic osteochondroma is characterized by mutations of the tumor suppressor genes EXT1 or EXT2 (Szuhai et al. 2011; Reijnders et al. 2010),

M.S. Benassi, PhD
Laboratory of Experimental Oncology, Istituto Ortopedico Rizzoli, Bologna, Italy
e-mail: mariaserena.benassi@ior.it

P. Picci et al. (eds.), *Atlas of Musculoskeletal Tumors and Tumorlike Lesions*, 
DOI 10.1007/978-3-319-01748-8_32, 

resulting in a loss of heparan sulfate (HS) synthesis and disturbance of hedgehog signaling pathways. EXT1 and EXT2 are located on the chromosome bands 8q24 and 11p11–12, respectively, and their germline mutations impair the HS biosynthesis, essential for the diffusion of hedgehog proteins. Reduced HS production can be reversed by the use of a DNA demethylating agent. In vitro and in vivo studies demonstrated that the reintroduction of EXT1 into cancer cell lines displaying methylation-dependent silencing of EXT1 induced reduction in colony formation density and tumor growth in nude mouse xenograft models (Ropero et al. 2004).

Progression from osteochondroma toward low-grade peripheral CHS requires reactivation of PTHLH signaling and antiapoptotic BCL2 expression. Further progression toward high-grade peripheral CHS involves additional genetic and structural changes including defective cell cycle checkpoints associated with mutation in p53; deregulation of cellular signaling pathways WINT, IHH, and TGFβ; loss of cellular organization with decrease in chondroid matrix; and increase in cellularity and vascularization (van Oosterwijk et al. 2012).

The lack of response to conventional chemotherapy or radiotherapy indicates that new treatments are needed to improve disease-free and overall survival in patients with chondrosarcoma, irrespective of the subtype (Schrage et al. 2010; van Oosterwijk et al. 2013).

Taking into account that molecular therapy effectiveness depends on biomarker activation state, potential targets for adjuvant treatment in CHS include monoclonal antibody to PTHLH, COX2, MMP, TGFβ inhibitors, BH-3 mimetics, small molecule kinase inhibitors, and Hh antagonist.

Extraskeletal myxoid chondrosarcoma, a soft tissue malignant tumor (see specific chapter), is characterized in 67 % of cases by recurrent translocation t(9;22)(q22;q12), resulting in CHNEWSR1 fusion product that may activate the PPARG nuclear receptor gene. Less frequently, translocations t(9;17)(q22.3;q12) and t(9;15)(q22.3;q21.3) are found, fusing RBP56 to CHN and TCF12 to CHN (Stacchiotti et al. 2012). The fusion transcripts interact with checkpoint cell cycle protein regulators such as p21, p27, and p16 inducing cell proliferation and survival.

## Bibliography

Bovée JV, Cleton-Jansen AM, Taminiau AH, Hogendoorn PC (2005) Emerging pathways in the development of chondrosarcoma of bone and implications for targeted treatment. Lancet Oncol 6:599–607

Bovée JV, Hogendoorn PC, Wunder JS, Alman BA (2010) Cartilage tumours and bone development: molecular pathology and possible therapeutic targets. Nat Rev Cancer 10:481–488

Grignani G, Palmerini E, Stacchiotti S, Boglione A, Ferraresi V, Frustaci S, Comandone A, Casali PG, Ferrari S, Aglietta M (2011) A phase 2 trial of imatinib mesylate in patients with recurrent nonresectable chondrosarcomas expressing platelet-derived growth factor receptor-alpha or -beta: an Italian Sarcoma Group study. Cancer 117:826–831

Reijnders CM, Waaijer CJ, Hamilton A, Buddingh EP, Dijkstra SP, Ham J, Bakker E, Szuhai K, Karperien M, Hogendoorn PC, Stringer SE, Bovée JV (2010) No haploinsufficiency but loss of heterozygosity for EXT in multiple osteochondromas. Am J Pathol 177:1946–1957

Ropero S, Setien F, Espada J, Fraga MF, Herranz M, Asp J, Benassi MS, Franchi A, Patino A, Ward LS, Bovee J, Cigudosa JC, Wim W, Esteller M (2004) Epigenetic loss of the familial tumor-suppressor gene exostosin-1(EXT-1) disrupts heparan sulfate synthesis in cancer cells. Hum Mol Genet 13:2753–2765

Schrage YM, Briaire-de Bruijn IH, de Miranda NFCC et al (2009) Kinome profiling of chondrosarcoma reveals Src-pathway activity and dasatinib as option for treatment. Cancer Res 69:6216–6222

Schrage YM, Machado I, Meijer D, Briaire-de Bruijn I, van den Akker BE, Taminiau AH, Kalinski T, Llombart-Bosch A, Bovée JV (2010) COX-2 expression in chondrosarcoma: a role for celecoxib treatment? Eur J Cancer 46:616–624

Stacchiotti S, Dagrada GP, Carlo Morosi C, Negri T, Romanini A, Silvana Pilotti S, Gronchi A, Casali PG (2012) Extraskeletal myxoid chondrosarcoma: tumor response to sunitinib. Clin Sarcoma Res 2:22

Szuhai K, Jennes I, de Jong D, Bovée JV, Wiweger M, Wuyts W, Hogendoorn PC (2011) Tiling resolution array-CGH shows that somatic mosaic deletion of the EXT gene is causative in EXT gene mutation negative multiple osteochondromas patients. Hum Mutat 32:E2036–E2049

Szuhai K, Cleton-Jansen AM, Hogendoorn PC, Bovée JV (2012) Molecular pathology and its diagnostic use in bone tumors. Cancer Genet 205:193–204

Tiet TD, Hopyan S, Nadesan P, Gokgoz N, Poon R, Lin AC, Yan T, Andrulis IL, Alman BA, Wunder JS (2006) Constitutive hedgehog signaling in chondrosarcoma up-regulates tumor cell proliferation. Am J Pathol 168:321–330

van Oosterwijk JG, Herpers B, Meijer D, Briaire-de Bruijn IH, Cleton-Jansen AM, Gelderblom H, van de Water B, Bovée JV (2012) Restoration of chemosensitivity for doxorubicin and cisplatin in chondrosarcoma in vitro: BCL-2 family members cause chemoresistance. Ann Oncol 23:1617–1626

van Oosterwijk JG, Meijer D, van Ruler MA, van den Akker BE, Oosting J, Krenács T, Picci P, Flanagan AM, Liegl-Atzwanger B, Leithner A, Athanasou N, Daugaard S, Hogendoorn PC, Bovée JV (2013) Screening for potential targets for therapy in mesenchymal, clear cell, and dedifferentiated chondrosarcoma reveals Bcl-2 family members and TGFβ as potential targets. Am J Pathol 182:1347–1356

# Chapter 33
# Osteosarcomas (OS)

**Piero Picci**

**Definition:** Malignant tumor composed of mesenchymal cells producing osteoid and immature bone. Almost constantly intramedullary, rarely it may originate at the bony surface. OS may sometimes present with skip or more distant skeletal mets, but it can be also skeletally multifocal at presentation. There are, consequently, several varieties of OS whose anatomo-clinical presentation, treatment and prognosis, however, are not as distinctive as to justify a separate classification. Other OS types, instead, are different in their clinical, pathologic, and therapeutic-prognostic features and are classified as separate entities (periosteal OS, parosteal OS, low-grade central OS).

P. Picci, MD
Laboratory of Experimental Oncology, Musculoskeletal Oncology,
Istituto Ortopedico Rizzoli, Bologna, Italy
e-mail: piero.picci@ior.it

P. Picci et al. (eds.), *Atlas of Musculoskeletal Tumors and Tumorlike Lesions*,
DOI 10.1007/978-3-319-01748-8_33, © Springer International Publishing Switzerland 2014

OS varieties and their prevalence on 100 cases

| High-grade OS varieties | 90 % |
|---|---|
| Classic osteosarcoma | 75 |
| Telangiectatic | 6/7 |
| Of jaw bones | 4 |
| Secondary | 4 |
| Small cell | <1 |
| Of bone surface | 1/2 |
| Multicentric | 0.5 |
| Intracortical | 0.05 |
| **Low-grade OS varieties** | **9/10 %** |
| Parosteal | 5 |
| Central | 2.5 |
| Periosteal | 1.5 |

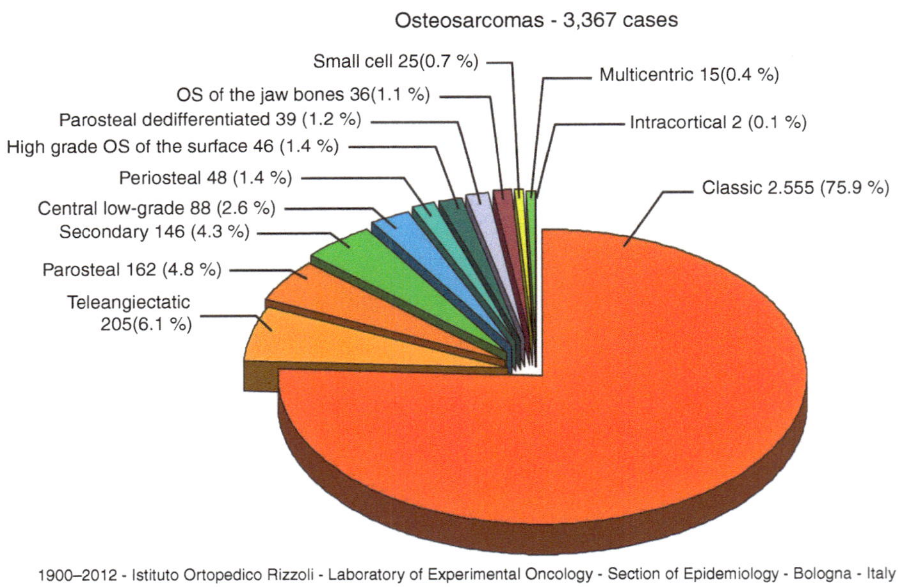

# Chapter 34
# Classic Osteosarcoma

**Piero Picci**

**Epidemiology:** It is the most frequent primary malignant tumor of bone, excluding myeloma, considered a systemic neoplasm. Its incidence is of 2–3 cases/million/year (only 0.2 % of all malignancies). Males are preferred (1.5:1). Most cases occur between 10 and 20 years of age. Older patients are usually secondary to radiation, Paget, chondrosarcoma, other primary bone tumors, or idiopathic.
**Localization:** Seventy percent of OS are localized around the knee or shoulder. Other locations are proximal or mid-femur, ilium, mid- and distal tibia, proximal fibula, and spine. Exceptional in the hand and foot. OS usually grows in the metaphysis or meta-diaphysis but tends to invade the epiphysis even in presence of a growth plate.
**Clinical:** Pain is usually the first symptom, often referred to trauma. In few weeks it increases and painful swelling appears. High temperature and limited joint motion are advanced signs. Pathologic fracture may occur in osteolytic forms. Alkaline phosphatase is frequently elevated. Less frequently, also LDH may be increased.

P. Picci, MD
Laboratory of Experimental Oncology, Musculoskeletal Oncology,
Istituto Ortopedico Rizzoli, Bologna, Italy
e-mail: piero.picci@ior.it

P. Picci et al. (eds.), *Atlas of Musculoskeletal Tumors and Tumorlike Lesions*,
DOI 10.1007/978-3-319-01748-8_34, © Springer International Publishing Switzerland 2014

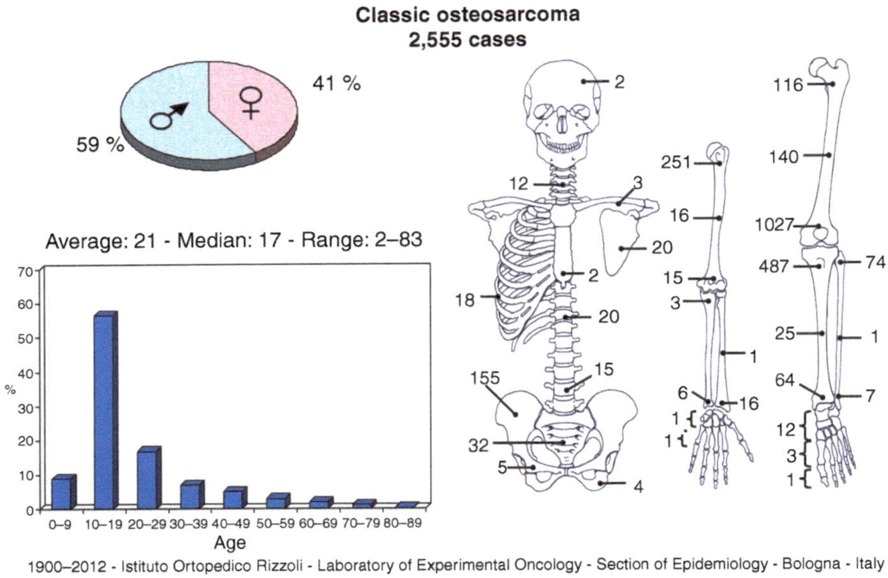


**Imaging:** The plain x-ray is usually diagnostic. Typically, OS starts intramedullary but breaches the cortex and expands in the soft tissues. It is usually a combination of radiolucency and radiodensity, sometimes it is entirely eburneous with edges always faded. The pure osteolytic form is typical of the telangiectatic variety. The tumor soft tissue extension shows irregular, cloud-like radiodensities and/or stripes of density perpendicular to the cortex (sunray image). Occasionally, it is purely radiolucent, and it can be appreciated only by CT and MRI. At the periphery of the area where the tumor breaches the cortex, a triangular buttress of immature bone (Codman's triangle) is seen. This is due to reactive bone acutely produced by the periosteum. Isotope bone scan is intensely hot even beyond the radiographic limits of the tumor. Rarely, it may reveal skips or distant bone mets. CT demonstrates intraosseous and extraosseous extension of the tumor and intratumoral radiodensities. With MRI, most OS exhibit the usual pattern of low T1 and high T2 signal. Sclerosing OS, however, have a low intensity on both T1 and T2 images. MRI is the best way to determine medullary tumor borders, epiphyseal invasion, and skip mets. Contrasted CT and MRI show the relationship with vessels, but in some cases, an angiogram may be more reliable. Patient workup always includes CT of the lungs where mets may appear as radiodense round nodules.

**Histopathology:** OS is fleshy soft in cellular areas with little matrix, firm and rubbery in fibroblastic areas with collagen production, gritty to stone-hard in osteogenic areas, and cartilaginous to myxoid in chondroblastic areas. Hemorrhage, necrosis, and cystic alterations are common. The most relevant diagnostic feature is constant permeation of marrow spaces, trapping host trabecular bone along its margins, predominantly. Cortex is also permeated and usually breached. An endosteal and periosteal production of reactive bone is associated. OS may invade the joint capsule and ligaments. Rarely, OS plugs are found in the adjacent veins. Skip mets, usually in the same but also in the adjacent cross-joint bone, may be detected in a small percentage of cases. Microscopically OS has a wide range of histological presentations but the characterizing feature is represented by high-grade sarcomatous cells producing osteoid and woven bone. The less osteogenic areas of the tumor (usually the periphery) are highly cellular and show more clear-cut features of high-grade malignancy as compared to more osteoid or bone-rich central areas of the tumor. Cells are large, with striking pleomorphism, hyperchromia, prominent nucleoli, and frequent atypical mitoses, although some 10 % of cases may show little anaplasia and lead to confusion with benign entities such as osteoblastoma, chondroblastoma, and giant cell tumors. Tumor osseous matrix varies from slender lacelike seams of osteoid to islands or dense sheets of woven bone. No regular trabeculae rimmed by osteoblasts are produced by OS cells. Where OS is intensely sclerotic, cells are scarce and small, with no mitoses; tissue is scarcely vascular and may be necrotic. In these areas, features of malignancy may be absent and diagnosis of OS is suggested by the permeative pattern of the tumor. Occasionally, OS is extensively chondroblastic (as a high-grade chondrosarcoma) or fibroblastic (similar to a fibrosarcoma). The osteoid-osseous production, which identifies the OS, may be found only in the microscopic study of the entire specimen. Reactive giant cells are seen, particularly in areas of hemorrhage. Particularly at the periphery of the tumor, reactive osteogenesis associates and should not be confused with tumorous osteogenesis. Histochemical stains show a high content of alkaline phosphatase in OS cells.

**Course and Staging:** OS has a rapid course. At presentation 80 % of OS are stage II-B; only 5 % are stage II-A. About 15 % of OS are stage III. Because, without the use of chemotherapy, 80–90 % of OS patients die of metastases, notwithstanding ablation of the primary tumor, it can be said that in 80–90 % of cases occult micromets are present at the start. Metastatic spread occurs primarily to the lungs.

**Treatment:** Treatment of OS is based on chemotherapy (ctx) and surgery of both the primary tumor and the metastases. Postop ctx (adjuvant) started in 1971, pre- and postop ctx (neoadjuvant) was introduced in 1978. Presently, the most effective drugs are adriamycin (ADM), high-dose methotrexate (HDMTX), cisdiaminoplatinum (CDP), and ifosfamide (IFO). Preop ctx is started immediately after diagnosis. After a few cycles (approximately 2 months), the tumor is restaged, evaluating response: clinical (regression of pain, reduction and hardening of the tumor mass), laboratory (<serum alkaline phosphatase), and imaging (arrest of growth, ossification and capsulation of the tumor, regression of vascularity and edema, decreased isotope uptake). A precise method for evaluating response to ctx preoperatively does not exist. Dynamic isotope scan and contrasted MRI, however, are comparatively good indexes of response. Based on postctx staging study, surgery is performed. The entire tumor is sampled histologically and examined to quantify tumor necrosis. Good response is indicated by necrosis from 90 to 100 %. Postop ctx is continued, starting 1–2 weeks after surgery and lasting from 4 to 6 months. Usually, the same drugs as used preop are given in good responders. In poor responders, different drugs are administered or added in more prolonged trials.

**Prognosis:** Without chemotherapy, the 10-year survival rate was around 10–15 %. With current ctx, the same figure is about 70 %, for OS nonmetastatic at presentation and involving the appendicular skeleton. Local recurrence was 2–3 % after amputation. Presently, it is around 5 % after conservative surgery. Metastases (mainly to the lungs) occur usually in the first 2–3 years. There are, however, rare cases of mets occurring even 5 to 10 years after treatment. The second most frequent site for mets is the skeleton.

**Key Points**

| | |
|---|---|
| Clinical | Young age, pain, and swelling |
| Radiological | Central lesion, aggressive with radio opacities |
| Histological | High grade with osteoid/bone production |
| Differential diagnosis | None |

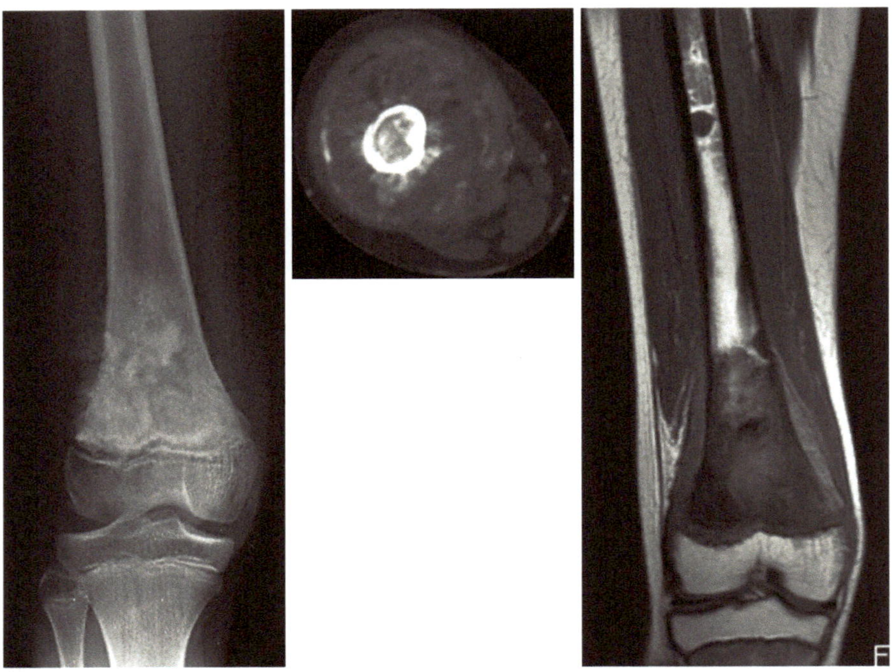

Radiograph, CT, and coronal T1 MR image. The tumor is metaphyseal and heterogeneous, forms bone, destroys part of the cortex, and invades soft tissues. A skip metastasis is detected on MR in the medullary cavity

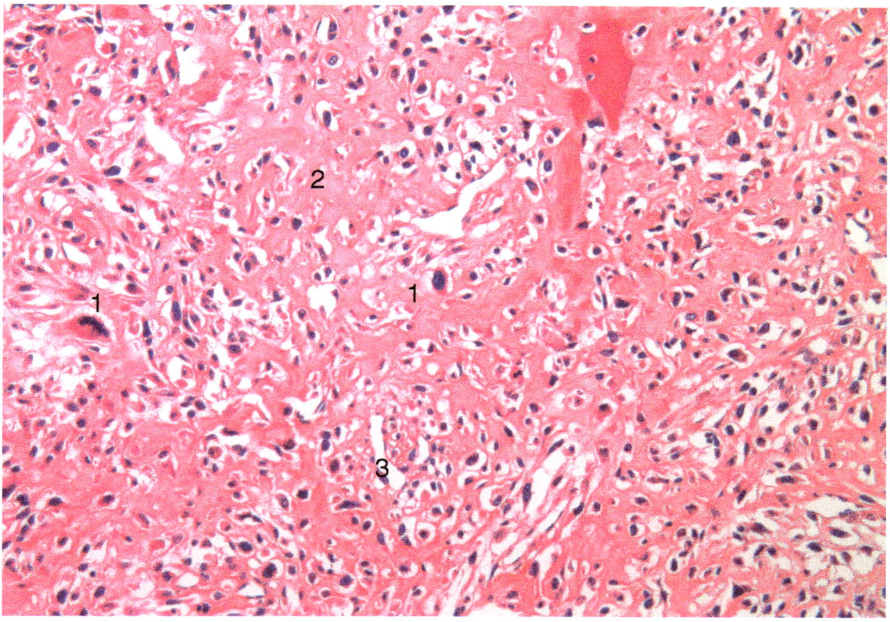


Sarcomatous tissue with cells producing osteoid and bone. (*1*) Sarcomatous tissue. The aspects of the high-grade malignancy are fairly evident: large, pleomorphic, and hyperchromic cells are seen. (*2*) Neoplastic osteoid and osseous material, shaped with an absolutely anarchical architecture. It is nearly impossible to find trabeculae bordered by a regular row of osteoblasts. (*3*) Abundant blood vessels. They do not have their own well-formed and continuous wall. In some areas, they are directly walled by sarcomatous cells

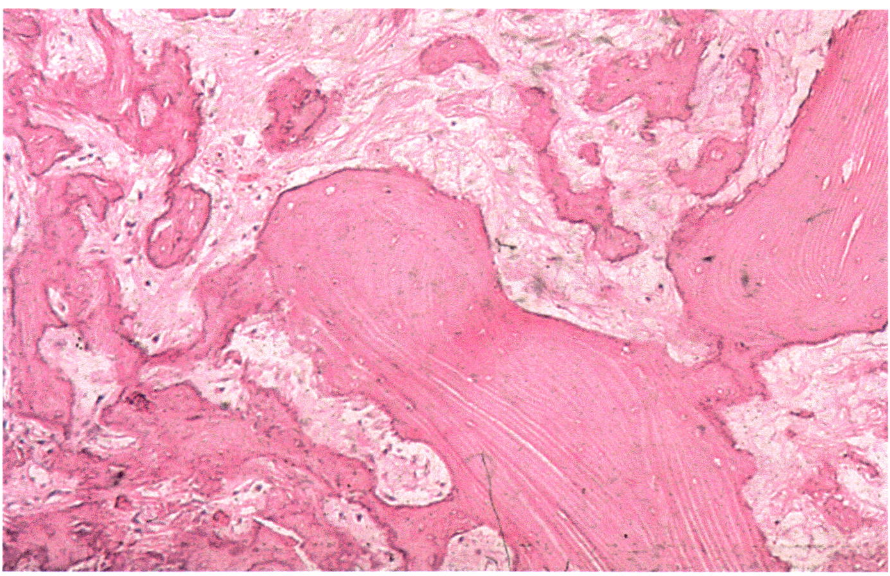

Histology patterns of a good response to preoperative chemotherapy

## Selected Bibliography

Klein MJ, Siegal GP (2006) Osteosarcoma: anatomic and histologic variants. Am J Clin Pathol 125(4):555–581. Review

Picci P (2007) Osteosarcoma (osteogenic sarcoma). Orphanet J Rare Dis 2:6. Review

Picci P, Mercuri M, Ferrari S, Alberghini M, Briccoli A, Ferrari C, Pignotti E, Bacci G (2010) Survival in high-grade osteosarcoma: improvement over 21 years at a single institution. Ann Oncol 21(6):1366–1373

Raymond AK, Jaffe N (2009a) Osteosarcoma multidisciplinary approach to the management from the pathologist's perspective. Cancer Treat Res 152:63–84. Review

Raymond AK, Jaffe N (2009b) Conditions that mimic osteosarcoma. Cancer Treat Res 152: 85–121. Review

# Chapter 35
# Telangiectatic Osteosarcoma

**Piero Picci**

**Definition:** This variety (about 6 % of all OS) is a completely osteolytic sarcoma, with a spongelike structure filled of blood, a scarce, immature osteogenesis.
**Epidemiology:** Sex, age, and localization are the same as in classic OS.

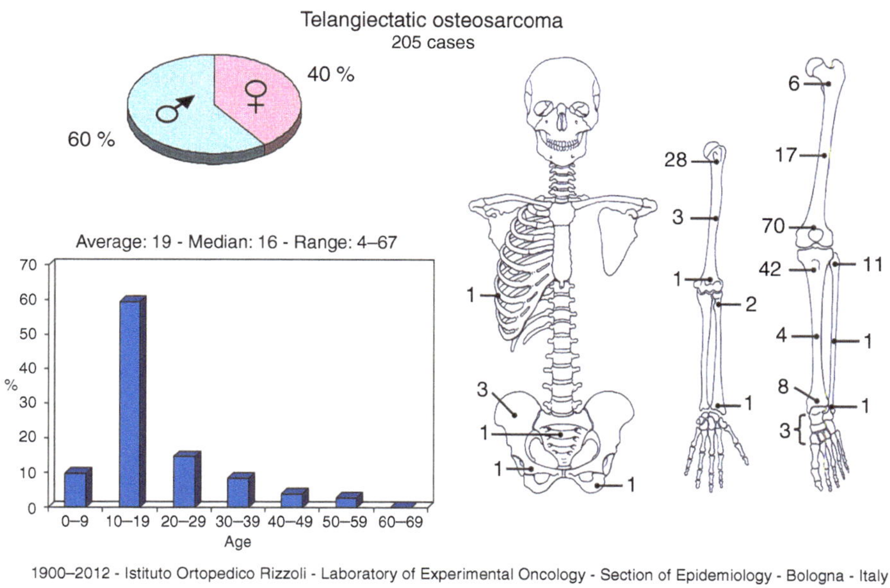


P. Picci, MD
Laboratory of Experimental Oncology, Musculoskeletal Oncology,
Istituto Ortopedico Rizzoli, Bologna, Italy
e-mail: piero.picci@ior.it

P. Picci et al. (eds.), *Atlas of Musculoskeletal Tumors and Tumorlike Lesions*,
DOI 10.1007/978-3-319-01748-8_35, 

**Clinical:** Clinically, the tumor has an aggressive course, with an expanding soft tissue mass, soft and tender on palpation, with increased local temperature. Pathologic fracture is not uncommon.

**Imaging:** Imaging shows a purely osteolytic lesion with ill-defined limits. Isotope scan shows an increased uptake. CT and MRI show the multicystic pattern (high signal in both T1 and T2 because of fluid and hemosiderin content) with fluid levels.

**Histopathology:** Grossly, the tumor is constituted by large cavities filled with blood and clots. Neoplastic tissue constitutes the septa; it is very soft and can be very scarce, overwhelmed by hemorrhage. The permeative pattern is particularly evident, and the cortex and periosteum are often extensively destroyed. Histologically under low power, the pattern is similar or identical to an ABC. Only at higher power, the sarcomatous nature of cells becomes apparent, because all telangiectatic OS are high-grade malignant. Sometimes, however, the hemorrhage and necrosis are so massive that it is difficult to find viable cells and to appreciate anaplasia. Osteogenesis is usually focal, and often it must be searched for in numerous sections and rarely may not be found at all, yet with the overall clinical, gross, and histological patterns are entirely consistent with telangiectatic OS, minus finding of osteoid or bone production by the sarcoma cells within the aneurysmal bone-cyst-like walls. Atypical mitoses are easily identified.

**Course:** The course is rapid, and the stage is almost regularly II-B or III.

**Treatment and Prognosis:** Treatment and prognosis are the same as for classic OS. Some data suggest that response to preoperative CTx is particularly good, which can be explained by the rich vascularity of the tumor, enhancing the ability of the chemotherapeutic agents to reach the malignant cells.

**Key Points**

| | |
|---|---|
| Clinical | Young age, pain and swelling, pathologic fracture possible |
| Radiological | Very aggressive and completely lytic |
| Histological | High-grade with lacunae |
| Differential diagnosis | Aneurysmal bone cyst |

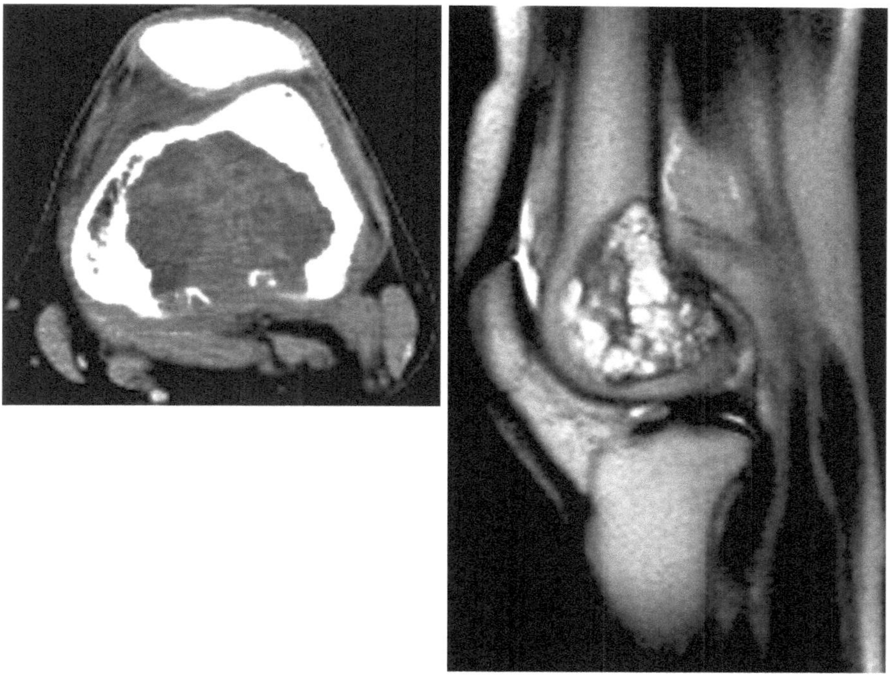

CT and sagittal T2 MR image. Aggressive metaphyseal tumor, destroying the cortex, invading the soft tissues, heterogeneous. Fluid-fluid levels in multiple small cavities are easily detected on MR

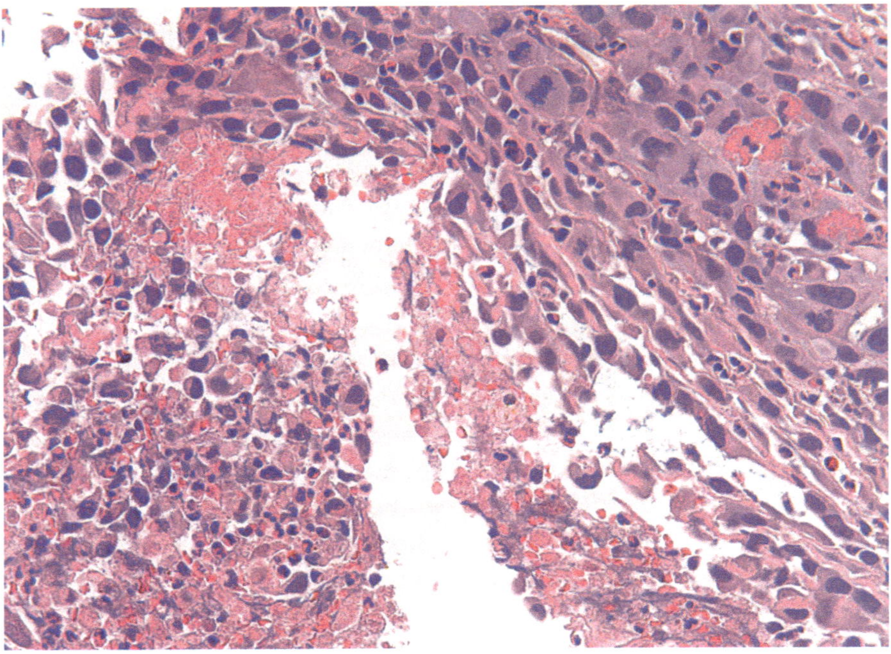

Blood-filled cavities rimmed by malignant mesenchymal cells only focally producing osteoid

## Selected Bibliography

Bacci G, Picci P, Ferrari S, Sangiorgi L, Zanone A, Brach del Prever A (1994) Primary chemotherapy and delayed surgery for non-metastatic telangiectatic osteosarcoma of the extremities. Results in 28 patients. Eur J Cancer 30A(5):620–626

Mervak TR, Unni KK, Pritchard DJ, McLeod RA (1991) Telangiectatic osteosarcoma. Clin Orthop Relat Res 270:135–139

Murphey MD, Robbin MR, McRae GA, Flemming DJ, Temple HT, Kransdorf MJ (1997) The many faces of osteosarcoma. Radiographics 17(5):1205–1231

Murphey MD, wan Jaovisidha S, Temple HT, Gannon FH, Jelinek JS, Malawer MM (2003) Telangiectatic osteosarcoma: radiologic-pathologic comparison. Radiology 229(2):545–53

Weiss A, Khoury JD, Hoffer FA, Wu J, Billups CA, Heck RK, Quintana J, Poe D, Rao BN, Daw NC (2007) Telangiectatic osteosarcoma: the St. Jude Children's Research Hospital's experience. Cancer 109(8):1627–1637

# Chapter 36
# Secondary Osteosarcoma

**Piero Picci**

**Definition:** These OS occur in cartilage tumors (dedifferentiated chondrosarcoma), radiated bone, Paget, fibrous dysplasia, bone infarct, and chronic osteomyelitis.

Most of these cases are observed in advanced age. Treatment is the same as for usual OS, except for the fact that patient age may contraindicate the use of the same CTx used in youngsters. Prognosis of secondary OS is generally worse as compared to primary OS, since the response to chemotherapy and advanced age of the patients are adverse parameters.

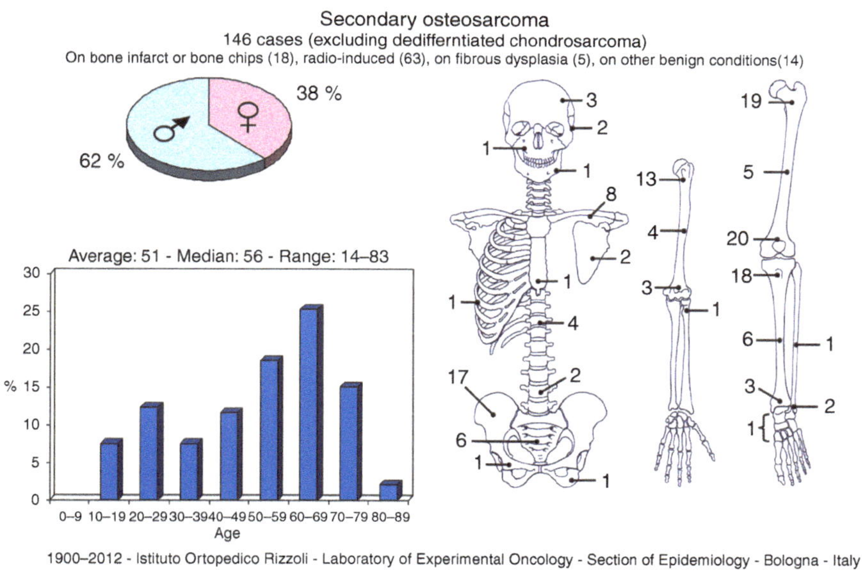


P. Picci, MD
Laboratory of Experimental Oncology, Musculoskeletal Oncology,
Istituto Ortopedico Rizzoli, Bologna, Italy
e-mail: piero.picci@ior.it

P. Picci et al. (eds.), *Atlas of Musculoskeletal Tumors and Tumorlike Lesions*,
DOI 10.1007/978-3-319-01748-8_36, © Springer International Publishing Switzerland 2014

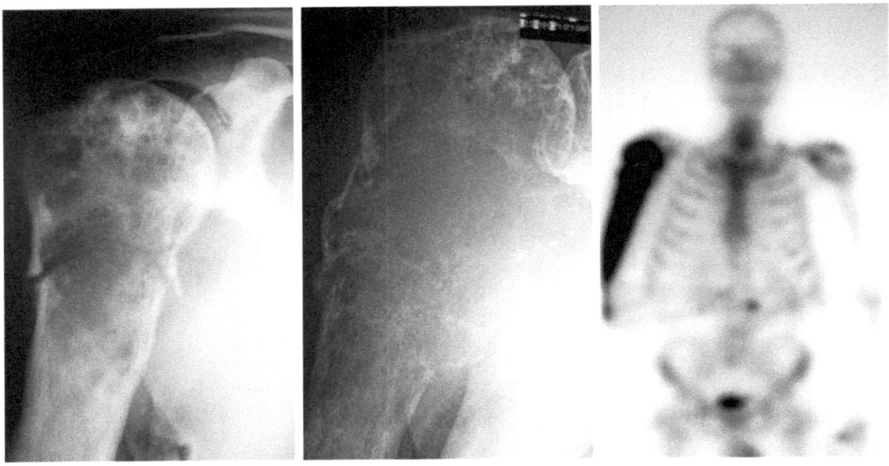

Radiograph at onset and two months later, bone scintigraphy. Lytic heterogeneous broken lesion involving the epiphysis and metaphysis. The thick lamellar cortex indicates a preexisting Paget disease, confirmed by widespread intense fixation on bone scintigraphy. Two months later, very fast progression of the tumor

## Selected Bibliography

Dray MS, Miller MV (2008) Paget's osteosarcoma and post-radiation osteosarcoma: secondary osteosarcoma at Middlemore Hospital, New Zealand. Pathology 40(6):604–610

Hamre MR, Severson RK, Chuba P, Lucas DR, Thomas RL, Mott MP (2002) Osteosarcoma as a second malignant neoplasm. Radiother Oncol 65(3):153–157

Okada A, Hatori M, Hosaka M, Watanuki M, Itoi E (2009) Secondary osteosarcoma arising after treatment for childhood hematologic malignancies. Ups J Med Sci 114(4):249–255

Picci P, Sieberova G, Alberghini M, Balladelli A, Vanel D, Hogendoorn PC, Mercuri M (2011) Late sarcoma development after curettage and bone grafting of benign bone tumors. Eur J Radiol 77(1):19–25

Tabone MD, Terrier P, Pacquement H, Brunat-Mentigny M, Schmitt C, Babin-Boilletot A, Mahmoud HH, Kalifa C (1999) Outcome of radiation-related osteosarcoma after treatment of childhood and adolescent cancer: a study of 23 cases. J Clin Oncol 17(9):2789–2795

# Chapter 37
# Small Cell Osteosarcoma

**Piero Picci**

**Definition:** Tumor composed by small cells, similar to Ewing sarcoma, but producing bone matrix.

This is a rare variety (1–2 % of all OS). Epidemiology, clinical presentation, and imaging are not consistently different from usual OS. Prognosis in the comparatively small number of reported cases has been worse as compared to usual OS. This tumor responds best to an Ewing sarcoma treatment protocol.

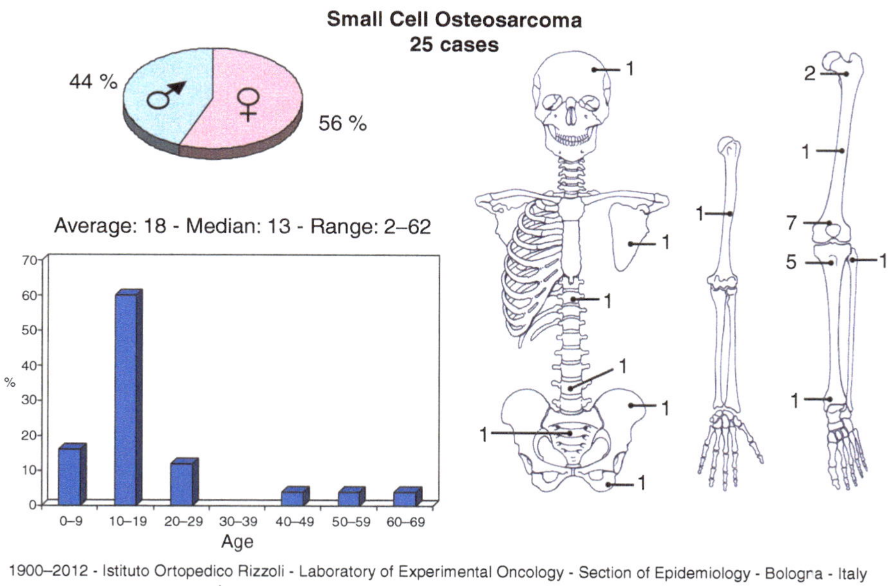


P. Picci, MD
Laboratory of Experimental Oncology, Musculoskeletal Oncology,
Istituto Ortopedico Rizzoli, Bologna, Italy
e-mail: piero.picci@ior.it

P. Picci et al. (eds.), *Atlas of Musculoskeletal Tumors and Tumorlike Lesions*,
DOI 10.1007/978-3-319-01748-8_37, © Springer International Publishing Switzerland 2014

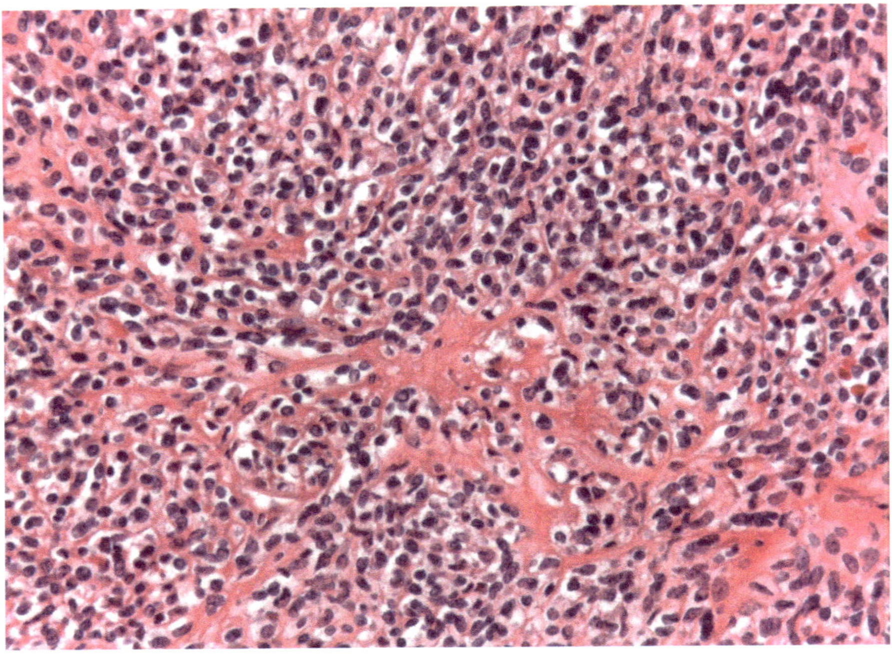

Round- to ovoid-shaped cells, similar to Ewing sarcoma cells, organized in sheets and nests, with osteoid matrix

## Selected Bibliography

Bertoni F, Present D, Bacchini P, Pignatti G, Picci P, Campanacci M (1989) The Istituto Rizzoli experience with small cell osteosarcoma. Cancer 64(12):2591–2599

Bishop JA, Shum CH, Sheth S, Wakely PE Jr, Ali SZ (2010) Small cell osteosarcoma: cytopathologic characteristics and differential diagnosis. Am J Clin Pathol 133(5):756–761

Devaney K, Vinh TN, Sweet DE (1993) Small cell osteosarcoma of bone: an immunohistochemical study with differential diagnostic considerations. Hum Pathol 24(11):1211–1225

Lee AF, Hayes MM, Lebrun D, Espinosa I, Nielsen GP, Rosenberg AE, Lee CH (2011) FLI-1 distinguishes Ewing sarcoma from small cell osteosarcoma and mesenchymal chondrosarcoma. Appl Immunohistochem Mol Morphol 19:233–238

Nakajima H, Sim FH, Bond JR, Unni KK (1997) Small cell osteosarcoma of bone. Review of 72 cases. Cancer 79(11):2095–2106

# Chapter 38
# High-Grade Osteosarcoma of the Surface

**Piero Picci**

**Definition:** A high-grade OS arising on the bony surface, with minimal involvement of the underlying cortex.

Apart from its site, it does not differ from the usual intramedullary OS in age, site, and histology. Imaging shows mixed non-mineralized and mineralized tumor matrix and some degree of periosteal reactive osteogenesis, but not the features specific of periosteal or parosteal OS. Histology is the same as in conventional OS. Treatment and prognosis also do not differ.

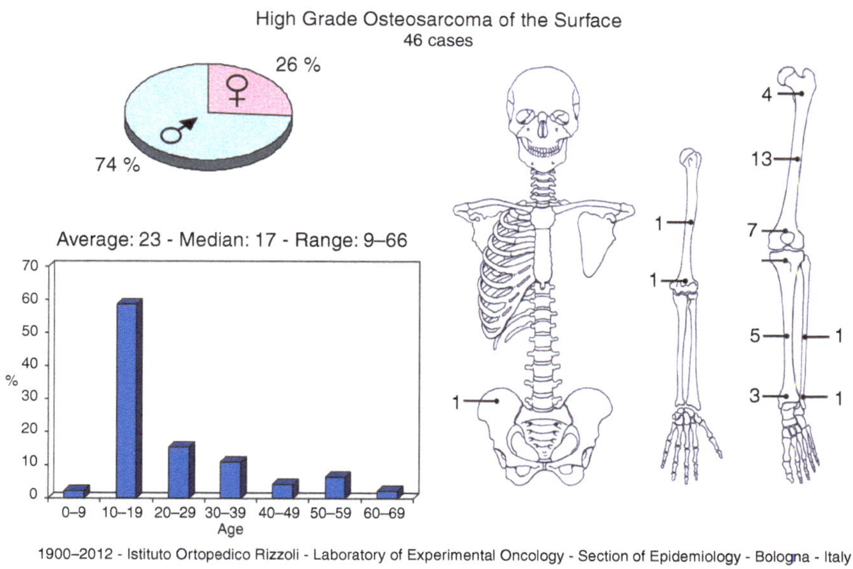


P. Picci, MD
Laboratory of Experimental Oncology, Musculoskeletal Oncology,
Istituto Ortopedico Rizzoli, Bologna, Italy
e-mail: piero.picci@ior.it

P. Picci et al. (eds.), *Atlas of Musculoskeletal Tumors and Tumorlike Lesions*,
DOI 10.1007/978-3-319-01748-8_38, © Springer International Publishing Switzerland 2014

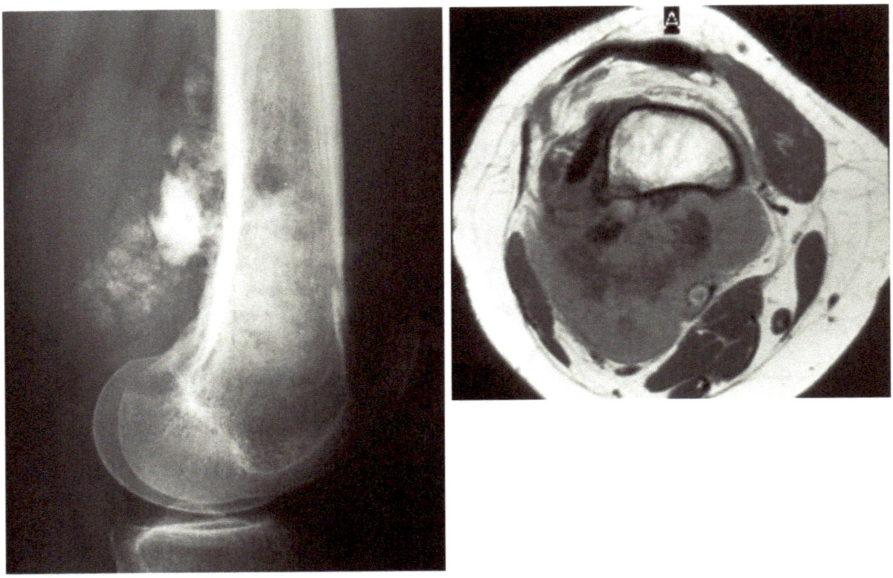

Lateral radiograph of the knee and axial T1 MR image. The tumor forms bone, mainly develops in the soft tissues, but starts from the cortex, which has a limited lysis. Minimal bone extension is visible on MR

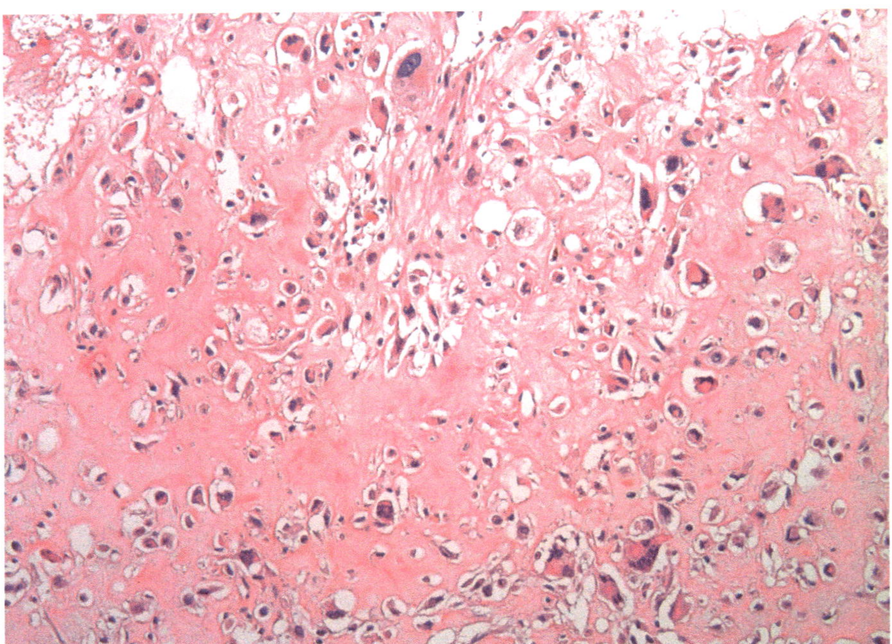

High-grade osteosarcoma identical to conventional central osteosarcoma, situated on the surface of bone. The majority are osteoblastic variants

## Selected Bibliography

Okada K, Unni KK, Swee RG, Sim FH (1999) High grade surface osteosarcoma: a clinicopathologic study of 46 cases. Cancer 85(5):1044–1054

Raymond AK (1991) Surface osteosarcoma. Clin Orthop Relat Res (270):140–148. Review

Schajowicz F, McGuire MH, Santini Araujo E, Muscolo DL, Gitelis S (1988) Osteosarcomas arising on the surfaces of long bones. J Bone Joint Surg Am 70(4):555–564

Staals EL, Bacchini P, Bertoni F (2008) High-grade surface osteosarcoma: a review of 25 cases from the Rizzoli Institute. Cancer 112(7):1592–1599

Vanel D, Picci P, De Paolis M, Mercuri M (2001) Radiological study of 12 high-grade surface osteosarcomas. Skeletal Radiol 30(12):667–671

# Chapter 39
# Multicentric Osteosarcoma

**Piero Picci**

**Definition:** Almost synchronous appearance of multiple OS in the skeleton usually without pulmonary mets.

It is very rare (0.5 % of all high-grade OS). Preferred age is from 5 to 15 years. (lower than usual OS). The number of foci varies from few to many, being distributed in the long bones but also in the trunk. They may involve locations, such as the vertebrae, ribs and sternum, skull, hand and foot, and epiphyses, where OS occurs very rarely. Usually one larger lesion is considered primary in the hypothesis that the others are metastatic deposits. All lesions are extensively sclerotic and do not easily breach the cortex. Histology is that of a usual osteoblastic generally extensively sclerotic high-grade OS. Whether it represents an early metastatic spread or a multicentric origin is unknown. The course is rapid and prognosis is regularly bad.

P. Picci, MD
Laboratory of Experimental Oncology, Musculoskeletal Oncology,
Istituto Ortopedico Rizzoli, Bologna, Italy
e-mail: piero.picci@ior.it

P. Picci et al. (eds.), *Atlas of Musculoskeletal Tumors and Tumorlike Lesions*,
DOI 10.1007/978-3-319-01748-8_39, © Springer International Publishing Switzerland 2014

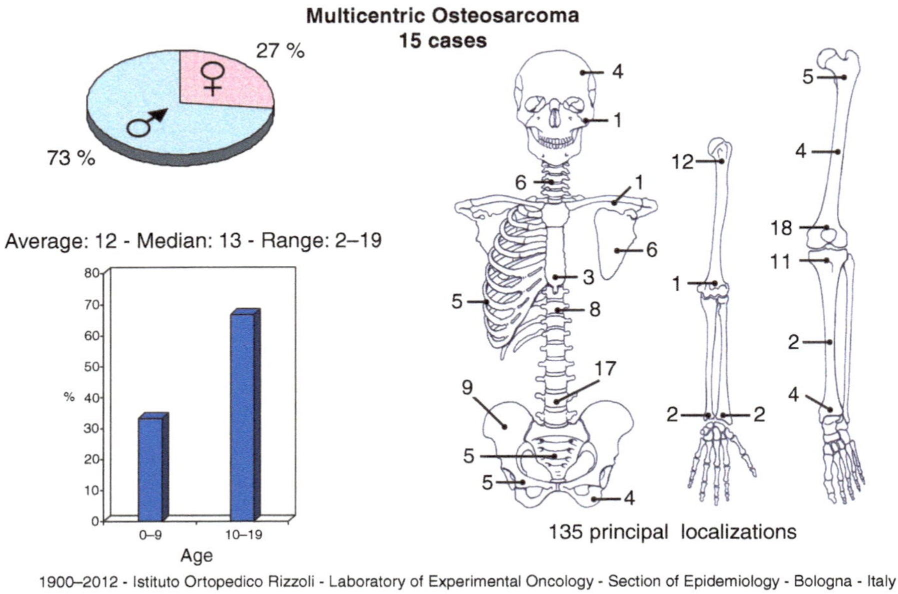
Multicentric Osteosarcoma
15 cases
27 %
73 %
Average: 12 - Median: 13 - Range: 2–19
%
0
10
20
30
40
50
60
70
80
0–9
10–19
Age
4
1
6
1
6
3
5
8
17
9
5
5
4
12
1
2
2
5
4
18
11
2
4
135 principal localizations
1900–2012 - Istituto Ortopedico Rizzoli - Laboratory of Experimental Oncology - Section of Epidemiology - Bologna - Italy

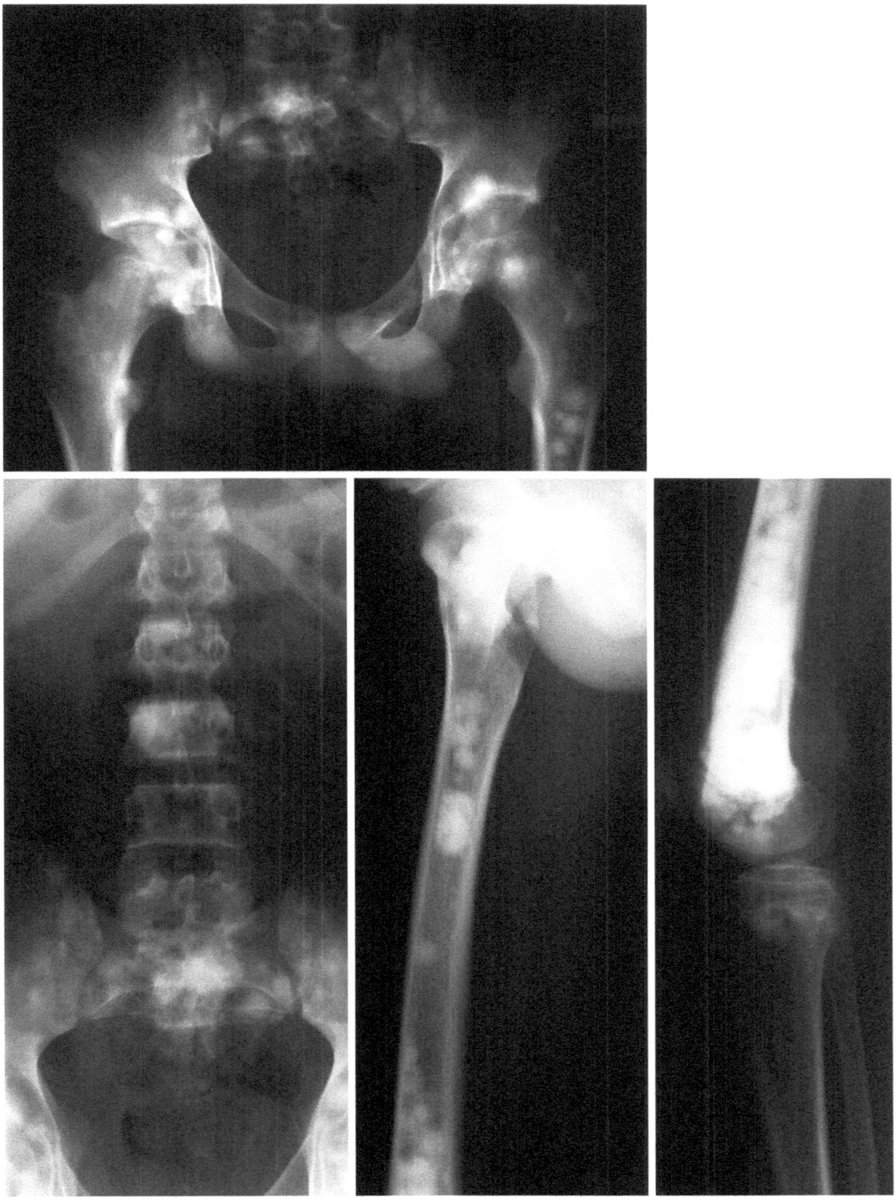

On radiographs multiple mainly sclerotic lesions are easily detected. The epiphyseal involvement is frequent in this lesion

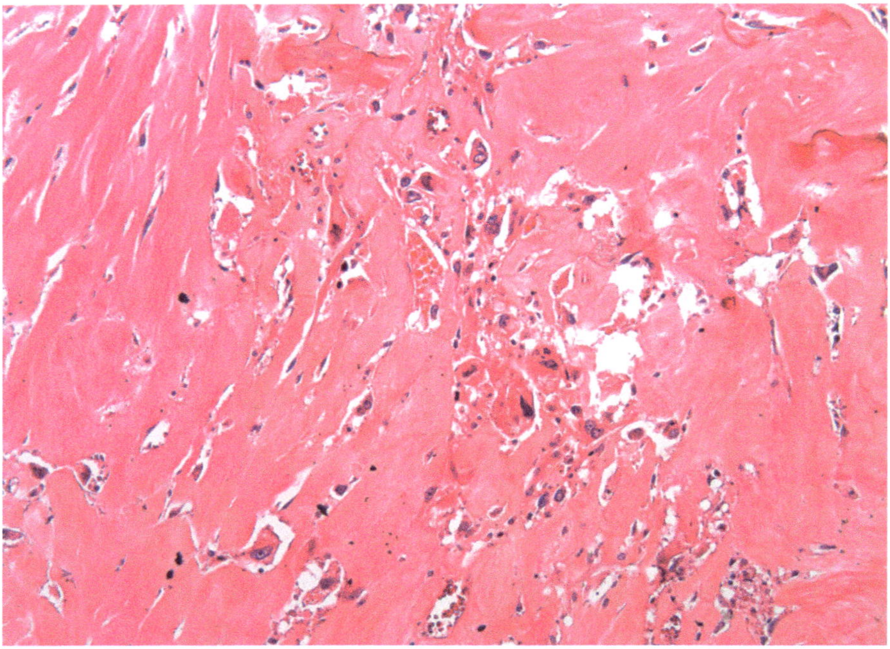

High-grade osteoblastic osteosarcoma usually extensively sclerotic with low cellularity and diminished cytological anaplasia when compared to the less sclerotic areas of the tumor

## Selected Bibliography

Bacci G, Fabbri N, Balladelli A, Forni C, Palmerini E, Picci P (2006) Treatment and prognosis for synchronous multifocal osteosarcoma in 42 patients. J Bone Joint Surg Br 88(8):1071–1075
Biagini R, Ruggieri P, De Cristofaro R, Torricelli P, Picci P, Bacci G (1991) Multicentric osteosarcoma. Report of 5 cases. Chir Organi Mov 76(2):113–122
Cermik TF, Salan A, Firat MF, Berkarda S (2003) Tc-99m MDP bone scintigraphy in a patient with multicentric osteosarcoma. Clin Nucl Med 28(7):599–600
Daffner RH, Kennedy SL, Fox KR, Crowley JJ, Sauser DD, Cooperstein LA (1997) Synchronous multicentric osteosarcoma: the case for metastases. Skeletal Radiol 26(10):569–578
Hopper KD, Eggli KD, Haseman DB, Moser RP Jr (1993) Osteosarcomatosis and metastatic osteosarcoma. Cancer Treat Res 62:163–171. Review

# Chapter 40
# Periosteal Osteosarcoma

**Piero Picci**

**Definition:** It is a predominantly chondroblastic OS of intermediate malignancy, originating from the periosteum usually in long bone diaphyses.
**Epidemiology:** Periosteal OS is rare (1/2 % of all OS). It has a preference for males and a wide range of age distribution, with prevalence in the second decade of life.

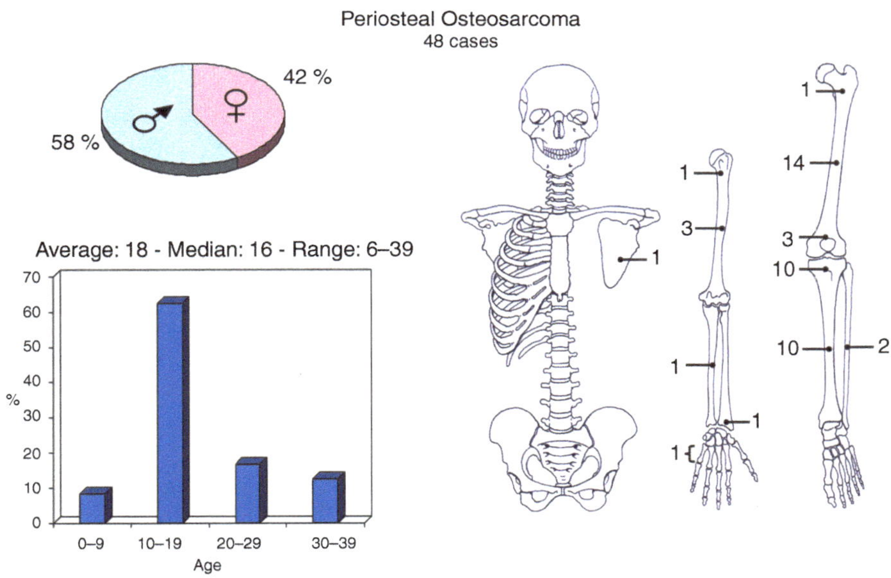


1900–2012 - Istituto Ortopedico Rizzoli - Laboratory of Experimental Oncology - Section of Epidemiology - Bologna - Italy

P. Picci, MD
Laboratory of Experimental Oncology, Musculoskeletal Oncology,
Istituto Ortopedico Rizzoli, Bologna, Italy
e-mail: piero.picci@ior.it

P. Picci et al. (eds.), *Atlas of Musculoskeletal Tumors and Tumorlike Lesions*,
DOI 10.1007/978-3-319-01748-8_40, © Springer International Publishing Switzerland 2014

**Localization:** Tibia and femur. Less frequent in the humerus, rare in the fibula and other long bones. Usually located in the diaphysis, occasionally in the metaphysis.
**Clinical:** Palpable mass and moderate pain. Slower growth rate than in usual OS.
**Imaging:** Periosteal, fusiform, radiolucent mass, with well-defined borders. It usually contains spiculated radiodensities perpendicular to the cortex that is either intact or superficially eroded, sometimes with a Codman triangle. Typically, the medullary canal is uninvolved. This is better demonstrated by CT and MRI.
**Histopathology:** The cut surface of the tumor is soft to rubbery with a translucent aura. The tumor forms large lobules, separated by streaks of ossification perpendicular to the cortex. Tissue is mainly chondroblastic, being more cellular at the periphery, with spindle cells and lacelike osteoid, more chondroid at the center of the lobuli. Lacelike osteoid among malignant osteoblasts is present and defines this tumor as OS. Histological grade of malignancy is 1–3. At the bony interface, the tumor can permeate for a small thickness into the cortex.
**Course and Staging:** It generally grows slower than usual OS, but much less slow than parosteal OS. The stage is I-A or I-B if there is intramedullary involvement.
**Treatment and Prognosis**: Due to the rather typical imaging presentation, frozen section intraoperative biopsy is usually adequate. Treatment consists of en bloc resection with wide margins. Mets (lung) have been observed in about 15 % of cases. Therefore, prognosis is good, after wide surgery and without chemotherapy.

**Key Points**

| | |
|---|---|
| Clinical | Young age, mild symptoms |
| Radiological | Surface lesion, diaphyseal with radio opacities |
| Histological | Cartilaginous with osteoid and low grade |
| Differential diagnosis | Periosteal chondrosarcoma, high-grade surface osteosarcoma |

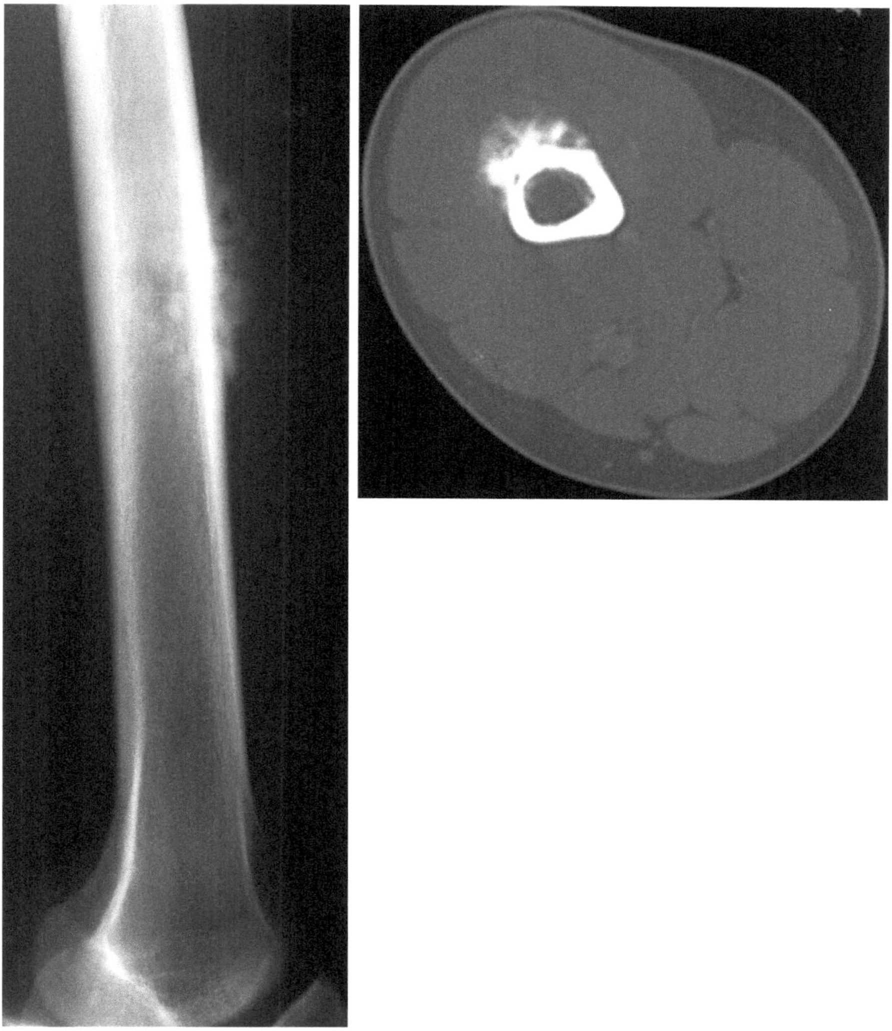

Radiograph and CT. Diaphyseal lesion of the bone surface, forming bone, with thick perpendicular periosteal bone formations. The cortex is eroded. Bone marrow is not involved

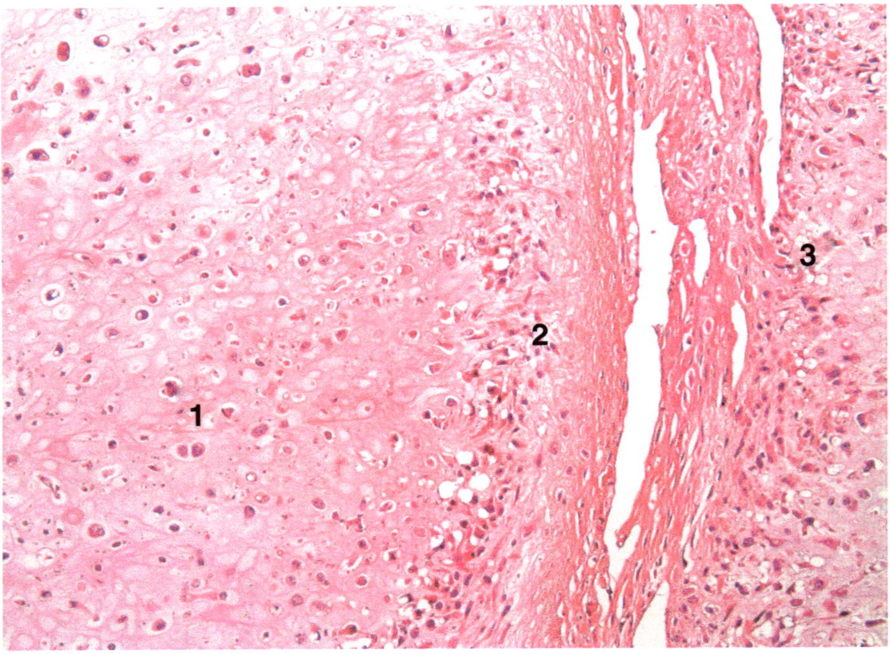


Predominantly chondroblastic tissue of intermediate malignancy that forms large lobules. (*1*) Chondroid cells. (*2*) Lacelike osteoid among malignant cells. (*3*) Spindle cells with minimal matrix at the periphery of the lobules

## Selected Bibliography

Cesari M, Alberghini M, Vanel D, Palmerini E, Staals EL, Longhi A, Abate M, Ferrari C, Balladelli A, Ferrari S (2011) Periosteal osteosarcoma: a single-institution experience. Cancer 117: 1731–1735

Grimer RJ, Bielack S, Flege S, Cannon SR, Foleras G, Andreeff I, Sokolov T, Taminiau A, Dominkus M, San-Julian M, Kollender Y, Gosheger G (2005) European Musculo Skeletal Oncology Society. Periosteal osteosarcoma – a European review of outcome. Eur J Cancer 41(18):2806–2811

Murphey MD, Jelinek JS, Temple HT, Flemming DJ, Gannon FH (2004) Imaging of periosteal osteosarcoma: radiologic-pathologic comparison. Radiology 233(1):129–138

Revell MP, Deshmukh N, Grimer RJ, Carter SR, Tillman RM (2002) Periosteal osteosarcoma: a review of 17 cases with mean follow-up of 52 months. Sarcoma 6(4):123–130

Rose PS, Dickey ID, Wenger DE, Unni KK, Sim FH (2006) Periosteal osteosarcoma: long-term outcome and risk of late recurrence. Clin Orthop Relat Res 453:314–317

# Chapter 41
# Parosteal Osteosarcoma

**Piero Picci**

**Definition:** This OS originates at the surface of the bone, with abundant production of dense bone and low-grade anaplasia (1–2). Progression of malignancy may occur in about 10 % of cases, particularly those with multiple recurrences or those previously mistaken for benign bone tumor.

**Epidemiology:** It is infrequent (5/6 % of all OS). Slight preference for females. Usually it appears between ages 20 and 50, very rare before the end of growth.

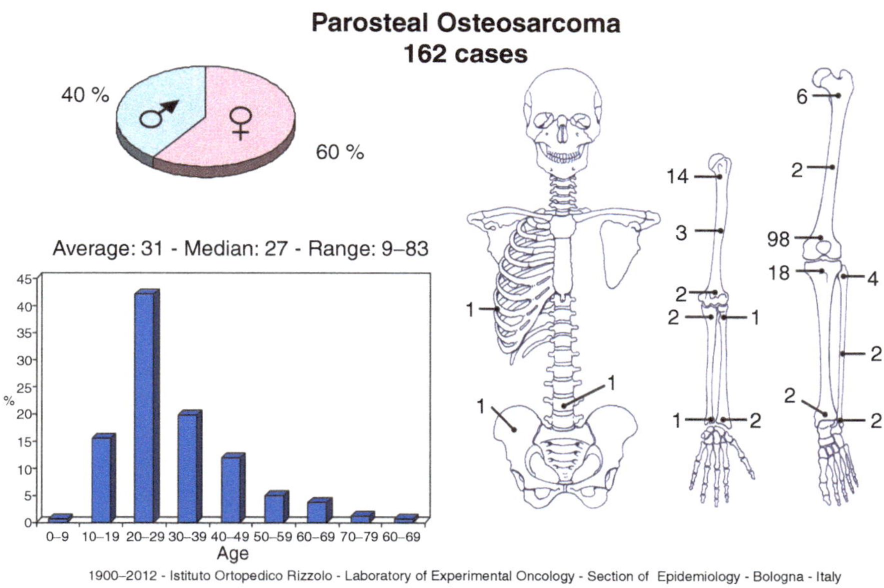


P. Picci, MD
Laboratory of Experimental Oncology, Musculoskeletal Oncology,
Istituto Ortopedico Rizzoli, Bologna, Italy
e-mail: piero.picci@ior.it

P. Picci et al. (eds.), *Atlas of Musculoskeletal Tumors and Tumorlike Lesions*, 173
DOI 10.1007/978-3-319-01748-8_41, © Springer International Publishing Switzerland 2014

**Localization:** Almost exclusive of the long bones, it originates from the metaphysis. Rare in the diaphyses, exceptional in flat bones. The most typical site is the distal metaphysis of the femur in its posterior aspect (60 % of cases).
**Clinical:** No or slight pain. There is a bony hard mass. Duration of symptoms not infrequently exceeds 1, 2, or even 5 years. It is not uncommon to see patients having undergone one or more surgical excisions and presenting with a local recurrence.
**Imaging:** Due to slow growth, parosteal OS is usually seen when it is large. It is a lobulated mass of osseous radiodensity, fused to the cortex with a broad base that tends to wrap around the bone. Radiodensity is maximal near the implant base, while the outer margin tends to be blurred. Detailed radiographies show a mesh of trabeculae ("steel-wool" pattern), from ground glass to ivory density. Toward the diaphysis, the medullary canal is usually not involved. In metaepiphyses, the thin cortex is frequently breached with extension to the cancellous bone. This involvement increases with malignancy and with the standing duration of the tumor and is appreciated by CT in 60 % of cases. CT scan, MRI, and angiography show the tumor adhering to or enwrapping the vascular bundle. Isotope scan is very hot.
**Histopathology:** Parosteal OS is composed by spindle cells and collagen fibers embedding osseous trabeculae. Occasionally cartilage may be associated. Cells form long sweeping fascicles with parallel nuclei. Trabeculae appear also as parallel streamers with broken cement lines as pagetic mosaic. Bone is formed by metaplasia from tumor cells (fibro-osseous metaplasia) and trabeculae may not show osteoblastic rimming. Parosteal OS is grade 1 or 2 but may progress in malignancy and transform into a high-grade OS. It is indeed not rare to find areas of different histological grade in the same case. Mets (lung, skeleton) have usually the same pattern as the most malignant part of the primary and are often densely sclerotic.
**Course and Staging:** The course is usually slow. Parosteal OS is usually staged I-A or B. When histological grade is 3 or 4 (dedifferentiated), the stage is II.
**Treatment and Prognosis:** Surgery must aim to wide margins. Hemicylindric resection is usually possible in the popliteal region, through a double approach (medial and lateral), elevating the quadriceps. More often, a complete segmental resection of the affected bone is needed. When the tumor impinges on the vascular bundle, wide margins may often require vessel resection and reconstruction. Chemotherapy is only indicated in grade 3 and 4 lesions and it is the same as that used in classic OS. Local recurrence is the rule after intralesional surgery. Mets may be seen in grade 2 and mainly when the underlying bone is invaded. Mets are less common as compared to conventional OS, about 2–10 % in grades 1–2 parosteal OS, but rising to about 60–70 % in grades 3–4 to dedifferentiated parosteal OS.

### Key Points

| | |
|---|---|
| Clinical | Long history of mild symptoms, pain, and swelling |
| Radiological | Surface, eburneus aspect |
| Histological | Low-grade bone production |
| Differential diagnosis | Dedifferentiated parosteal osteosarcoma, high-grade surface osteosarcoma |

### Amplicons

| | | |
|---|---|---|
| MDM2 12q15 | Gene amplification | FISH analysis: 47.7 ± 21.9 |
| CDK4 12q13-14 | Gene amplification | FISH analysis: 31.5 ± 13.9 |

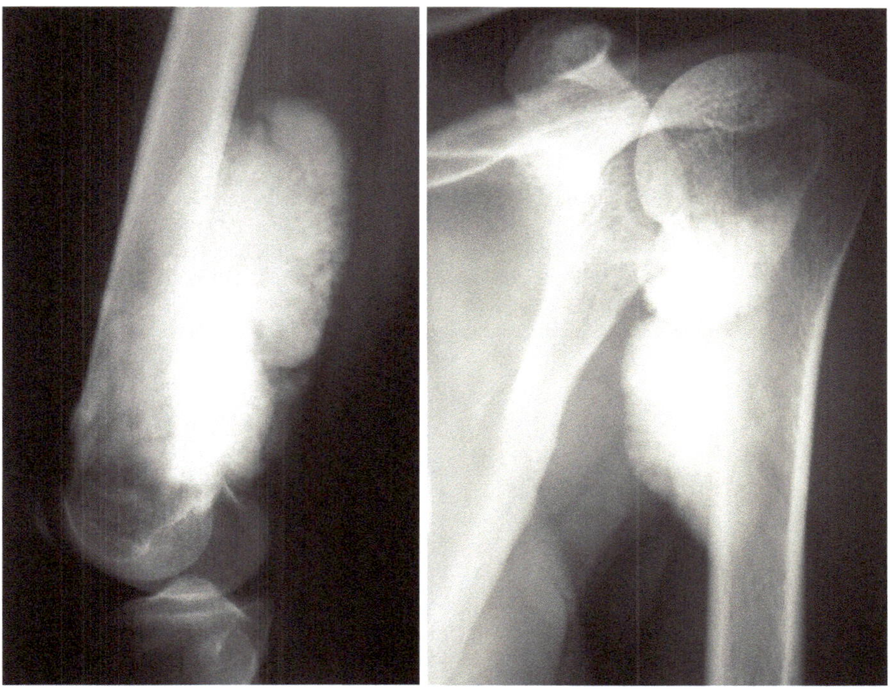

Radiographs. The tumors are homogeneous and sclerotic. They are attached to the bone

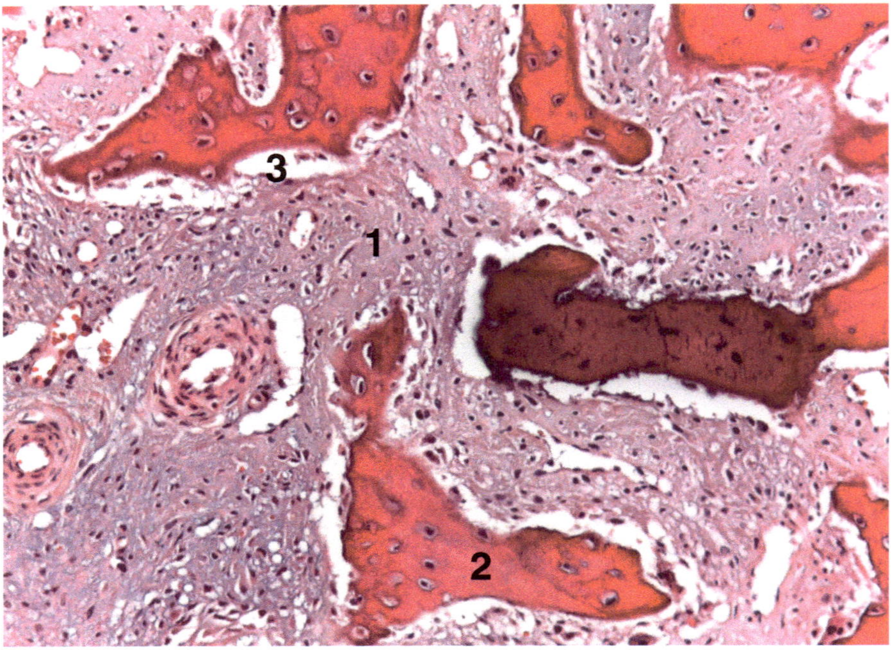


Histopathologic features are substantially represented by bony trabeculae enmeshed in a sarcomatous spindle cell and collagen stroma. (*1*) Numerous and atypical spindle cells, like those of a fibrosarcoma. Nuclei are large, oval, and pleomorphic. The tumor is graded from 1 to 2 based on the aspect of the spindle cells. (*2*) Bone trabeculae arranged rather regularly. (*3*) Osteocytes do not present aspects of malignancy. The best place to find anaplasia is in the fibroblastic stroma and sometimes in the cartilage that is produced as caps over the bony portions of the tumor

In rare cases, within a typical parosteal osteosarcoma, lytic areas (at imaging) can be found corresponding histologically to high-grade areas, showing therefore a dedifferentiation. Dedifferentiated parosteal OS is as malignant as conventional OS.

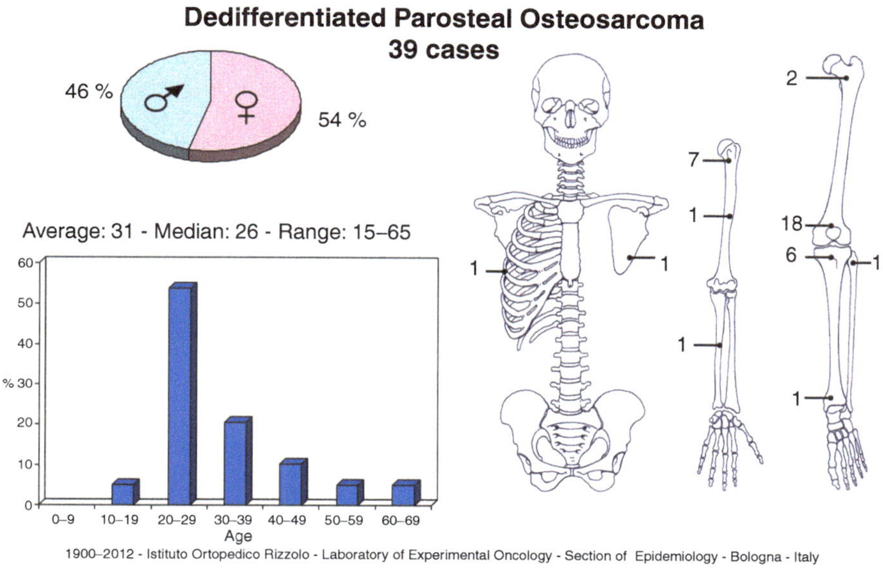

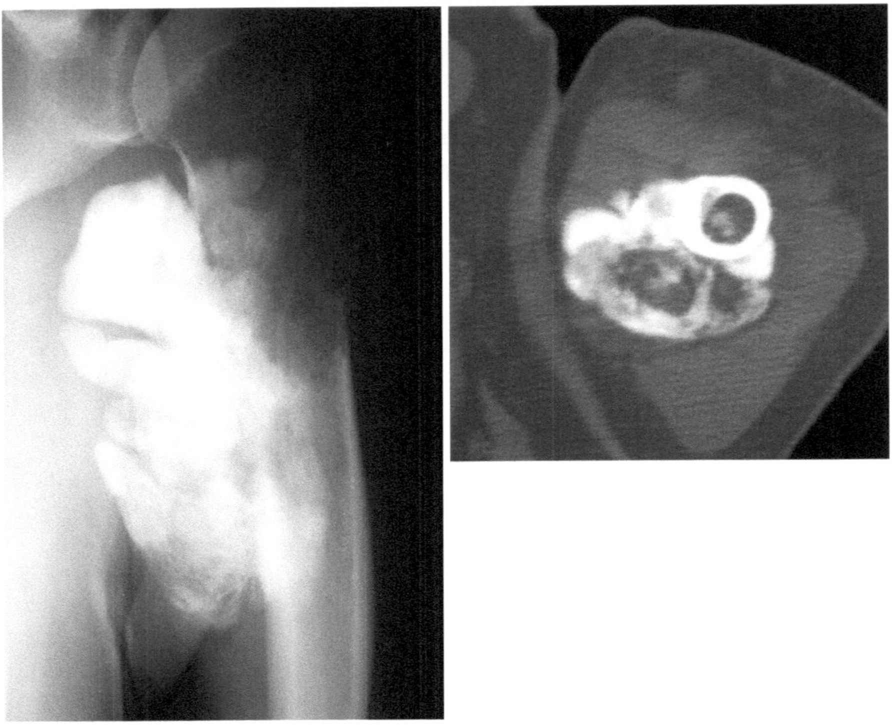

Radiograph and CT of the humerus. Typical parosteal osteosarcoma on radiograph. The lytic central areas on CT are very suggestive of dedifferentiation

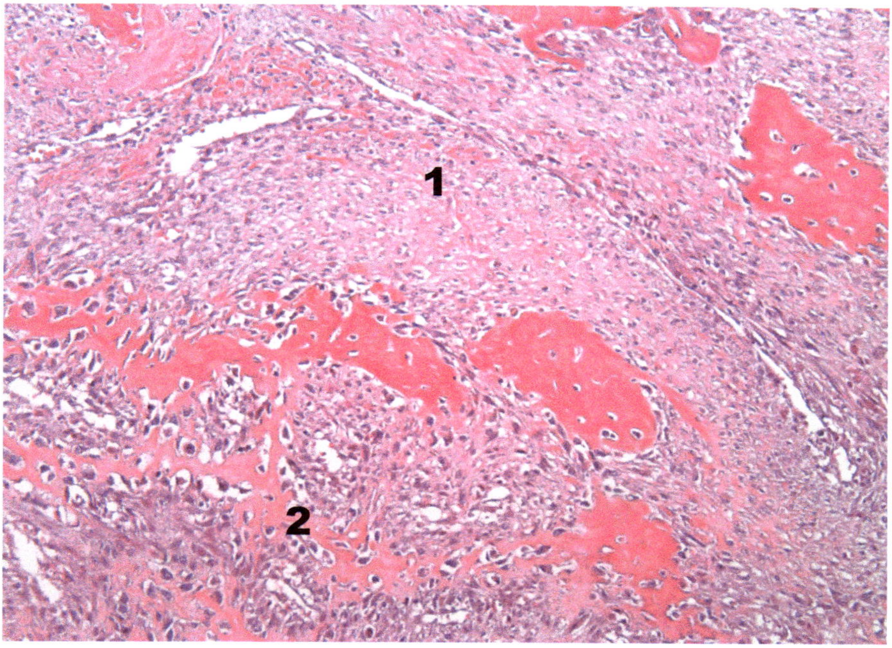


(*1*) Parosteal grade 2 osteosarcoma. (*2*) Dedifferentiated grade 3 osteoblastic osteosarcoma

## Selected Bibliography

Azura M, Vanel D, Alberghini M, Picci P, Staals E, Mercuri M (2009) Parosteal osteosarcoma dedifferentiating into telangiectatic osteosarcoma: importance of lytic changes and fluid cavities at imaging. Skeletal Radiol 38(7):685–690

Bertoni F, Bacchini P, Staals EL, Davidovitz P (2005) Dedifferentiated parosteal osteosarcoma: the experience of the Rizzoli Institute. Cancer 103(11):2373–2382

Campanacci M, Picci P, Gherlinzoni F, Guerra A, Bertoni F, Neff JR (1984) Parosteal osteosarcoma. J Bone Joint Surg Br 66(3):313–321

Mejia-Guerrero S, Quejada M, Gokgoz N, Gill M, Parkes RK, Wunder JS, Andrulis IL (2010) Characterization of the 12q15 MDM2 and 12q13-14 CDK4 amplicons and clinical correlations in osteosarcoma. Genes Chromosomes Cancer 49(6):518–525

Ritschl P, Wurnig C, Lechner G, Roessner A (1991) Parosteal osteosarcoma. 2–23-year follow-up of 33 patients. Acta Orthop Scand 62(3):195–200

Sheth DS, Yasko AW, Raymond AK, Ayala AG, Carrasco CH, Benjamin RS, Jaffe N, Murray JA (1996) Conventional and dedifferentiated parosteal osteosarcoma. Diagnosis, treatment, and outcome. Cancer 78(10):2136–2145

# Chapter 42
# Central Low-Grade Osteosarcoma

**Piero Picci**

**Definition:** It is an intramedullary low-grade (1–2) bone-producing tumor. It has been suggested to represent the central counterpart of parosteal OS.
**Epidemiology:** Rare (2/3 % of all OS). There is no preference for sex. Age ranges from 10 to 70 years, the mean being around 30.

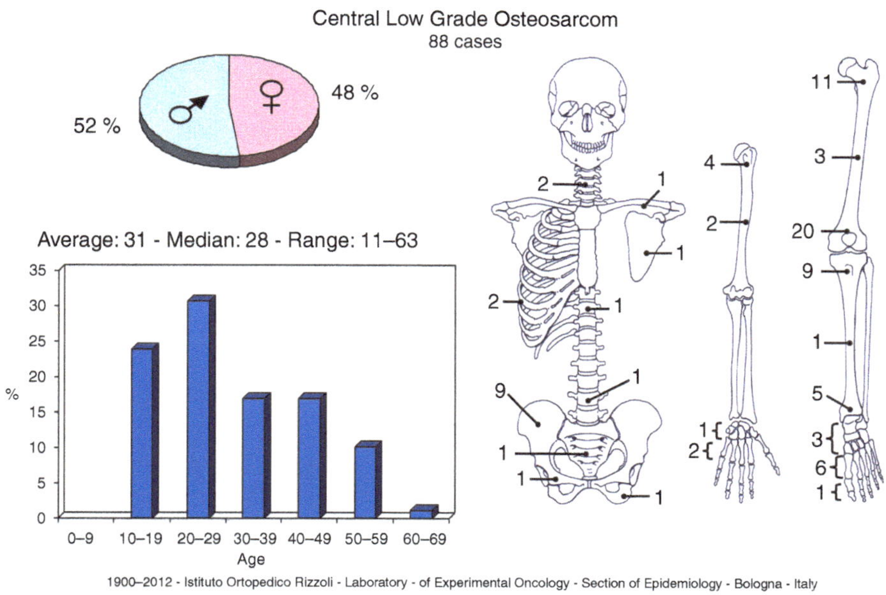


P. Picci, MD
Laboratory of Experimental Oncology, Musculoskeletal Oncology,
Istituto Ortopedico Rizzoli, Bologna, Italy
e-mail: piero.picci@ior.it

P. Picci et al. (eds.), *Atlas of Musculoskeletal Tumors and Tumorlike Lesions*,
DOI 10.1007/978-3-319-01748-8_42, © Springer International Publishing Switzerland 2014

**Localization:** Preferred sites are long bones, particularly the distal femur and proximal tibia. Rare in flat bones, exceptional in the hands and feet. It is usually centered in the metaphysis and, as it occurs mostly in the adults, it usually involves the epiphysis too.
**Clinical:** Mild to moderate pain and, less frequently, a moderate swelling. Clinical history is often of long duration. Pathologic fracture is rare.
**Imaging:** Usually, due to the slow course, tumors are large. The lesion appears mixed, lytic, and blastic. Even when lytic, some ground-glass density can usually be found. Conversely, when it is sclerotic, some radiolucent areas are usually present. The bone contour may be expanded with a thin cortex that may present some discontinuity. In half of the cases, a soft tissue mass can be found, which can be either radiolucent or densely mineralized. Frequently, the tumor presents a coarse trabeculation similar to desmoid tumor of the bone. A chronic periosteal reaction and thickening of the cortex are occasionally observed. Isotope scan is hot. CT and MRI are essential to define the extensions of the tumor.
**Histopathology:** Spindle cells producing collagen and bone. The tumor closely mimics the histological pattern of parosteal OS. When osteogenesis is scarce, the tumor is similar to desmoid tumor or low-grade fibrosarcoma. Less frequently, it has the "Chinese characters" appearance of fibrous dysplasia. Occasionally foci of cartilage can be seen. The cells demonstrate slight atypia with few mitotic figures. Osteocytes of the bone trabeculae are normal. The grade is either 1 or 2. Almost regularly, the tumor has a permeative pattern.
**Course and Staging:** The course is slow. In about 15 % of cases, the tumor shows a progression of malignancy, transforming into a high-grade OS. The stage of the tumor is I-A or I-B. Transformed tumors become stage II.
**Treatment:** Intralesional or marginal excision is almost regularly followed by local recurrence. Thus, surgery must obtain wide margins, which is usually feasible with conservative resection. Chemotherapy is not indicated, except in case of progression in malignancy.
**Prognosis:** Mets (lungs) are reported in 10 % of cases and mostly in tumors that progressed in malignancy. They can occur many years after the onset of symptoms. Thus, prognosis is good, unless the tumor becomes a high-grade OS.

**KeyPoints**

| | |
|---|---|
| Clinical | Long history of mild symptoms, usually pain |
| Radiological | Central, mixed aspect |
| Histological | Low-grade bone production |
| Differential diagnosis | Osteoblastoma, high-grade osteosarcoma |

**Amplicons**

| | | |
|---|---|---|
| MDM2 12q15 | Gene amplification | CGH analysis: 100 % |
| CDK4 12q13-14 | Gene amplification | CGH analysis: 100 % |

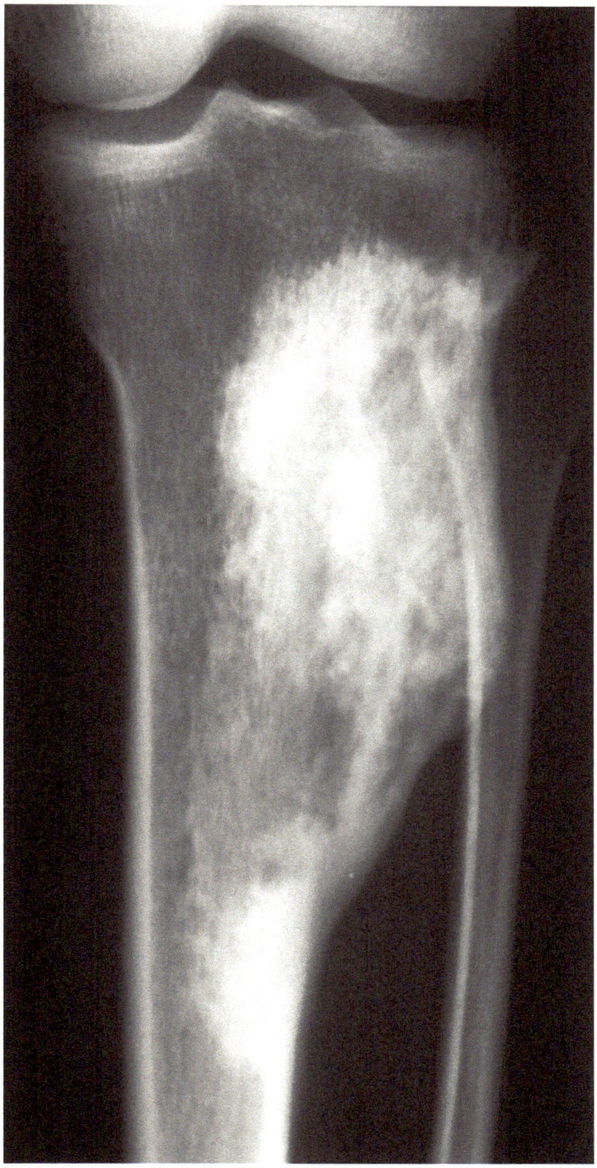

Radiograph. Heterogeneous lytic and sclerotic metadiaphyseal tumor. The lesion is well limited, but destroys the cortex

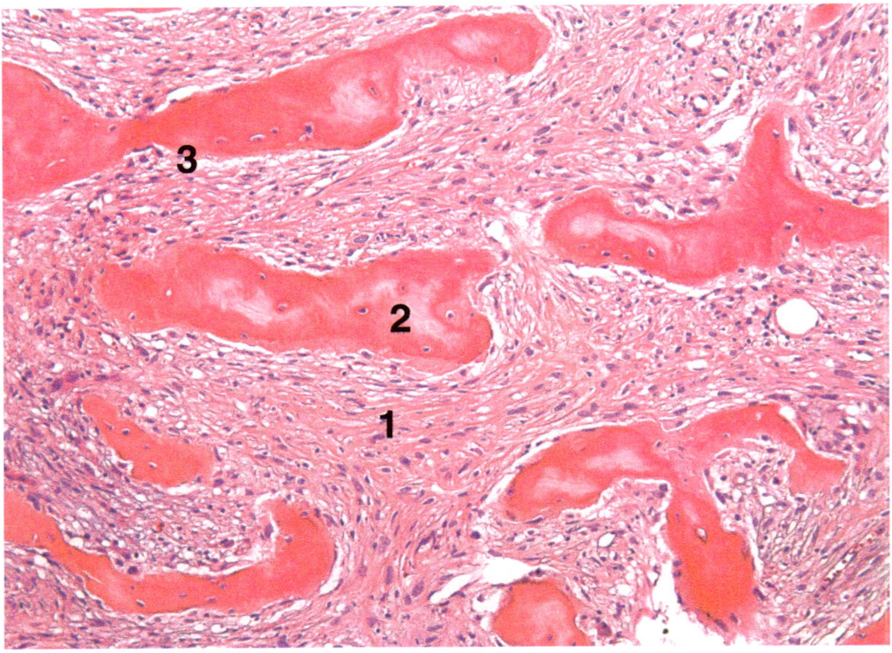


It histologically appears to be identical to parosteal osteosarcoma. (1) Cells surrounding the bony trabeculae are prevalently spindle shaped, with aspects of minimal atypia. (2) Neoplastic trabeculae. A regular row of osteoblasts lining the trabeculae can be absent

## Selected Bibliography

Bertoni F, Bacchini P, Donati D, Martini A, Picci P, Campanacci M (1993a) Osteoblastoma-like osteosarcoma. The Rizzoli Institute experience. Mod Pathol 6(6):707–716

Bertoni F, Bacchini P, Fabbri N, Mercuri M, Picci P, Ruggieri P, Campanacci M (1993b) Osteosarcoma. Low-grade intraosseous-type osteosarcoma, histologically resembling parosteal osteosarcoma, fibrous dysplasia, and desmoplastic fibroma. Cancer 71(2):338–345

Choong PF, Pritchard DJ, Rock MG, Sim FH, McLeod RA, Unni KK (1996) Low grade central osteogenic sarcoma. A long-term followup of 20 patients. Clin Orthop Relat Res (322):198–206

Dujardin F, Binh MB, Bouvier C, Gomez-Brouchet A, Larousserie F, Muret AD, Louis-Brennetot C, Aurias A, Coindre JM, Guillou L, Pedeutour F, Duval H, Collin C, de Pinieux G (2011) MDM2 and CDK4 immunohistochemistry is a valuable tool in the differential diagnosis of low-grade osteosarcomas and other primary fibro-osseous lesions of the bone. Mod Pathol 24(5):624–637

Yoshida A, Ushiku T, Motoi T, Shibata T, Beppu Y, Fukayama M, Tsuda H (2010) Immunohistochemical analysis of MDM2 and CDK4 distinguishes low-grade osteosarcoma from benign mimics. Mod Pathol 23(9):1279–1288

# Chapter 43
# Biology of Osteosarcoma

**Massimo Serra**

Unlike other sarcomas, osteosarcoma (OS) is characterized by complex, unbalanced karyotypes and alterations in multiple genes and pathways. This genetic background originates from the high genetic instability of OS cells, which can lead to multiple malignant cell populations within the same tumor (Hattinger et al. 2010; Atiye et al. 2005).

The pathways governed by the tumor suppressor genes P53 and retinoblastoma 1 (RB1) are those that have been most consistently found to be involved OS pathogenesis. In fact, the majority (around 80 %) of OS patients have alterations of one or both pathways (Hattinger et al. 2010). The P53 gene product plays a major role in the cell response to DNA damage and RB1 gene regulates cell cycle progression: alteration of pathways governed by these two genes may allow cells to proliferate and become malignant after the acquisition of additional genetic aberrations. This is the reason why children affected by the Li-Fraumeni syndrome (carrying germline deletion/mutations of P53) or familial retinoblastoma (carrying germline mutations of RB1) have a dramatically higher risk of developing OS.

The study of the natural history of OS has clearly shown that during development and progression tumor cells acquire several genetic changes, which may account for not only the aggressive behavior of this neoplasm but can also be responsible for the development of resistance to chemotherapeutic drugs (Hattinger et al. 2003, 2009). Taking these features into consideration, research on new drugs for novel treatment modalities of high-grade OS has been devoted to identify and validate agents against new candidate therapeutic targets, which have proved or appeared to be relevant for OS pathogenesis, treatment response, or clinical outcome. The current research goal of drug development for OS consists in the identification and validation of agents that can be administered as adjuvant to conventional

M. Serra, PhD
Laboratory of Experimental Oncology, Istituto Ortopedico Rizzoli,
Bologna, Italy
e-mail: massimo.serra@ior.it

P. Picci et al. (eds.), *Atlas of Musculoskeletal Tumors and Tumorlike Lesions*, 185
DOI 10.1007/978-3-319-01748-8_43, 

chemotherapeutics to better control the local and metastatic disease, as well as to improve the efficacy of standard chemotherapy regimens without increasing their collateral adverse toxicity. These facts offer the hope for not only an increased survival probability but also for an improved quality of life of cured patients, which is particularly relevant for tumors mainly affecting young people like OS.

As a complement to these goals, the validation of predictive, prognostic markers for high-grade OS is highly needed in order to allow a patient stratification based on specific characteristics of each tumor and on a precise risk evaluation aimed to identify those subgroups of patients with the highest probability to benefit from each innovative treatment.

Several studies on high-grade OS have drawn attention to clinical parameters like age, tumor size, anatomical location, histological type, or tumor necrosis, but their association with clinical outcome significantly varied across different reports, often resulting contradictory or inconsistent (Hattinger et al. 2010).

An information that clearly emerged from clinical studies is that the major cause of failure of the current treatment protocols for high-grade OS is the natural or acquired drug resistance, which occurs in 35–45 % of patients. Therefore, the identification and validation of drug resistance-related markers as prognostic factors and potential new therapeutic targets are highly warranted. Although several evidences in this field have been reported in the past 10–15 years, a clear picture of molecular determinants of drug unresponsiveness in OS is still far to be completed (Hattinger et al. 2010).

Several studies have indicated that ABCB1 transporter (also named as MDR1 or P-glycoprotein) plays an important role in drug resistance and treatment response of high-grade OS patients. Therefore, targeting this molecule appears to be an interesting therapeutic option to improve treatment results in OS patients which are unresponsive to conventional regimens (Hattinger et al. 2010; Serra et al. 1995, 2003, 2006; Baldini et al. 1995; Scionti et al. 2008).

In the past 30 years, several ABC transporter modulators or inhibitors have been described and entered clinical Phase I–III trials for different human tumors. The clinical use of such modulators has however been limited by the severe collateral toxicity which has been encountered at the concentrations required to enable these drugs to significantly inhibit the ABC transporter activity. More recently, a new generation of ABC transporter inhibitors has been developed and some of them have undergone clinical evaluation in different human tumors. Among these, some curcuminoids have appeared to be particularly promising for a possible clinical use, and curcumin has been recently included in Phase I–II clinical trials (Clinical Trial Identifier: NCT00689195; http://www.clinicaltrial.gov/) which are currently recruiting patients (Hattinger et al. 2010). These trials are planned to study the efficacy of curcumin in high-grade relapsed or metastatic OS patients, who are not treated with other second-line chemotherapies. In the next few years, on the basis of the results which will be obtained from the ongoing trials in OS and other human tumors, we should have more information about the possibility to include ABC transporter inhibitors in association with conventional chemotherapeutics in the treatment of conventional drug-unresponsive high-grade OS patients.

One important challenge that has emerged in the past few years as a possibility to improve the clinical results of conventional treatments in several human cancers is to target not only the tumor features that are directly associated with treatment unresponsiveness but also the biological systems that support tumor growth and progression. The scientific rationale of these different therapeutic approaches is the potentiation of the classical chemotherapeutic drugs' efficacy, without increasing their adverse collateral toxicities.

Scientific advances in understanding the genetic and molecular bases of OS, together with the identification of markers associated with a reduced response to conventional treatments and a poor outcome, have revealed several new potential targets for intervention to treat high-grade OS patients (Hattinger et al. 2010; Pasello et al. 2008; Toffoli et al. 2009; Buddingh et al. 2011; Wolf et al. 2000).

Presently, several Phase I/II/III clinical trials for high-grade OS are assessing the efficacy and collateral toxicities of agents targeting specific pathways (IGF/IGF-IR, Src, mTOR, VEGF) and bone resorption (RANK/RANK-L) or enhancing the immune system activity (muramyl tripeptide-phosphatidyl ethanolamine). Although several preclinical and clinical evidences have been already reported about the efficacy of some of these drugs and agents in high-grade OS, most of these findings must still be considered as preliminary and need to be further confirmed before including any of these agents in standard treatment protocols (Hattinger et al. 2010).

On the basis of the few information which has been reported so far, it can be predicted that, in the next 5–10 years, there is a concrete possibility to identify agents with efficacy and safety profiles superior or complementary to those of conventional drugs, which may be considered for innovative treatment strategies for selected groups of high-grade OS patients.

## Bibliography

Atiye J, Wolf M, Kaur S, Monni O, Bohling T, Kivioja A, Tas E, Serra M, Tarkkanen M, Knuutila S (2005) Gene amplifications in osteosarcoma-CGH microarray analysis. Genes Chromosomes Cancer 42:158–163

Baldini N, Scotlandi K, Barbanti-Brodano G, Manara MC, Maurici D, Bacci G, Bertoni F, Picci P, Sottili S, Campanacci M, Serra M (1995) Expression of P-glycoprotein in high-grade osteosarcomas in relation to clinical outcome. N Engl J Med 333:1380–1385

Buddingh EP, Kuijjer ML, Duim RA, Bürger H, Agelopoulos K, Myklebost O, Serra M, Mertens F, Hogendoorn PC, Lankester AC, Cleton-Jansen AM (2011) Tumor-infiltrating macrophages are associated with metastasis suppression in high-grade osteosarcoma: a rationale for treatment with macrophage activating agents. Clin Cancer Res 17:2110–2119

Hattinger CM, Reverter-Branchat G, Remondini D, Castellani G, Benini S, Pasello M, Manara MC, Scotlandi K, Picci P, Serra M (2003) Genomic imbalances associated with methotrexate resistance in human osteosarcoma cell lines detected by comparative genomic hybridization-based techniques. Eur J Cell Biol 82:483–493

Hattinger CM, Stoico G, Michelacci F, Pasello M, Scionti I, Remondini D, Castellani GC, Fanelli M, Scotlandi K, Picci P, Serra M (2009) Mechanisms of gene amplification and evidence of coamplification in drug-resistant human osteosarcoma cell lines. Genes Chromosomes Cancer 48:289–309

Hattinger CM, Pasello M, Ferrari S, Picci P, Serra M (2010) Emerging drugs for high-grade osteosarcoma. Expert Opin Emerg Drugs 15:615–634

Pasello M, Michelacci F, Scionti I, Hattinger CM, Zuntini M, Caccuri AM, Scotlandi K, Picci P, Serra M (2008) Overcoming glutathione S-transferase P1-related cisplatin resistance in osteosarcoma. Cancer Res 68:6661–6668

Scionti I, Michelacci F, Pasello M, Hattinger CM, Alberghini M, Manara MC, Bacci G, Ferrari S, Scotlandi K, Picci P, Serra M (2008) Clinical impact of the methotrexate resistance-associated genes C-MYC and dihydrofolate reductase (DHFR) in high-grade osteosarcoma. Ann Oncol 19:1500–1508

Serra M, Scotlandi K, Manara MC, Maurici D, Benini S, Sarti M, Campanacci M, Baldini N (1995) Analysis of P-glycoprotein expression in osteosarcoma. Eur J Cancer 12:1998–2002

Serra M, Scotlandi K, Reverter-Branchat G, Ferrari S, Manara MC, Benini S, Incaprera M, Bertoni F, Mercuri M, Briccoli A, Bacci G, Picci P (2003) Value of P-glycoprotein and clinicopathologic factors as the basis for new treatment strategies in high-grade osteosarcoma of the extremities. J Clin Oncol 21:536–542

Serra M, Pasello M, Manara MC, Scotlandi K, Ferrari S, Bertoni F, Mercuri M, Alvegard TA, Picci P, Bacci G, Smelalnd S (2006) May P-glycoprotein status be used to stratify high-grade osteosarcoma patients? Results from the Italian/Scandinavian Sarcoma Group 1 treatment protocol. Int J Oncol 29:1459–1468

Toffoli G, Biason P, Russo A, De Mattia E, Cecchin E, Hattinger CM, Pasello M, Alberghini M, Ferrari C, Scotlandi K, Picci P, Serra M (2009) Impact of TP53 Arg72Pro and MDM2 SNP309 polymorphisms on the risk of high grade osteosarcoma development and survival. Clin Cancer Res 15:3550–3556

Wolf M, El-Rifai W, Tarkkanen M, Kononen J, Serra M, Eriksen EF, Elomaa I, Kallionemi A, Kallionemi O-P, Knuutila S (2000) Novel findings in gene expression detected in human osteosarcoma by cDNA microarray. Cancer Genet Cytogenet 123:128–132

# Chapter 44
# Chemotherapy in Osteosarcoma

**Stefano Ferrari**

Treatment of osteosarcoma requires a combined approach of surgery of the primary tumor and systemic chemotherapy.

The use of chemotherapy has significantly changed prognosis of patients with osteogenic sarcoma. In the past, when patients received only surgery, survival rate was lower than 20 % (Campanacci 1999).

Standard strategy of chemotherapy for osteosarcoma is based on primary chemotherapy and delayed surgery followed by adjuvant chemotherapy.

The use of chemotherapy before surgery (neoadjuvant) was introduced in the 1970s and was widely adopted since it offers the opportunity of evaluating chemotherapy-induced tumor necrosis by histological examination of the resected surgical specimen (Rosen et al. 1982).

The degree of tumor necrosis used as a marker of chemosensitivity proved to be an important factor predictive of survival. Also in a recent paper exploring several prognostic factors of survival in nonmetastatic osteosarcoma, chemotherapy-induced tumor necrosis still retains prognostic significance (Bacci et al. 2006). Imaging techniques such as PET and dynamic magnetic resonance have been assessed to predict the pathological response (Denecke et al. 2010; Kong et al. 2013).

Microarray technology has been recently investigated to predict chemotherapy response, and a multigene predictive model was developed to classify good and poor responders to preoperative chemotherapy (Man et al. 2005).

S. Ferrari, MD
Chemotherapy Ward of Musculoskeletal Tumors,
Istituto Ortopedico Rizzoli, Bologna, Italy
e-mail: stefano.ferrari@ior.it

P. Picci et al. (eds.), *Atlas of Musculoskeletal Tumors and Tumorlike Lesions*, 189
DOI 10.1007/978-3-319-01748-8_44, © Springer International Publishing Switzerland 2014

## 44.1 Choice of Chemotherapy

Most chemotherapy regimens adopted for osteogenic sarcoma are based on methotrexate, cisplatin, doxorubicin, and ifosfamide.

It is well known that when only CDP and ADM are used, a probability of DFS around 45 % can be expected (Whelan et al. 2012).

Strategies of chemotherapy based on the use of the four active drugs lead to a probability of DFS around 60–65 % (Bacci et al. 2000; Fuchs et al. 1998; Ferrari et al. 2005; Meyers et al. 2005).

It is a matter of discussion whether all patients require intensive and prolonged treatment based on all four active drugs (Anninga et al. 2011).

In a recent study, the Italian Sarcoma Group (ISG) demonstrated that IFO can be given postoperatively and only to patients with a poor pathological response to primary chemotherapy based on MTX, CDP, and ADM (Ferrari et al. 2012). This approach allowed the same results with lower toxicity than a chemotherapy regimen that used all four drugs since the primary phase and for all patients.

A notable study from the Children's Cancer Group and Pediatric Oncology Group has been recently reported in which classic cytotoxic chemotherapy was combined to immunotherapy. The study evaluated whether the addition of ifosfamide and/or muramyl tripeptide-phosphatidylethanolamine (MTP-PE) to methotrexate, cisplatin, and doxorubicin could improve prognosis (Meyers et al. 2005). MTP-PE is a component of the bacterial cell wall conjugated to phosphatidylethanolamine and encapsulated in liposomes with immunostimulating activity. Overall the probability of survival was higher in those patients treated with MTP-PE and a trend was reported in terms of EFS (Meyers).

Based on these data, EMA licensed the drug for the commercial use in Europe.

A new interest is growing around immunotherapy in an adjuvant setting in osteosarcoma patients. The EURAMOS 1 trial randomized patients with a good pathological response to the MAP regimen to receive a maintenance treatment with interferon. Unfortunately the results recently reported at the ASCO meeting did not show advantages for patients assigned to the interferon arm (Bielack et al. 2013).

## 44.2 Salvage Chemotherapy

While no doubt can be raised on the importance of chemotherapy in the first-line treatment of osteosarcoma, the use of chemotherapy after relapse is still under discussion. Complete surgical remission of all sites of recurrence is pivotal for survival after relapse, but the role of chemotherapy has been recently emphasized in patients with a relapse-free interval shorter than 24 months (Chou et al. 2005). In particular, the use of high-dose ifosfamide was recommended to prevent or delay a second recurrence (Chou et al. 2005). In a paper on the same topic, reporting a monoinstitutional experience, only complete surgical removal of metastatic disease was reported as important for long-term survival (Crompton et al. 2006; Ferrari et al. 2003).

The results obtained in patients with synchronous metastases, in spite of an aggressive surgical and chemotherapeutical approach, are worse than those obtained in patients without evident metastases at diagnosis (Daw et al. 2006; Bacci et al. 2003).

A recent paper reported improved results in patients metastatic at presentation who received MTP-PE added to the classical chemotherapy (Chou et al. 2009).

A possible different strategy is suggested in a recent article (Grignani et al. 2012) describing the use of multikinase inhibitor sorafenib. These data make it worthwhile to test the use of sorafenib in a larger population of patients with osteosarcoma and possibly in combination with classic antineoplastic agents.

## 44.3 Parosteal, Periosteal, and Low-Grade Central Osteosarcoma

Wide resection is the standard treatment for these low-grade osteosarcomas. Chemotherapy as used in classic high-grade osteosarcoma is indicated in case of dedifferentiated parosteal osteosarcoma or in case of progression of malignancy in recurrences.

## 44.4 Small-Cell Osteosarcoma

Small-cell osteosarcoma is a rare variant, for which standard treatment is the same as for classic osteosarcoma. There is some controversy over the type of chemotherapy, since small-cell osteosarcoma seems to have a somewhat different chemosensitivity and a worse prognosis. Some authors have recommended treatment with the same chemotherapy as that used for Ewing sarcoma, but this is still investigational, while good results have also been reported with chemotherapy used for conventional osteosarcoma.

## 44.5 Multicentric Osteosarcoma

Multicentric osteosarcoma is extremely rare and has a poor prognosis, whatever the treatment. Aggressive neoadjuvant chemotherapy, followed by, or combined with, radiation therapy to involved sites, and possibly surgical excision of selected lesions, may be considered standard treatment. Nonetheless, the course of the disease is usually rapid and fatal.

## 44.6 Secondary Osteosarcoma

Secondary osteosarcoma may occur in several benign diseases of the bone: Paget's disease, fibrous dysplasia, bone infarcts, chronic osteomyelitis, irradiated bone, as well as in dedifferentiated chondrosarcoma. Poor prognosis has usually been reported, but the issue now raised is that in some cases, this could be due to the older age of these patients, resulting in a less aggressive treatment, and thus that prognosis does not necessarily differ from that of conventional osteosarcoma provided adequate treatment is given. Consequently, aggressive neoadjuvant chemotherapy and wide surgery are strongly recommended also for these rare osteosarcomas.

## Bibliography

Anninga JK, Gelderblom H, Fiocco M, Kroep JR, Taminiau AH, Hogendoorn PC, Egeler RM (2011) Chemotherapeutic adjuvant treatment for osteosarcoma: where do we stand? Eur J Cancer 47(16):2431–2445

Bacci G, Ferrari S, Bertoni F et al (2000) Long-term outcome for patients with nonmetastatic osteosarcoma of the extremity treated at the istituto ortopedico rizzoli according to the istituto ortopedico rizzoli/osteosarcoma-2 protocol: an updated report. J Clin Oncol 18:4016–4027

Bacci G, Briccoli A, Rocca M et al (2003) Neoadjuvant chemotherapy for osteosarcoma of the extremities with metastases at presentation: recent experience at the Rizzoli Institute in 57 patients treated with cisplatin, doxorubicin, and a high dose of methotrexate and ifosfamide. Ann Oncol 14(7):1126–1134

Bacci G, Longhi A, Versari M et al (2006) Prognostic factors for osteosarcoma of the extremity treated with neoadjuvant chemotherapy: 15-year experience in 789 patients treated at a single institution. Cancer 106:1154–1161

Bielack SS, Smeland S, Whelan J et al (2013) MAP plus maintenance pegylated interferon α-2b (MAPIfn) versus MAP alone in patients with resectable high-grade osteosarcoma and good histologic response to preoperative MAP: first results of the EURAMOS-1 "good response" randomization. J Clin Oncol 31 (suppl; abstr LBA10504)

Campanacci M (1999) High grade osteosarcomas. In: Campanacci M (ed) Bone and soft tissue tumors, 2nd edn. Springer, Wien/New York, pp 463–515

Chou AJ, Merola PR, Wexler LH et al (2005) Treatment of osteosarcoma at first recurrence after contemporary therapy: the Memorial Sloan-Kettering Cancer Center experience. Cancer 104:2214–2221

Chou AJ, Kleinerman ES, Krailo M et al (2009) Addition of muramyl tripeptide to chemotherapy for patients with newly diagnosed metastatic osteosarcoma: a report from the Children's Oncology Group. Cancer 115(22):5339–5348

Crompton BD, Re G, Weinberg VK et al (2006) Survival after recurrence of osteosarcoma: a 20-year experience at a single institution. Pediatr Blood Cancer 47:255–259

Daw NC, Billups CA, Rodriguez-Galindo C et al (2006) Metastatic osteosarcoma. Cancer 106:403–412

Denecke T, Hundsdörfer P, Misch D et al (2010) Assessment of histological response of paediatric bone sarcomas using FDG PET in comparison to morphological volume measurement and standardized MRI parameters. Eur J Nucl Med Mol Imaging 37(10):1842–1853

Ferrari S, Briccoli A, Mercuri M et al (2003) Postrelapse survival in osteosarcoma of the extremities: prognostic factors for long-term survival. J Clin Oncol 21(4):710–715

Ferrari S, Smeland S, Mercuri M et al (2005) Neoadjuvant chemotherapy with high-dose ifosfamide, high-dose methotrexate, cisplatin, and doxorubicin for patients with localized osteosarcoma of the extremity: a joint study by the Italian and Scandinavian Sarcoma Groups. J Clin Oncol 23:8845–8852

Ferrari S, Ruggieri P, Cefalo G et al (2012) Neoadjuvant chemotherapy with methotrexate, cisplatin, and doxorubicin with or without ifosfamide in nonmetastatic osteosarcoma of the extremity: an Italian sarcoma group trial ISG/OS-1. J Clin Oncol 30(17):2112–2118

Fuchs N, Bielack SS, Epler D et al (1998) Long-term results of the co-operative German-Austrian-Swiss osteosarcoma study group's protocol COSS-86 of intensive multidrug chemotherapy and surgery for osteosarcoma of the limbs. Ann Oncol 9:893–899

Grignani G, Palmerini E, Dileo P et al (2012) A phase II trial of sorafenib in relapsed and unresectable high-grade osteosarcoma after failure of standard multimodal therapy: an Italian Sarcoma Group study. Ann Oncol 23(2):508–516

Kong CB, Byun BH, Lim I et al (2013) $^{18}$F-FDG PET SUVmax as an indicator of histopathologic response after neoadjuvant chemotherapy in extremity osteosarcoma. Eur J Nucl Med Mol Imaging 40(5):728–736

Man TK, Chintagumpala M, Visvanathan J et al (2005) Expression profiles of osteosarcoma that can predict response to chemotherapy. Cancer Res 65:8142–8150

Meyers PA, Schwartz CL, Krailo M et al (2005) Osteosarcoma: a randomized, prospective trial of the addition of ifosfamide and/or muramyl tripeptide to cisplatin, doxorubicin, and high-dose methotrexate. J Clin Oncol 23:2004–2011

Rosen G, Caparros B, Huvos A et al (1982) Preoperative chemotherapy for osteogenic sarcoma: Selection of postoperative adjuvant chemotherapy based on the response of the primary tumor to preoperative chemotherapy. Cancer 49:1221–1230

Whelan JS, Jinks RC, McTiernan A et al (2012) Survival from high-grade localised extremity osteosarcoma: combined results and prognostic factors from three European Osteosarcoma Intergroup randomised controlled trials. Ann Oncol 23(6):1607–1616

# Chapter 45
# Undifferentiated Pleomorphic Sarcoma (UPS)

**Piero Picci**

**Definition**: A high-grade malignant neoplasm characterized by tumor cells with diffuse pleomorphism in the absence of a specific line of differentiation.
**Epidemiology**: It is less common than OS, but more frequent than fibrosarcoma. It is the most frequent among secondary sarcomas of bone. It is distributed in all age groups. Secondary UPS are observed in more advanced age groups.

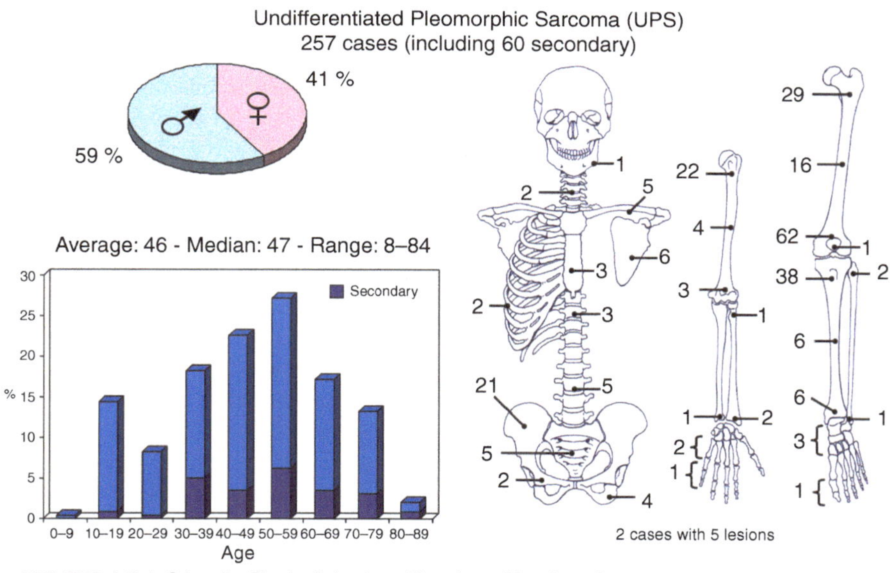


P. Picci, MD
Laboratory of Experimental Oncology, Musculoskeletal Oncology,
Istituto Ortopedico Rizzoli, Via di Barbiano 1/10, Bologna, 40136, Italy
e-mail: piero.picci@ior.it

P. Picci et al. (eds.), *Atlas of Musculoskeletal Tumors and Tumorlike Lesions*,
DOI 10.1007/978-3-319-01748-8_45, 

**Localization**: It is most commonly localized in the long bones, femur, tibia, and humerus. It prevails in the metaphyses and metaepiphyses, although a purely diaphyseal location may be seen. There is prevalence for the distal femur and proximal tibia, but not as pronounced as it is in OS.
**Clinical**: Pain and swelling usually of short duration. Pathologic fracture is common.
**Imaging**: It is a purely osteolytic tumor, centrally or eccentrically located. Osteolysis may be uniform with geographic ill-defined borders. Sometimes scattered or confluent radiolucent areas give a moth-eaten appearance. Infrequently some osteosclerosis is present at the margins of the tumor. The cortex is usually permeated or destroyed by the tumor. Periosteal reaction is not frequent. As in all purely osteolytic tumors, tumor extension in bone and in the surrounding soft tissues is often more pronounced than radiographs indicate. CT, MRI, isotope scan, and angiography, as well as gross and microscopic study, will show the real extension.
**Histopathology**: UPS usually breaches the cortex and extends to soft tissues. The tumor is soft, parenchymatous or encephaloid, with frequent yellowish or tan areas. It may appear more firm and whitish with frequent zones of necrosis or hemorrhage. The margins of the tumor show permeation both in bone and soft tissue. UPS is usually a pleomorphic, high-grade malignancy, with storiform pattern. Large and roundish histiocyte-like cells may be sometimes more spindle, with increasing collagenogenesis. In these fibroblastic areas, the storiform pattern is more evident. Cells have an extensive cytoplasm and are deeply stained, eosinophilic, or granular to foamy and thus clearer. Their nuclei are also large, round, or oval, with a thick nuclear membrane, chromatin lumps, and a prominent nucleolus. Some cells may have monstrous, multiple nuclei. Atypical mitoses are usually abundant. Benign sparse giant cells may be seen. The "fibrous" differentiation associates with a "normalizing" effect on cells, so that cells may be sparse, with small nuclei, and only subtle signs of anaplasia. Permeative patterns and coexistence of more cellular and anaplastic areas however will allow diagnosis. UPS of bone is frequently secondary (bone infarcts, old chondromas and chondrosarcomas, radiated bone, Paget, giant-cell tumor, osteomyelitis).
**Course and Stage**: Most UPS are stage II-B, occasionally stage II-A. As compared to OS, UPS has a higher tendency to multifocal involvement and to bone mets.
**Treatment and Prognosis**: UPS moderately responds to chemotherapy. After preoperative chemotherapy, the same used for OS, the percentage of good responders (necrosis >90 %) is less than in OS, but higher to that of fibrosarcoma. Postoperative chemotherapy is also indicated with the same schedules used for OS. With this aggressive treatment, the percentage of long-term survivors seems to be similar to that obtained in OS, around 60–70 %. UPS is also relatively responsive to radiation therapy, which can be used rarely as adjuvant to surgery or in inoperable cases.

**Key Points**

| | |
|---|---|
| Clinical | Adults, pain and swelling |
| Radiological | Pure lytic lesion |
| Histological | High-grade spindle and pleomorphic cells |
| Differential diagnosis | Bone metastasis and all other primary purely osteolytic lesions of the adults |

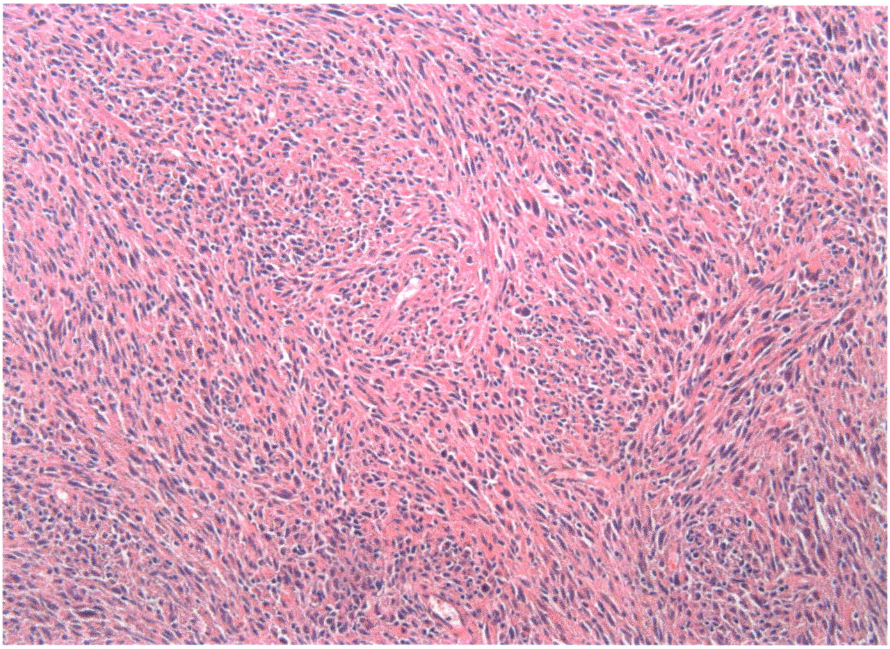

Cytologically high-grade malignant proliferation of spindle and pleomorphic cells arranged in a storiform growth pattern. Fine filamentous collagen is detected between tumor cells. Osteoid production must be absent

## Selected Bibliography

Bielack SS, Schroeders A, Fuchs N, Bacci G, Bauer HC, Mapeli S, Tomeno B, Winkler K (1999) Malignant fibrous histiocytoma of bone: a retrospective EMSOS study of 125 cases. European Musculo-Skeletal Oncology Society. Acta Orthop Scand 70(4):353–360

Hoekstra HJ, Ham SJ, van der Graaf WT, Kamps WA, Molenaar WM, Schraffordt KH (1998) Malignant fibrous histiocytoma of bone: a clinicopathologic study of 81 patients. Cancer 82(5):993–994

Koplas MC, Lefkowitz RA, Bauer TW, Joyce MJ, Ilaslan H, Landa J, Sundaram M (2010) Imaging findings, prevalence and outcome of de novo and secondary malignant fibrous histiocytoma of bone. Skeletal Radiol 39(8):791–798

Niini T, Lahti L, Michelacci F, Ninomiya S, Hattinger CM, Guled M, Böhling T, Picci P, Serra M, Knuutila S (2011) Array comparative genomic hybridization reveals frequent alterations of G1/S checkpoint genes in undifferentiated pleomorphic sarcoma of bone. Genes Chromosomes Cancer 50:291–306

Tarkkanen M, Larramendy ML, Böhling T, Serra M, Hattinger CM, Kivioja A, Elomaa I, Picci P, Knuutila S (2006) Malignant fibrous histiocytoma of bone: analysis of genomic imbalances by comparative genomic hybridisation and C-MYC expression by immunohistochemistry. Eur J Cancer 42(8):1172–1180

# Chapter 46
# Fibrosarcoma

**Piero Picci**

**Definition**: Intermediated to high-grade spindle cell malignant neoplasm lacking any line of differentiation other than fibroblastic.
**Epidemiology**: Rare tumor. Slight preference for males. It is a tumor of adult age.

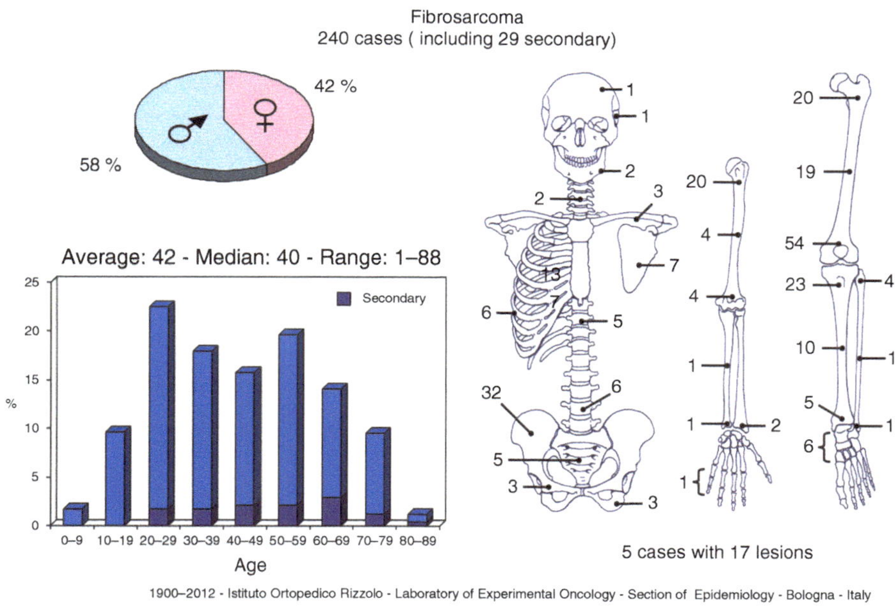


P. Picci, MD
Laboratory of Experimental Oncology, Musculoskeletal Oncology,
Istituto Ortopedico Rizzoli, Via di Barbiano 1/10, Bologna 40136, Italy
e-mail: piero.picci@ior.it

P. Picci et al. (eds.), *Atlas of Musculoskeletal Tumors and Tumorlike Lesions*,
DOI 10.1007/978-3-319-01748-8_46, © Springer International Publishing Switzerland 2014

**Localization**: The most frequent locations are femur, tibia, humerus, and pelvis. Usually located in the meta-diaphysis, it may invade the epiphysis or be multicentric.
**Clinical**: In low-grade tumors, there is mild pain, no soft tissue mass, and a slow course. In high-grade fibrosarcoma, pain may be severe with a soft tissue mass, joint involvement, the course is rapidly progressive, and pathologic fracture is common.
**Imaging**: It is a purely osteolytic tumor with ill-defined limits, interruption of the cortex, and soft tissue mass. Periosteal reaction is scarce or absent. Low-grade tumors may present better-defined borders.
**Histopathology**: Fibrosarcoma tends to be firm and whitish when containing more collagen (low grade) and pink gray and soft to encephaloid, when the cells prevail (high grade). It is typically a spindle cell tumor with a rather uniform aspect throughout. The tumor is composed by spindle cells and collagen fibers. These form fascicles or streams, where cells are all oriented in the same direction (as a school of fish). Grading (from 1 to 4) is based on differentiation, number of cells, hyperchromatism, pleomorphism of nuclei, prominence of nucleoli, and frequency of mitoses. This grading has a clear bearing on treatment and prognosis.
**Course and Staging**: Generally, the course is slower as compared to OS. Grade 1 and 2 fibrosarcomas are staged I (low grade); grades 3–4 are staged II.
**Treatment**: Fibrosarcoma is resistant to radiotherapy. Preoperative chemotherapy is scarcely effective and not feasible in elderly patients. Often surgery cannot be as conservative as it is nowadays in OS.
**Prognosis**: Low-grade fibrosarcomas have a good prognosis with late metastases after a wide surgery. Stage II tumors demonstrate a high incidence of recurrences and metastases. The average survival rate at 10 years is around 30 %.

**Key Points**

| | |
|---|---|
| Clinical | Adults, pain and swelling |
| Radiological | Pure lytic lesion |
| Histological | High-grade spindle cells |
| Differential diagnosis | Bone metastasis and all other primary purely osteolytic lesions of the adults |

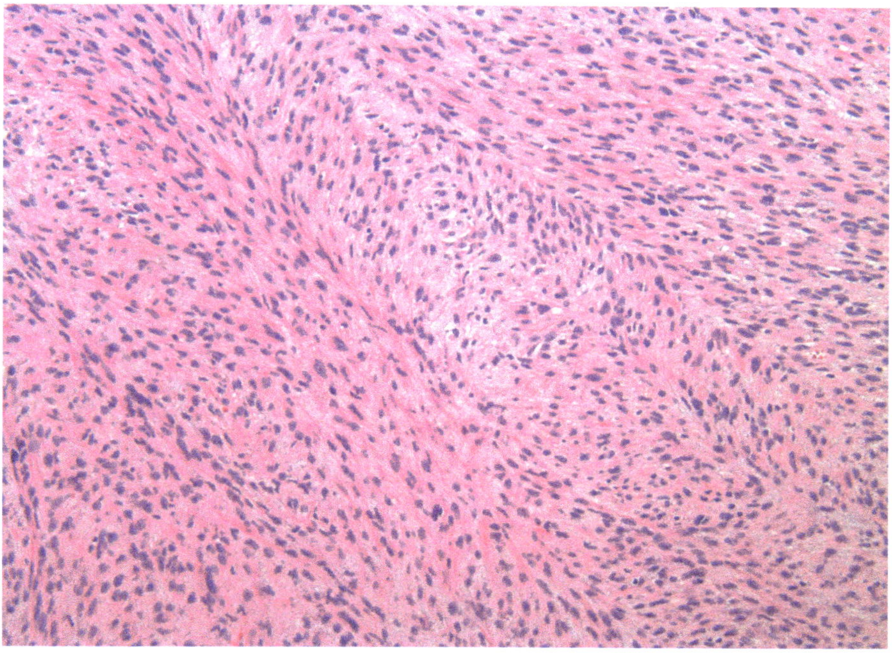

Proliferation of atypical spindle cells with slender and tapered nuclei, arranged in a herringbone growth pattern; slight to abundant collagen is present between cells

## Selected Bibliography

Bertoni F, Capanna R, Calderoni P, Patrizia B, Campanacci M (1984) Primary central (medullary) fibrosarcoma of bone. Semin Diagn Pathol 1(3):185–198

Hattinger CM, Tarkkanen M, Benini S, Pasello M, Stoico G, Bacchini P, Knuutila S, Scotlandi K, Picci P, Serra M (2004) Genetic analysis of fibrosarcoma of bone, a rare tumour entity closely related to osteosarcoma and malignant fibrous histiocytoma of bone. Eur J Cell Biol 83(9):483–491

Niini T, López-Guerrero JA, Ninomiya S, Guled M, Hattinger CM, Michelacci F, Böhling T, Llombart-Bosch A, Picci P, Serra M, Knuutila S (2010) Frequent deletion of CDKN2A and recurrent coamplification of KIT, PDGFRA, and KDR in fibrosarcoma of bone–an array comparative genomic hybridization study. Genes Chromosomes Cancer 49(2):132–143

Papagelopoulos PJ, Galanis EC, Trantafyllidis P, Boscainos PJ, Sim FH, Unni KK (2002) Clinicopathologic features, diagnosis, and treatment of fibrosarcoma of bone. Am J Orthop (Belle Mead NJ) 31(5):253–257. Review

Sanerkin NG (1980) Definitions of osteosarcoma, chondrosarcoma, and fibrosarcoma of bone. Cancer 46(1):178–185

# Chapter 47
# Ewing Sarcoma (ES)

**Nicola Fabbri and Marco Manfrini**

**Definition**: Highly malignant poorly differentiated tumors composed by small round cells. The current definition includes the historical entities of Ewing sarcoma, peripheral primitive neuroectodermal tumor, and Askin tumor, all of which share the same oncogenic fusions and similar biologic behavior. Some present cytologic neurogenic features (PNET): otherwise, they do not differ from ES, and therefore they are considered as a unique entity.
**Epidemiology**: ES is 2/3 times less frequent than OS. Males are preferred by 1.5 to 1. It is more common in Caucasians and rarely appears in individuals of African heritage: around 90 % of cases occur from age 5 to 25; ES represents the second most common bone malignancy affecting children and adolescents.

N. Fabbri, MD (✉)
Department of Surgery, Orthopaedic Service,
Memorial Sloan-Kettering Cancer Center,
York Avenue 1275, New York, NY 10065, USA
e-mail: fabbrin@mskcc.org

M. Manfrini, MD (✉)
Department of Surgery, 3rd Orthopaedic and Traumatologic
Clinic Prevalently Oncologic, Istituto Ortopedico Rizzoli,
Via Pupilli 1, Bologna 40136, Italy
e-mail: marco.manfrini@ior.it

P. Picci et al. (eds.), *Atlas of Musculoskeletal Tumors and Tumorlike Lesions*,
DOI 10.1007/978-3-319-01748-8_47, 

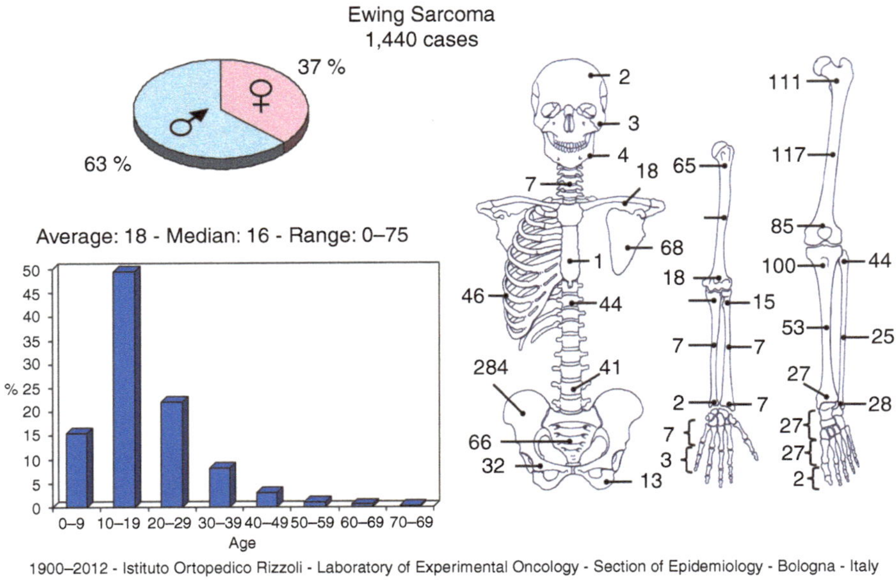


**Localization**: ES is as frequent in flat and short bones as it is in long bones. Among the appendicular skeleton, the most common localization is the femur, followed by the tibia, humerus, fibula, and forearm bones. It is rare in hands and feet. In the trunk, the most frequent localization is the pelvis, followed by vertebrae and sacrum, scapula, ribs, and clavicle. It is rare in the skull. In long bones, tumor is common in the midshaft, but it may involve a larger portion or even the entire bone.

**Clinical**: Pain is the earliest symptom (intense and radiating if nerve impingement is present, as in sacral and vertebral locations). Swelling is also a common and early sign. Occasionally, however, no soft tissue mass is appreciated, even for a long time. Low-grade fever may be associated. An increase of serum HDL and erythrocyte sedimentation rate is frequently seen, sometimes associated with leukocytosis and anemia.

**Imaging**: The imaging presentation is variable: the most frequent is a permeative, infiltrative, poorly defined osteolysis that can also be more destructive, uniform, or vaguely trabeculated. Increased bone radiodensities may associate (bone necrosis and/or deposition of reactive bone). The cortex is almost regularly breached by the tumor or destroyed, and a radiolucent extraosseous tumor mass is usually present. Particularly in flat bones, this mass may be prominent, in contrast with a minimal cortical osteolysis. Rarely, however, the tumor remains intramedullary for a long time. Periosteal reaction is common in the diaphyses of younger patients. The most typical, but infrequent, and unspecific sign is an onionskin appearance, leading to a fusiform enlargement of the shaft. Rarely ES is predominantly periosteal, with erosion of the outer cortex. CT defines alterations of bone and of the extraosseous mass (density similar to muscle). MRI is the best way to study tumor extension in marrow spaces (better demonstrated by STIR technique). Isotope scan and PET scan are hot and may occasionally show skeletal mets at presentation. Typically, extension of ES, as shown by MRI, CT, and isotope scan is much more than what appears on X-rays.

**Histopathology**: ES is very soft and grayish, and when incised it produces a milky fluid. It may be frankly hyperemic or hemorrhagic. Necrosis is common, particularly in the center with a semiliquid appearance, similar to that of purulence. Microscopically the tumor is uniformly made of sheets of small round cells closely packed and without matrix. Cells have a rather monotonous appearance. Cytoplasm is scarce, pale, granular, and clear to eosinophilic, with poorly defined limits. The nuclei are round/oval, with a distinct nuclear membrane and powder-like chromatin, with one or more tiny nucleoli. Mitotic figures are usually rare. Reticulin fibers are lacking between cells but may enwrap cell islands. Blood vessels are numerous and thin. PAS diastase staining (particularly after alcohol fixation) shows intracytoplasmatic glycogen, even very abundant, in the vast majority of cases. Immunohistochemistry shows positivity to vimentin and specific antigens. Electron microscopy shows glycogen in the cytoplasm. Sometimes cells form a rosette pattern (with nuclei at the periphery and cytoplasmatic projections toward the center). Pseudorosette appearance can also be related to necrosis in the center of a cellular nest. When immunohistochemistry shows positivity to S-100 and to NSE, and electron microscopy shows dendritic processes, neurotubes, and dense-core neurosecretory granules, the tumor is labeled as PNET. Necrosis is very frequent and extensive. In areas of necrosis, leukocytic infiltration may be prominent, and it should not be confused with osteomyelitis. The tumor diffusely permeates the marrow spaces, before initiating bone destruction. There is a prominent neovascular and inflammatory response, while fibro-osseous reaction is not able to provide a significant encapsulation. At the edges of the tumor, there are frequent satellites in and well beyond the reactive zone, and even skips.

**Course and Staging**: The tumor usually displays an aggressive growth. There are, however, cases where ES remains intraosseous causing only discontinuous pain for a time ranging from 1 to several years. ES frequently has a multiple metastatic spread to the lungs, skeleton, lymph nodes, and brain, and even a multicentric origin has been postulated. Because of this, Enneking proposed a modified special staging for this tumor: EW I, solitary intraosseous; EW II, solitary extraosseous; EW III, multicentric, skeletal; and EW IV, distant metastases. The most frequent stage at presentation is EW II (70 % of cases). Metastases at presentation are about 20 %.

**Treatment**: ES is sensitive to chemotherapy and radiotherapy, but surgery plays an essential role in treatment of the primary tumor. Thus, treatment is based on a combination of chemotherapy (always), surgery (whenever feasible), and radiotherapy (alternative to surgery or associated to less than wide or radical surgery). Chemotherapy is given pre- and postoperatively and is used as a combination of at least 4 drugs, the most effective being cyclophosphamide, ifosfamide, adriamycin, vincristine, dactinomycin D, and etoposide. Preoperative ctx, in responsive patients, which are the vast majority, causes an extensive necrosis of tumor cells. Because ES cells do not produce any matrix and disappear short after necrosis, effective chemotherapy determines a remarkable shrinkage of the tumor, plus recession of inflammation. Local treatment (surgery alone, surgery and radiotherapy, or radiotherapy alone) is planned according to stage and several individual factors (age, locations, radiological response to ctx) and by balancing the risk of local recurrence against functional loss secondary to local treatment.

**Prognosis**: Before the use of chemotherapy, long-term survival was less than 10 %, also in localized cases. Presently, with multimodal treatment, the same figure is around 60 %. Local recurrence occurs in about 20 % of cases treated with radiation (without surgery) plus chemotherapy. It is rare after wide surgery plus chemotherapy (without radiation). Metastases, usually multiple or disseminated, appear most frequently in the lungs, followed by the skeleton and the lymph nodes. Treatment with aggressive chemotherapy and bone marrow transplantation or with whole-body irradiation is still experimental. A more favorable prognosis can be expected in distal appendicular localizations, small tumor volume, normal ESR and HDL values, use of adequate surgery, and good to complete response to preoperative chemotherapy.

**Key Points**

| | |
|---|---|
| Clinical | Young patients, pain, swelling, and general symptoms |
| Radiological | Variable, usually aggressive, with infiltrative pattern |
| Histological | Monomorphic blue round cell tumor |
| Differential diagnosis | Langerhans cell histiocytosis, osteomyelitis, lymphoma |

**Immunohistochemical Panel**

| | |
|---|---|
| VIM | + |
| CD99 | + |
| Caveolin-1 | + |
| Fli1 | + |
| CAM 5.2 | ± |
| S100 | ± |
| LAC | – |
| TdT | – |

**Chromosomal Translocations**

| | | |
|---|---|---|
| t (11,22)(q24;q12) | EWSR1-FLI1 | 90–95 % |
| t (21;22)(q22;q12) | EWSR1-ERG | 5–10 % |
| t (2;22)(q35;q12) | EWSR1-FEV | Rare |
| t (7;22)(p22;q12) | EWSR1-ETV1 | Rare |
| t (17;22)(q21;q12) | EWSR1-E1AF | Rare |
| t (2;16)(q35;p11) | FUS-FEV | Rare |
| t (16;21)(q11;p22) | FUS-ERG | Rare |
| t (20;22)(q13;q12) | EWSR1-NFACT2 | Rare |
| t (4;22)(q31;q12) | EWSR1-SMARCA5 | Rare |
| t (6;22)(p21;q12) | EWSR1-POU5F1 | Rare |
| t (2;22)(q31;q12) | EWSR1-SP3 | Rare |
| t (4;19)(q35;q13) | EWSR1-POU5F1 | Rare |
| Submicroscopic inv(22) in t(1;22)(p36.1;q12) | EWSR1-PATZ | Rare |

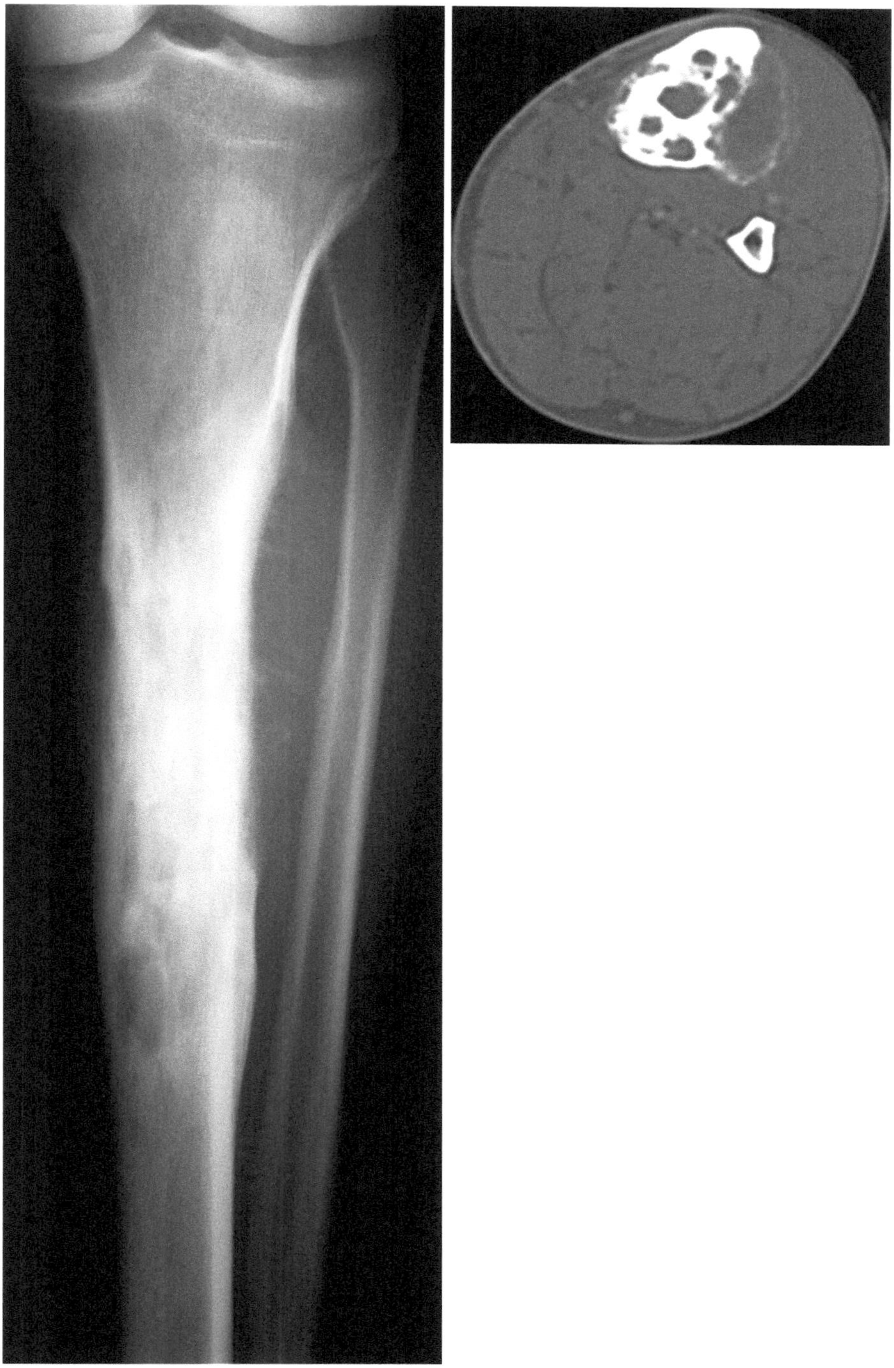

Radiograph and CT. Metaphyseal tumor of the tibia. The lesion is poorly limited and partially sclerotic (not rare in Ewing sarcoma because of the reaction of the remaining bone), destroys the cortex, and invades soft tissues

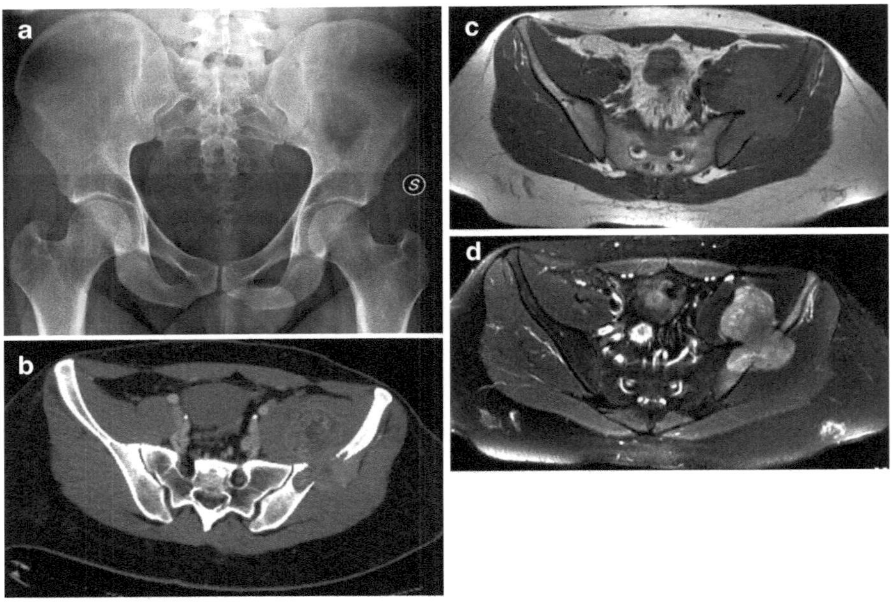

Radiograph (**a**), CT with contrast injection (**b**), and T1 (**c**) and T2 (**d**) W axial MR images

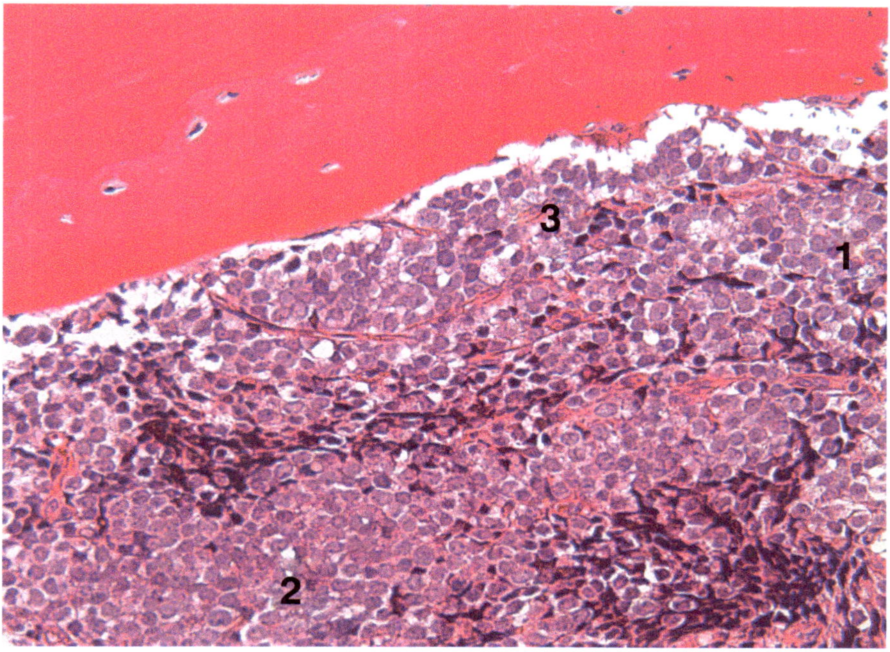


It is a very cellular tissue where, in viable areas, a uniform mat of small round nuclei intensely stained is observed. (*1*) Cytoplasms are rather scarce, pale, vacuolated, and characterized by faded boundaries. The nuclei, instead, are clearly visualized because of their intense color; as such, they constitute the diagnostic element of the tumor. (*2*) Nuclei are twice as large as those of lymphocytes, round oval in shape, with pulverulent chromatin and one or more very small nucleoli. Nearly all nuclei are the same size and characterized by a rather regular shape. Mitotic figures are usually rare. (*3*) Blood vessels, usually scarce, are thin and provided with their own walls

## Selected Bibliography

Balamuth NJ, Womer RB (2010) Ewing's sarcoma. Lancet Oncol 11(2):184–192. Review

Gamberi G, Cocchi S, Benini S, Magagnoli G, Morandi L, Kreshak J, Gambarotti M, Picci P, Zanella L, Alberghini M (2011) Molecular diagnosis in Ewing family tumors: the Rizzoli experience – 222 consecutive cases in four years. J Mol Diagn 13(3):314–324. doi:10.1016/j.jmoldx.2011.01.004. Epub 2011 Mar 31

Li S, Siegal GP (2010) Small cell tumors of bone. Adv Anat Pathol 17(1):1–11. Review

Longhi A, Ferrari S, Tamburini A, Luksch R, Fagioli F, Bacci G, Ferrari C (2012) Late effects of chemotherapy and radiotherapy in osteosarcoma and Ewing sarcoma patients: the Italian Sarcoma Group Experience (1983–2006). Cancer 118(20):5050–5059

Maheshwari AV, Cheng EY (2010) Ewing sarcoma family of tumors. J Am Acad Orthop Surg 18(2):94–107. Review

McCarville MB (2009) The child with bone pain: malignancies and mimickers. Cancer Imaging 9(Spec No A):S115–S121. Review

Paulussen M, Bielack S, Jürgens H, Casali PG, ESMO Guidelines Working Group (2009) Ewing's sarcoma of the bone: ESMO clinical recommendations for diagnosis, treatment and follow-up. Ann Oncol 20(Suppl 4):140–142. Review

# Chapter 48
# Biology of Ewing Sarcoma

**Katia Scotlandi**

Sarcomas are a heterogeneous group of malignant tumors that are derived from mesenchymal tissues, including bone, muscle, and cartilage. Unlike carcinomas, which affect the elderly and are the final result of a long history of progressively accumulating preneoplastic lesions that allowed the molecular definition of multiple carcinogenic events, mostly sarcomas arise abruptly in infants, and their natural history of sarcomas is still mostly unknown. However, in the last decade, we have gained significant new insights into the genetic abnormalities that underlie the pathogenesis of these tumors. Specific molecular alterations have been associated with specific histological subtypes of sarcomas, leading to a new classification of many sarcomas. Conventionally grouped in either soft tissue or bone sarcomas according to the site of their origin, these tumors can now be genetically distinguished in two main groups: those carrying a tumor-specific recurrent chromosome aberrations that appear to be central to the pathogenesis of the tumor and are therefore included among diagnostic criteria and those with complex karyotypes and variable genetic alterations (Helman and Meltzer 2003; Wunder et al. 2007). Sarcomas with recurrent molecular changes include, among others, Ewing sarcoma family tumors, synovial sarcoma, alveolar rhabdomyosarcoma, myxoid liposarcoma, and myxoid chondrosarcoma. These tumors typically carry disease-specific chromosome translocations that frequently result in the expression of an oncogenic chimeric transcription factor, such as EWS-FLI1 in Ewing sarcoma. EWS-FLI1, which is present in around 85 % of Ewing sarcoma, derives specifically from a chromosomal translocation between chromosomes 11 and 22 and is referred to as t(11;22). While other translocations have also been described in Ewing sarcoma, including t(21;22) and t(7;22), all of the translocations involve the fusion of the EWS gene with an ETS family gene. Survival rates of patients with the different translocations appear to be the same (Le Deley et al. 2010;

K. Scotlandi, PhD
Laboratory of Experimental Oncology, Istituto Ortopedico Rizzoli, Bologna, Italy
e-mail: katia.scotlandi@ior.it

P. Picci et al. (eds.), *Atlas of Musculoskeletal Tumors and Tumorlike Lesions*, 211
DOI 10.1007/978-3-319-01748-8_48, © Springer International Publishing Switzerland 2014

van Doorninck et al. 2010). Forced expression of EWS-FLI1 in normal cells can induce tumorigenesis. However, the effects of EWS-FLI1 expression are strongly dependent on cellular background (Kovar 2005). For example, EWS-FLI1 transforms immortalized murine NIH3T3 fibroblasts or mesenchymal stem cells and is required for the oncogenic phenotype of patient-derived EWS cells, but its introduction into primary human or murine fibroblasts leads to growth arrest or cell death, respectively. These data suggest that oncogenic transformation by EWS-FLI requires a permissive cellular background. The critical factors in the permissive background are largely unknown, but may include disruption of the p53 and RB pathways and the presence of an intact IGF pathway as well as of CD99, a 32 kD integral membrane glycoprotein that is highly expressed in most cases of Ewing sarcoma. Recently it was clearly shown how EWS-FLI1 can induce upregulation of IGF1, inducing autocrine activation of IGF-1R and/or of CD99, thus sustaining its transforming activity in mesenchymal stem cells (Cironi et al. 2008; Riggi et al. 2005; Herrero-Martín et al. 2009; McKinsey et al. 2011).

The presence of specific chimeric product is very attractive from a therapeutic point of view. Unfortunately the chimeric transcription factors that give rise to Ewing sarcoma are not druggable at the best of our current knowledge. Thus, the most interesting therapeutic options are druggable pathways regulated by EWS-FLI1, such as the IGF-1R-mediated signaling pathway. Antibodies or tyrosine-kinase inhibitors directed against the IGF-1 receptor protein have also been studied as a potential treatment for advanced Ewing sarcoma (Manara et al. 2007) and have implications for therapy. However, phase I–III clinical studies with anti-IGF-IR drugs have clearly indicated modest toxic effects, with mild and reversible hyperglycemia as the most common toxicity, but limited effectiveness. Particularly in Ewing's sarcoma (EWS), despite the presence of the target in all tumors and ample preclinical evidence supporting the potential value of anti-IGF-IR agents, less than 10 % of cases extraordinarily responded to this therapy (Olmos et al. 2010; Pappo et al. 2011). Evidences for a compensatory role of IR-A when IGF-1R is disrupted (Garofalo et al. 2011, 2012) have been provided, indicating the relationship between these two receptors as one mechanism responsible for acquired and intrinsic resistance to selective anti-IGF-IR therapy. However, further studies are clearly necessary to better define patients that may really benefit from an anti-IGF-IR therapy as well as to rationalize the use of this targeted therapy in combination treatments.

CD99 is another molecule being studied as a potential immunotherapy target for the treatment of Ewing sarcoma. Engagement by anti-CD99 monoclonal antibodies induces massive apoptosis and reduces malignant potential of Ewing sarcoma cells (Scotlandi et al. 2000; Cerisano et al. 2004). An increased antitumor effectiveness of the anti-CD99 monoclonal antibody has recently been demonstrated both in vitro and in vivo when combined with doxorubicin (Scotlandi et al. 2006). In addition recent research suggests that CD99 plays a role in preventing the normal neural differentiation of Ewing cells (Rocchi et al. 2010). Apoptosis also occurs following activation of cell surface receptors by the tumor necrosis factor-related apoptosis-inducing ligand (TRAIL), and the use of agonistic monoclonal antibodies activating the TRAIL receptors is one of the emerging targeted strategies in cancer management, either as monotherapy or in association with other treatment modalities. There is evidence that Ewing's sarcoma cells are exquisitely sensitive to TRAIL-mediated apoptosis. A preclinical study showing efficacy of TRAIL in animals was recently reported (Picarda et al. 2010). These drugs are still purely investigational at this point.

Finally EWS-FLI1 fusion genes were recently found to act in a positive feedback loop to maintain the expression of PARP1, which was required for EWS-FLI-mediated transcription, thereby enforcing oncogene-dependent sensitivity to PARP-1 inhibition (Brenner et al. 2012). Ewing sarcoma cells, primary tumor xenografts, and tumor metastases were all highly sensitive to PARP1 inhibition. Addition of a PARP1 inhibitor to the second-line chemotherapeutic agent temozolomide resulted in complete responses of all treated tumors in an EWS-FLI1-driven mouse xenograft model of Ewing sarcoma. These findings offer a strong preclinical rationale to target the EWS-FLI1:PARP1 intersection as a therapeutic strategy to improve the treatment of Ewing sarcoma.

## Bibliography

Brenner JC, Feng FY, Han S, Patel S, Goyal SV, Bou-Maroun LM, Liu M, Lonigro R, Prensner JR, Tomlins SA, Chinnaiyan AM (2012) PARP-1 inhibition as a targeted strategy to treat Ewing's sarcoma. Cancer Res 72(7):1608–1613. doi:10.1158/0008-5472.CAN-11-3648. Epub 2012 Jan 27

Cerisano V, Aalto Y, Perdichizzi S, Bernard G, Manara MC, Benini S, Cenacchi G, Preda P, Lattanzi G, Nagy B, Knuutila S, Colombo MP, Bernard A, Picci P, Scotlandi K (2004) Molecular mechanisms of CD99-induced caspase-independent cell death and cell-cell adhesion in Ewing's sarcoma cells: actin and zyxin as key intracellular mediators. Oncogene 23(33):5664–5674

Cironi L, Riggi N, Provero P, Wolf N, Suvà ML, Suvà D, Kindler V, Stamenkovic I (2008) IGF1 is a common target gene of Ewing's sarcoma fusion proteins in mesenchymal progenitor cells. PLoS One 3(7):e2634

Garofalo C, Manara MC, Nicoletti G, Marino MT, Lollini PL, Astolfi A, Pandini G, López-Guerrero JA, Schaefer KL, Belfiore A, Picci P, Scotlandi K (2011) Efficacy of and resistance to anti-IGF-1R therapies in Ewing's sarcoma is dependent on insulin receptor signaling. Oncogene 30(24):2730–2740. doi:10.1038/onc.2010.640. Epub 2011 Jan 31

Garofalo C, Mancarella C, Grilli A, Manara MC, Astolfi A, Marino MT, Conte A, Sigismund S, Carè A, Belfiore A, Picci P, Scotlandi K (2012) Identification of common and distinctive mechanisms of resistance to different anti-IGF-IR agents in Ewing's sarcoma. Mol Endocrinol 26(9):1603–1616. doi:10.1210/me.2012-1142. Epub 2012 Jul 13

Helman LJ, Meltzer P (2003) Mechanisms of sarcoma development. Nat Rev Cancer 3(9):685–694

Herrero-Martín D, Osuna D, Ordóñez JL, Sevillano V, Martins AS, Mackintosh C, Campos M, Madoz-Gúrpide J, Otero-Motta AP, Caballero G, Amaral AT, Wai DH, Braun Y, Eisenacher M, Schaefer KL, Poremba C, de Alava E (2009) Stable interference of EWS-FLI1 in an Ewing sarcoma cell line impairs IGF-1/IGF-1R signalling and reveals TOPK as a new target. Br J Cancer 101(1):80–90. Epub 2009 Jun 2

Kovar H (2005) Context matters: the hen or egg problem in Ewing's sarcoma. Semin Cancer Biol 15(3):189–196

Le Deley MC, Delattre O, Schaefer KL, Burchill SA, Koehler G, Hogendoorn PC, Lion T, Poremba C, Marandet J, Ballet S, Pierron G, Brownhill SC, Nesslböck M, Ranft A, Dirksen U, Oberlin O,

Lewis IJ, Craft AW, Jürgens H, Kovar H (2010) Impact of EWS-ETS fusion type on disease progression in Ewing's sarcoma/peripheral primitive neuroectodermal tumor: prospective results from the cooperative Euro-E.W.I.N.G. 99 trial. J Clin Oncol 28(12):1982–1988

Manara MC, Landuzzi L, Nanni P, Nicoletti G, Zambelli D, Lollini PL, Nanni C, Hofmann F, García-Echeverría C, Picci P, Scotlandi K (2007) Preclinical in vivo study of new insulin-like growth factor-I receptor–specific inhibitor in Ewing's sarcoma. Clin Cancer Res 13(4):1322–1330

McKinsey EL, Parrish JK, Irwin AE, Niemeyer BF, Kern HB, Birks DK, Jedlicka P (2011) A novel oncogenic mechanism in Ewing sarcoma involving IGF pathway targeting by EWS/Fli1-regulated microRNAs. Oncogene 30(49):4910–4920

Olmos D, Postel-Vinay S, Molife LR, Okuno SH, Schuetze SM, Paccagnella ML, Batzel GN, Yin D, Pritchard-Jones K, Judson I, Worden FP, Gualberto A, Scurr M, de Bono JS, Haluska P (2010) Safety, pharmacokinetics, and preliminary activity of the anti-IGF-1R antibody figitumumab (CP-751,871) in patients with sarcoma and Ewing's sarcoma: a phase 1 expansion cohort study. Lancet Oncol 11(2):129–135. Epub 2009 Dec 23

Pappo AS, Patel SR, Crowley J et al (2011) R1507, a monoclonal antibody to the insulin-like growth factor 1 receptor, in patients with recurrent or refractory Ewing sarcoma family of tumors: results of a phase II sarcoma alliance for research through collaboration study. J Clin Oncol 29:4541–4547

Picarda G, Lamoureux F, Geffroy L, Delepine P, Montier T, Laud K, Tirode F, Delattre O, Heymann D, Rédini F (2010) Preclinical evidence that use of TRAIL in Ewing's sarcoma and osteosarcoma therapy inhibits tumor growth, prevents osteolysis, and increases animal survival. Clin Cancer Res 16(8):2363–2374. Epub 2010 Apr 6

Riggi N, Cironi L, Provero P, Suvà ML, Kaloulis K, Garcia-Echeverria C, Hoffmann F, Trumpp A, Stamenkovic I (2005) Development of Ewing's sarcoma from primary bone marrow-derived mesenchymal progenitor cells. Cancer Res 65(24):11459–11468

Rocchi A, Manara MC, Sciandra M, Zambelli D, Nardi F, Nicoletti G, Garofalo C, Meschini S, Astolfi A, Colombo MP, Lessnick SL, Picci P, Scotlandi K (2010) CD99 inhibits neural differentiation of human Ewing sarcoma cells and thereby contributes to oncogenesis. J Clin Invest 120(3):668–680. doi:10.1172/JCI36667. Epub 2010 Feb 8

Scotlandi K, Baldini N, Cerisano V, Manara MC, Benini S, Serra M, Lollini PL, Nanni P, Nicoletti G, Bernard G, Bernard A, Picci P (2000) CD99 engagement: an effective therapeutic strategy for Ewing tumors. Cancer Res 60(18):5134–5142

Scotlandi K, Perdichizzi S, Bernard G, Nicoletti G, Nanni P, Lollini PL, Curti A, Manara MC, Benini S, Bernard A, Picci P (2006) Targeting CD99 in association with doxorubicin: an effective combined treatment for Ewing's sarcoma. Eur J Cancer 42(1):91–96. Epub 2005 Dec 2

van Doorninck JA, Ji L, Schaub B, Shimada H, Wing MR, Krailo MD, Lessnick SL, Marina N, Triche TJ, Sposto R, Womer RB, Lawlor ER (2010) Current treatment protocols have eliminated the prognostic advantage of type 1 fusions in Ewing sarcoma: a report from the Children's Oncology Group. J Clin Oncol 28(12):1989–1994. Epub 2010 Mar 22

Wunder JS, Nielsen TO, Maki RG, O'Sullivan B, Alman BA (2007) Opportunities for improving the therapeutic ratio for patients with sarcoma. Lancet Oncol 8(6):513–524

# Chapter 49
# Chemotherapy in Ewing Sarcoma

**Stefano Ferrari**

Neoadjuvant chemotherapy with the aim of reducing tumor mass, increasing the likelihood of local control, and facilitating conservative surgical approaches is standard (ESMO and European Sarcoma Network Working Group 2012).

The combination of the six drugs, vincristine, dactinomycin, Adriamycin, cyclophosphamide, ifosfamide, and etoposide, is the standard treatment for first-line chemotherapy of Ewing sarcoma (ES) (Grier et al. 2003).

In surgically treated patients, pathological evaluation of chemotherapy identifies patients with different probability of survival, higher in those patients with a good pathological response (Cotterill et al. 2000; Bacci et al. 2000).

Different strategies of dose intensification have been evaluated (Granowetter et al. 2009; Womer et al. 2012; Ferrari et al. 2011). The results of Protocol AEWS0031 suggest that a dose intensification obtained by an interval compression (every 2 weeks) of the chemotherapy courses may be more effective compared to a standard approach based on chemotherapy courses given every 3 weeks. This strategy seems to be beneficial in pediatric patients, but not in adults (Womer et al. 2012).

A dose intensification based on high-dose chemotherapy and peripheral blood stem cell rescue was used in poor responder patients to a six-drug primary chemotherapy in a joint study between Italian Sarcoma Group and Scandinavian Sarcoma Group. The results showed that poor responder patients had the same probability of event-free survival obtained in good responder patients (Ferrari et al. 2011).

To investigate the role of high-dose chemotherapy in ES, in 1999, a randomized clinical trial (EUROEWING) was activated, but unfortunately the results of the group of patients randomized for the high-dose chemotherapy are not yet available.

S. Ferrari, MD
Chemotherapy Ward of Musculoskeletal Tumors, Istituto Ortopedico Rizzoli, Bologna, Italy
e-mail: stefano.ferrari@ior.it

P. Picci et al. (eds.), *Atlas of Musculoskeletal Tumors and Tumorlike Lesions*,
DOI 10.1007/978-3-319-01748-8_49, © Springer International Publishing Switzerland 2014

The strategy of treatment for patients with synchronous metastases is based on chemotherapy combined with local and metastatic control by the use of aggressive surgical and/or radiotherapeutic approach.

The probability of survival of patients with metastatic disease is lower than 30 % even when intensive but standard chemotherapy regimens are used (Cangir et al. 1990; Kolb et al. 2003).

Use of high-dose therapy and autologous stem cell rescue has been adopted, and in a recent joint study of the Italian Sarcoma Group and Scandinavian Sarcoma Group, a survival higher than 50 % has been reported in patients with lung metastases and/or 1 bone metastasis (Luksch et al. 2012). In patients with very high risk, such as those with multivisceral involvement, multiple skeletal lesions, and/or bone marrow infiltration, the probability of survival reached 30 % when high-dose chemotherapy was delivered in addition to aggressive surgery and/or radiotherapy for all the sites of disease (Ladenstein et al. 2010).

In case of patients with only pulmonary recurrent disease, the relapse-free interval is the main prognostic factor (Bacci et al. 2006). Standard treatment of patients with isolated lung metastases is polychemotherapy, to be selected considering the previous chemotherapy treatment (ESMO and European Sarcoma Network Working Group 2012). High-dose treatments with autologous stem cell rescue are still investigational, but frequently adopted in several centers. In case of isolated lung metastases with a long (> 36 months) relapse-free survival, chemotherapy can be omitted, and surgical resection with radiation therapy of the lungs is recommended (Briccoli et al. 2004).

For patients with extrapulmonary metastases, prognosis is dismal, and also in this condition, chemotherapy is to be defined according to the previous treatment. High-dose treatments with peripheral stem cell rescue have been used by various investigators, with encouraging results. Generally, they are reserved to optimal responders after some kind of induction therapy willing to enter open clinical studies on such approaches.

Irinotecan and temozolomide, topotecan and cyclophosphamide, and high-dose ifosfamide have shown activity in recurrent ES of bone (Hunold et al. 2006; Wagner et al. 2007; Casey et al. 2009; Ferrari et al. 2009).

## Bibliography

Bacci G, Ferrari S, Bertoni F et al (2000) Prognostic factors in nonmetastatic Ewing's sarcoma of bone treated with adjuvant chemotherapy: Analysis of 359 patients at the Istituto Ortopedico Rizzoli. J Clin Oncol 18:4–11

Bacci G, Longhi A, Ferrari S et al (2006) Pattern of relapse in 290 patients with nonmetastatic Ewing's sarcoma family tumors treated at a single institution with adjuvant and neoadjuvant chemotherapy between 1972 and 1999. Eur J Surg Oncol 32(9):974–979

Briccoli A, Rocca M, Ferrari S, Mercuri M, Ferrari C, Bacci G (2004) Surgery for lung metastases in Ewing's sarcoma of bone. Eur J Surg Oncol 30(1):63–67

Cangir A, Vietti TJ, Gehan EA et al (1990) Ewing's sarcoma metastatic at diagnosis. Results and comparisons of two intergroup Ewing's sarcoma studies. Cancer 66:887–893

Casey DA, Wexler LH, Merchant MS et al (2009) Irinotecan and temozolomide for Ewing sarcoma: the Memorial Sloan-Kettering experience. Pediatr Blood Cancer 53(6):1029–1034

Cotterill SJ, Ahrens S, Paulussen M et al (2000) Prognostic factors in Ewing's tumor of bone: analysis of 975 patients from the European Intergroup Cooperative Ewing's Sarcoma Study Group. J Clin Oncol 18:3108–3114

ESMO, European Sarcoma Network Working Group (2012) Bone sarcomas: ESMO Clinical Practice Guidelines for diagnosis, treatment and follow-up. Ann Oncol 23(Suppl 7):vii100–vii109

Ferrari S, del Prever AB, Palmerini E et al (2009) Response to high-dose ifosfamide in patients with advanced/recurrent Ewing sarcoma. Pediatr Blood Cancer 52(5):581–584

Ferrari S, Sundby Hall K, Luksch R et al (2011) Nonmetastatic Ewing family tumors: high-dose chemotherapy with stem cell rescue in poor responder patients. Results of the Italian Sarcoma Group/Scandinavian Sarcoma Group III protocol. Ann Oncol 22(5):1221–1227

Granowetter L, Womer R, Devidas M et al (2009) Dose-intensified compared with standard chemotherapy for nonmetastatic Ewing sarcoma family of tumors: a Children's Oncology Group Study. J Clin Oncol 27(15):2536–2541

Grier HE, Krailo MD, Tarbell NJ et al (2003) Addition of ifosfamide and etoposide to standard chemotherapy for Ewing's sarcoma and primitive neuroectodermal tumor of bone. N Engl J Med 348(8):694–701

Hunold A, Weddeling N, Paulussen M, Ranft A, Liebscher C, Jürgens H (2006) Topotecan and cyclophosphamide in patients with refractory or relapsed Ewing tumors. Pediatr Blood Cancer 47(6):795–800

Kolb EA, Kushner BH, Gorlick R et al (2003) Long-term event-free survival after intensive chemotherapy for Ewing's family of tumors in children and young adults. J Clin Oncol 21(18):3423–3430

Ladenstein R, Pötschger U, Le Deley MC et al (2010) Primary disseminated multifocal Ewing sarcoma: results of the Euro-EWING 99 trial. J Clin Oncol 28(20):3284–3291

Luksch R, Tienghi A, Hall KS et al (2012) Primary metastatic Ewing's family tumors: results of the Italian Sarcoma Group and Scandinavian Sarcoma Group ISG/SSG IV Study including myeloablative chemotherapy and total-lung irradiation. Ann Oncol 23(11):2970–2976

Wagner LM, McAllister N, Goldsby RE et al (2007) Temozolomide and intravenous irinotecan for treatment of advanced Ewing sarcoma. Pediatr Blood Cancer 48(2):132–139

Womer RB, West DC, Krailo MD et al (2012) Randomized controlled trial of interval-compressed chemotherapy for the treatment of localized Ewing sarcoma: a report from the Children's Oncology Group. J Clin Oncol 30(33):4148–4154

# Chapter 50
# Adamantinoma

**Nicola Fabbri and Pietro Ruggieri**

**Definition**: Low-grade malignant tumor with peculiar epithelial-like component that sometimes is reminiscent of the ameloblastoma of jaw bones, hence its name.
**Epidemiology**: It is very rare, slightly prevailing in males, and affects all ages, but it is more common between 20 and 40. Exceptional in children below 11–12 years of age.

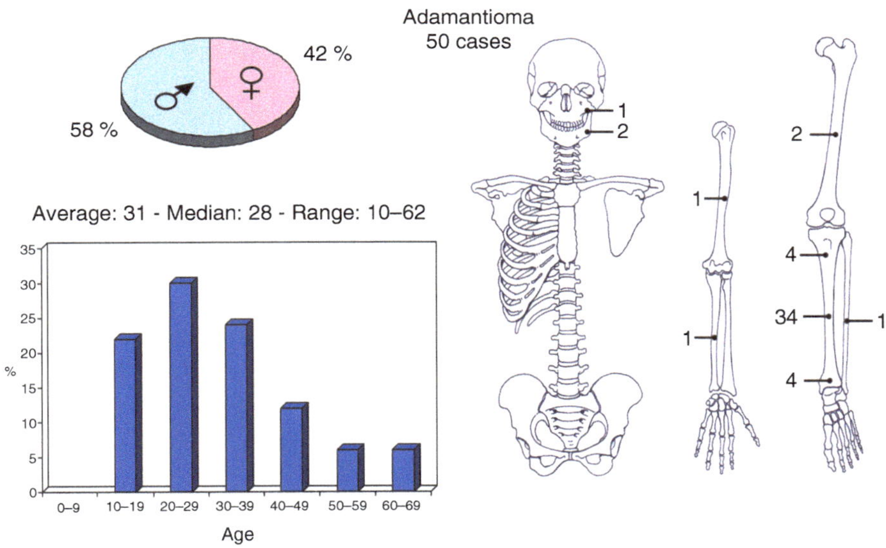


1900–2012 - Istituto Ortopedico Rizzoli - Laboratory of Experimental Oncology - Section of Epidemiology - Bologna - Italy

N. Fabbri, MD (✉)
Department of Surgery, Orthopaedic Service, Memorial Sloan-Kettering Cancer Center, 1275 York Avenue, New York, NY 10065, USA
e-mail: fabbrin@mskcc.org

P. Ruggieri, MD, PhD (✉)
2nd Orthopaedic and Traumatologic Clinic, Istituto Ortopedico Rizzoli, Bologna, Italy
e-mail: pietro.ruggieri@ior.it

P. Picci et al. (eds.), *Atlas of Musculoskeletal Tumors and Tumorlike Lesions*,
DOI 10.1007/978-3-319-01748-8_50, © Springer International Publishing Switzerland 2014

**Localization**: Diaphysis of the long bones. The tibia is affected in 80 % of cases. Followed by the fibula, femur, humerus, ulna, and radius. Exceptional in the soft tissues of the leg.

**Clinical**: Pain and swelling. Usually, trauma is recorded in clinical history.

**Imaging**: Eccentric osteolysis, mostly involving the anterior cortex of the tibia, with expansion and frequent cancellation of the cortex, with no or minimal periosteal reaction, frequent sclerotic borders, and soft tissue mass. CT and MRI indicate a solid fibrous tissue (low T1, high T2 signal), sharply demarcated from the surrounding tissues. Sometimes (mainly in younger patients), osteolyses are multifocal and extended in the diaphysis and associated with anterior bowing of the tibia, so that the imaging is, at least initially, identical to that of osteofibrous dysplasia.

**Histopathology**: Grossly, the tissue is solid, whitish, and fibrous-firm, with occasional hemorrhagic or cystic areas. Histologically, it is composed by epithelial-like aggregates, surrounded by a fibrous or fibro-osseous stroma. Different histological patterns can be observed. The more frequent and typical (basaloid) is represented by cords and islands of epithelial cells, similar to a basal cell carcinoma. Cells may be dispersed and spindling in the center of the islands, cubic, and palisading in the periphery, recalling the ameloblastoma of jaw bones. Reticulin fibers surround the islands, but are lacking between cells. Sometimes, the tumor is composed by spindle sarcomatous cells, in intersecting fascicles or in whorls, with reticulin fibers between cells. Otherwise, small ramified tubules or alveolar cavities are seen, matted by cubic or flattened cells, in one or more layers. This aspect may resemble an adenocarcinoma of the breast or a hemangioendothelioma. Rarely and focally is a squamocellular aspect seen, with prickles and pearls of squamous epithelium. The cells of adamantinoma, whatever their shape and aggregation, are of small to medium size, with rather hyperchromatic nuclei, small or absent nucleoli, but usually with scarce pleomorphism and rare mitosis. They regularly stain with keratin antibodies and, at the electron microscope, show tonofibrils, desmosomes, microfilaments, and basal membrane. All the aforementioned cellular aggregates are surrounded by a fibrous stroma, usually abundant and dense, whose cells are positive for vimentin, not for keratins. Frequently, and especially at the periphery of the tumor, the stroma is similar to the tissue of osteofibrous dysplasia.

**Course and Staging**: Adamantinoma has an indolent course. The interval between the first symptoms and surgical treatment may be of several years, up to 20. Mets occur in about 15 % of cases, to the lungs, lymph nodes, and skeleton. The stage of the tumor at presentation is I A or B.

**Treatment and Prognosis**: Treatment is only surgical and should aim at wide margins. Regional lymph nodes should be excised when showing some enlargement. An intralesional or marginal excision is regularly followed by local recurrence. With an adequate treatment, the cure rate should be around 90 %.

For differential diagnosis with osteofibrous dysplasia of long bones, see Chap. 9.

## Key Points

| | |
|---|---|
| Clinical | Young adults, pain and swelling |
| Radiological | Eccentric pure lytic lesion in the tibia diaphysis |
| Histological | Epithelial cells in a fibrous stroma |
| Differential diagnosis | Osteofibrous dysplasia of long bones |

## Immunohistochemical Panel

| | |
|---|---|
| CK | Groups of epithelial cells positive |

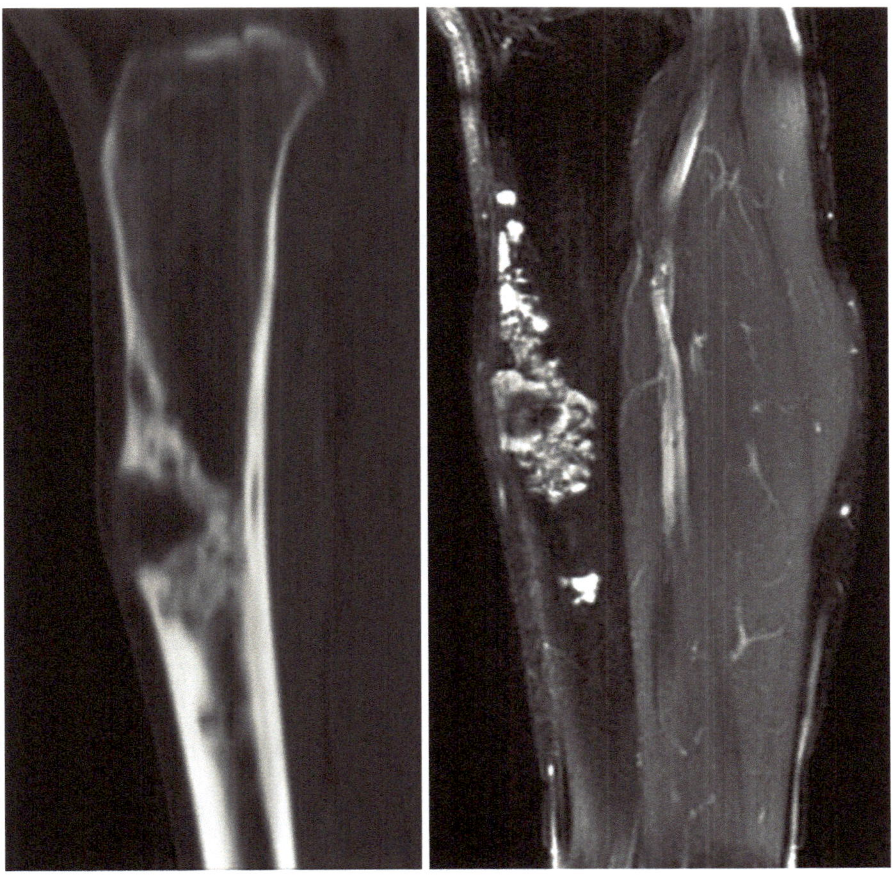

Sagittal CT reconstruction and T2 MR images. The lesion involves the anterior cortex of the shaft of the tibia and is heterogeneous. A satellite lesion is easily detected in the cortex

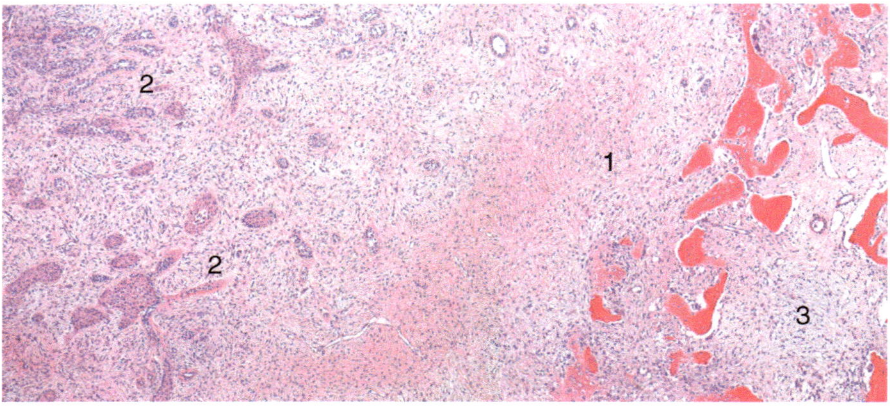


Different histological aspects may be observed, even within the same specimen. (*1*) Spindle cells arranged in whorls. (*2*) Group of epithelial cells sometimes recalling the aspect of tubular structures. (*3*) Abundant fibrous stroma; this collagenous connective tissue explains the sustained consistency and whitish color of the tumor

## Selected Bibliography

Campanacci M, Giunti A, Bertoni F, Laus M, Gitelis S (1981) Adamantinoma of the long bones. The experience at the Istituto Ortopedico Rizzoli. Am J Surg Pathol 5(6):533–542

Khanna M, Delaney D, Tirabosco R, Saifuddin A (2008) Osteofibrous dysplasia, osteofibrous dysplasia-like adamantinoma and adamantinoma: correlation of radiological imaging features with surgical histology and assessment of the use of radiology in contributing to needle biopsy diagnosis. Skeletal Radiol 37(12):1077–1084

Most MJ, Sim FH, Inwards CY (2010) Osteofibrous dysplasia and adamantinoma. J Am Acad Orthop Surg 18(6):358–366. Review

Roque P, Mankin HJ, Rosenberg A (2008) Adamantinoma: an unusual bone tumour. Chir Organi Mov 92(3):149–154

# Chapter 51
# Vascular Tumors

**Nicola Fabbri and Pietro Ruggieri**

In the human body, endothelial cells account for a 6–7 $m^2$ surface and weight approximately 1.2 kg. They are involved in several important processes, such as oxygen transportation, inflammation, and wound repair. Many regulatory factors are necessary to the functions of the endothelial cells.

**Course and Staging**: Benign: stage 1 or 2. Low grade: stage I seem to behave unpredictably. Some have a slow indolent course; very few metastasize. Whether these can progress to a high-grade hemangioendothelioma is still unknown. High grade: stage II, usually B, are most aggressive and even fulminating malignancies.

**Treatment**: Benign: curettage plus local adjuvants (phenol, cement) and radiation therapy in inoperable situations. Low grade: almost the same as benign, en bloc resection is often preferred to curettage. High grade: treated very aggressively, aiming to very wide or radical surgical margins. Survival rate at 5 years is around 20 %. It is not known whether chemotherapy can improve such figure.

N. Fabbri, MD (✉)
Department of Surgery, Orthopaedic Service, Memorial Sloan-Kettering Cancer Center, 1275 York Avenue, New York, NY 10065, USA
e-mail: fabbrin@mskcc.org

P. Ruggieri, MD, PhD (✉)
2nd Orthopaedic and Traumatologic Clinic, Istituto Ortopedico Rizzoli, Bologna, Italy
e-mail: pietro.ruggieri@ior.it

P. Picci et al. (eds.), *Atlas of Musculoskeletal Tumors and Tumorlike Lesions*,
DOI 10.1007/978-3-319-01748-8_51, © Springer International Publishing Switzerland 2014

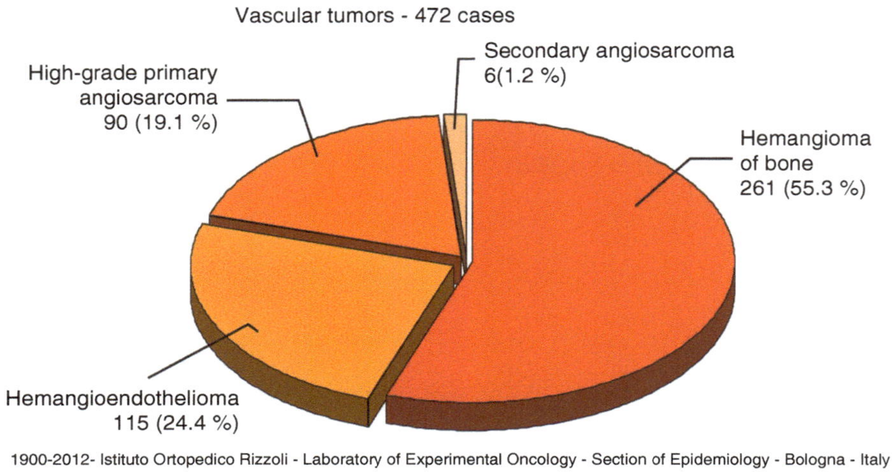


**Prognosis**

Hemangiomas → benign lesions
Hemangioendothelioma → good prognosis
Angiosarcoma → poor prognosis

## 51.1 Characteristics of the Single Entities

### *51.1.1 Hemangioma of Bone*

**Definition**: Benign solitary tumor composed of newly formed vessels.
**Epidemiology**: Rare tumors, accounting for less than 1 % of all bone tumors, even though at least 10 % of patients from autopsy series showed vertebral hemangiomas as occasional finding. About 70 % of the cases are diagnosed in patients between 30 and 60 years. The most frequent locations are the craniofacial bones, followed by the vertebrae, the ribs, and the appendicular skeleton.
**Clinical**: Frequently asymptomatic sometimes pain when pathologic fracture occurs.
**Histology**: Hemangiomas can be classified in four groups: capillary, cavernous, venous, and mixed. In the bone they almost all belong to the fourth category. The cells can be quite plump and composed of many interesting vessels and can be diagnosed as cellular or epithelioid variants of hemangioma and must be distinguished

from low-grade malignant (hemangioendothelioma) to high-grade (angiosarcoma) vascular tumors. One of the key factors is age. If the patient is younger than 25 years of age, the vascular tumor is almost always a hemangioma variant. Most patients with true angiosarcoma are over 50–60 years of age.

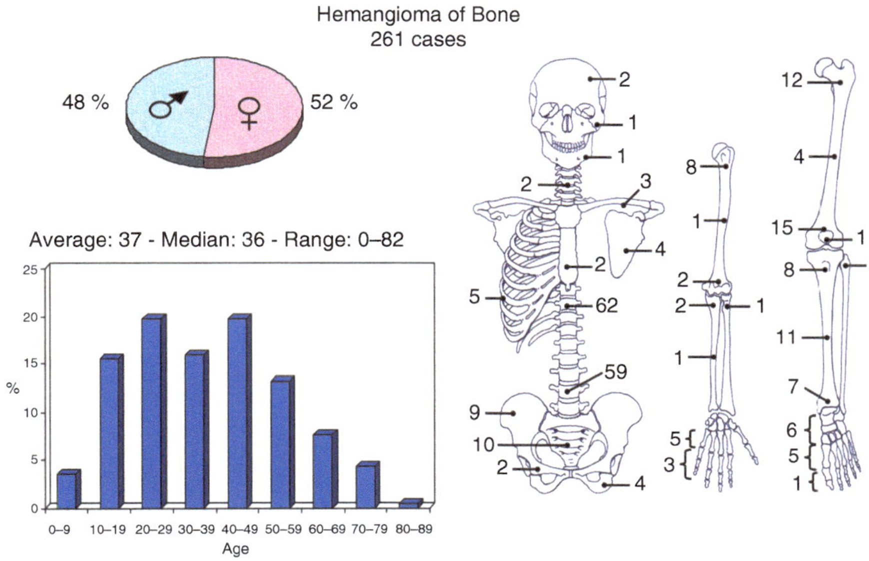


1900–2012 - Istituto Ortopedico Rizzoli - Laboratory of Experimental Oncology - Section of Epidemiology - Bologna - Italy

### *51.1.2 Hemangioendothelioma*

**Definition**: Borderline to low-grade malignant vascular tumors, capable of aggressive local growth, recurrence, or both, and some variants, such as epithelioid hemangioendothelioma, can also give distant metastases. Epithelioid hemangioendothelioma was defined as a peculiar entity in 1982 by S. Weiss and F. Enzinger.

**Incidence**: Extremely rare, they frequently occur during the second to third decade of life.

**Location**: Bones of the lower extremities are the most frequently involved location, followed by the spine. Separated synchronous foci are frequently present in different anatomic locations.

**Imaging**: Lytic frequently multiple lesions with variable peripheral sclerosis.

**Histology**: Solid nests or anastomosing chords of epithelioid cells that occasionally form narrow vascular channels. The supporting connective tissue may exhibit myxoid change.

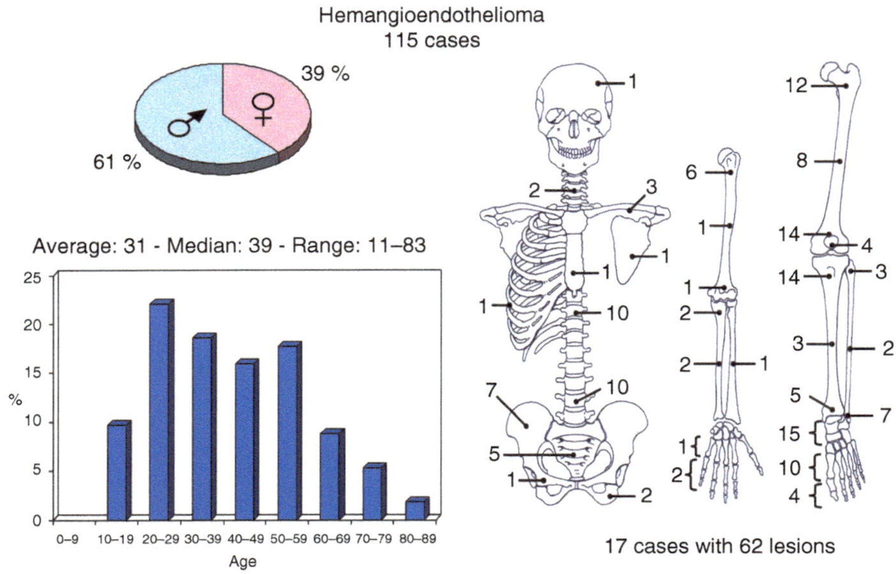


1900–2012 - Istituto Ortopedico Rizzoli - Laboratory of Experimental Oncology - Section of Epidemiology - Bologna - Italy

### *51.1.3 Angiosarcoma*

**Definition**: Malignant intermediate to high-grade vascular tumor, most frequently occurring in the soft tissues, rarely occurring in the bones. Primary angiosarcomas of bone are very rare and account for less than 1 % of all angiosarcomas.
**Location**: Usually occurring in the long tubular bones, rarely can be multicentric, most frequently occurring in the lower extremities.
**Imaging**: Destructive lytic mass with infiltrative borders.
**Histology**: Atypical endotheliod cell with vasoformative features. Intermediate grade lesions can be easily recognizable, while high-grade lesions exhibit prominent clear atypia and brisk mitotic activity.

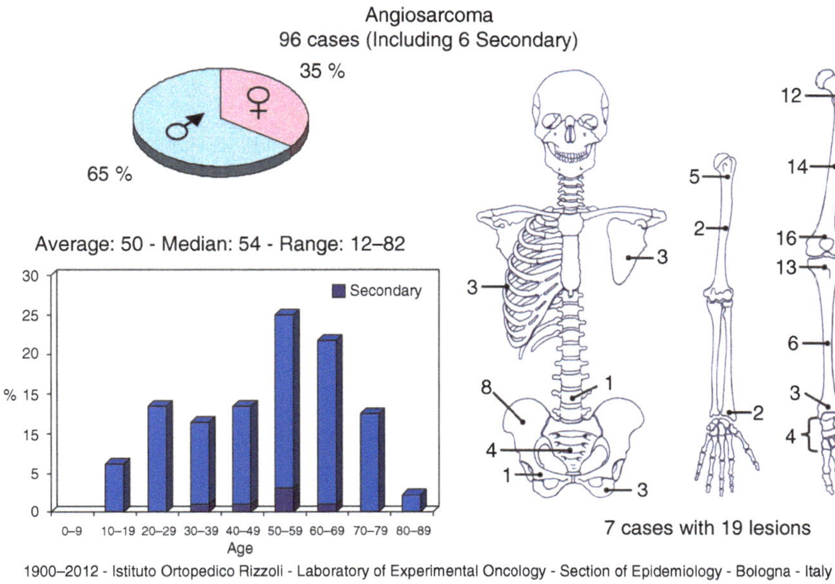


## Prognosis

| |
|---|
| Hemangiomas → benign lesions |
| Hemangioendothelioma → good prognosis |
| Angiosarcoma → poor prognosis |

## Key Points

| | |
|---|---|
| Clinical | Adults, pain and swelling |
| Radiological | Frequently multifocal, pure lytic lesion |
| Histological | Proliferation of endothelial cells at various grade of differentiation |
| Differential diagnosis | Bone metastasis, myeloma, and all other primary purely osteolytic lesions of adults |

## Immunohistochemical Panel

| | |
|---|---|
| VIM | + |
| CD31 | + |
| CD34 | +/– |
| Fli-1 | + |
| CK | +/– |
| ERG | + (Am J Surg Pathol 35; 3: 432–441) |

**Chromosomal Translocations**

| | | |
|---|---|---|
| Epithelioid hemangioendothelioma | | |
| t(1;3)(p36.3;q25) | WWTR1-CAMTA1 | 100 % |

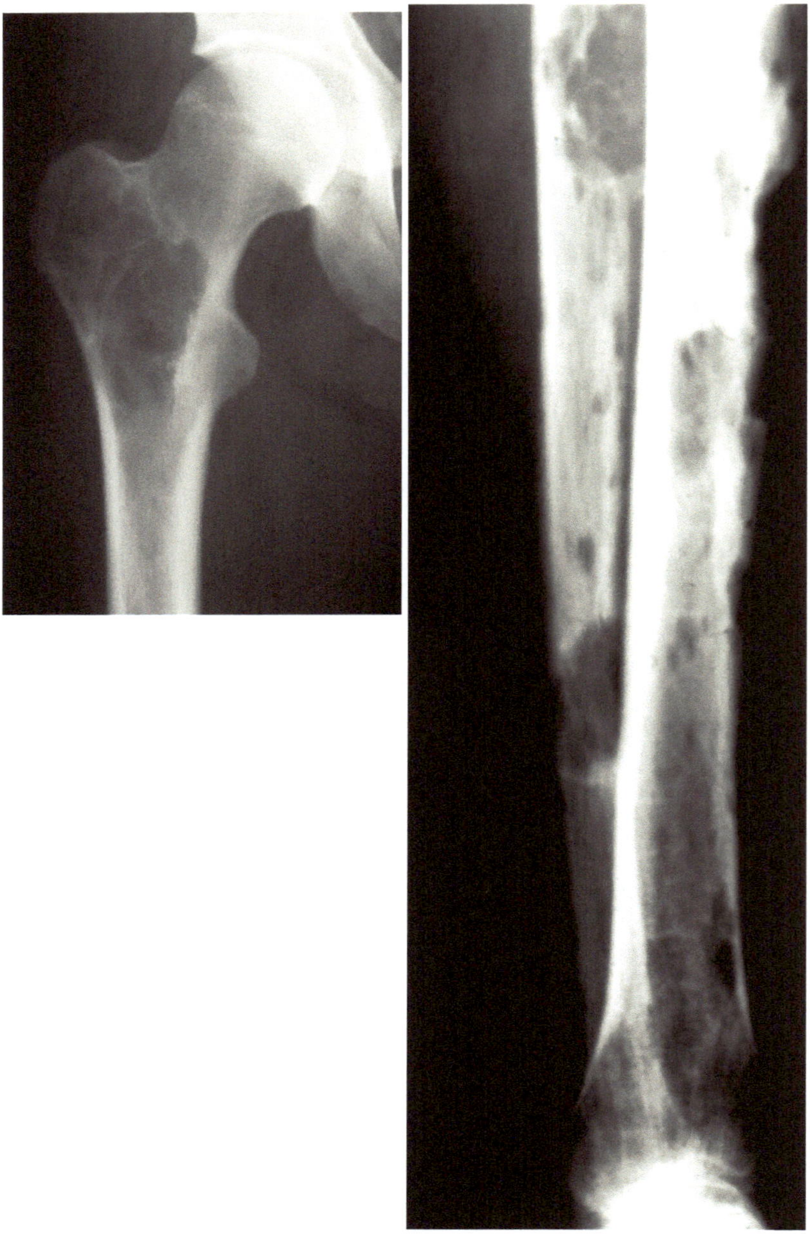

Radiograph. Purely lytic lesions. They can be unique or multiple usually in the same limb

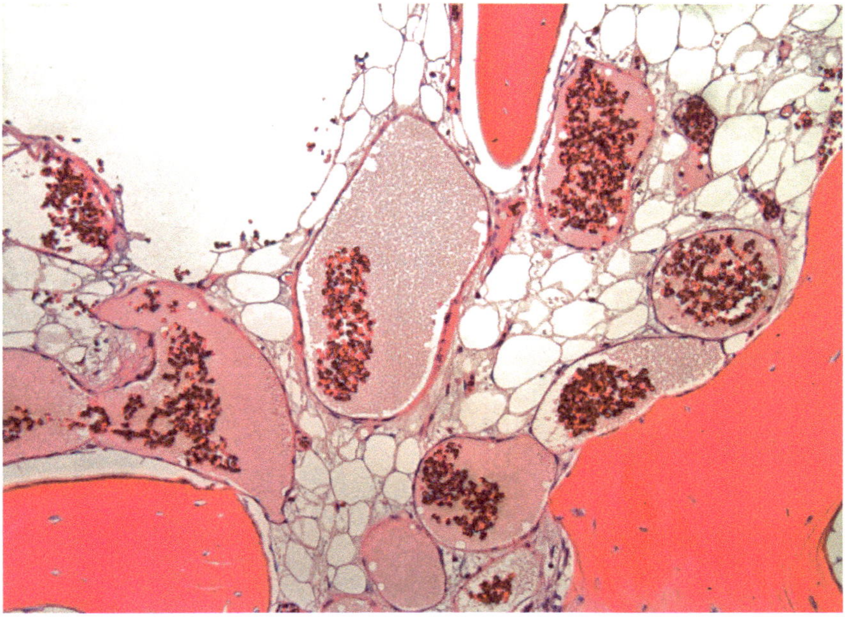

Hemangioma. Numerous blood-filled cavities rimmed by flat endothelium are present in between the host bone trabeculae

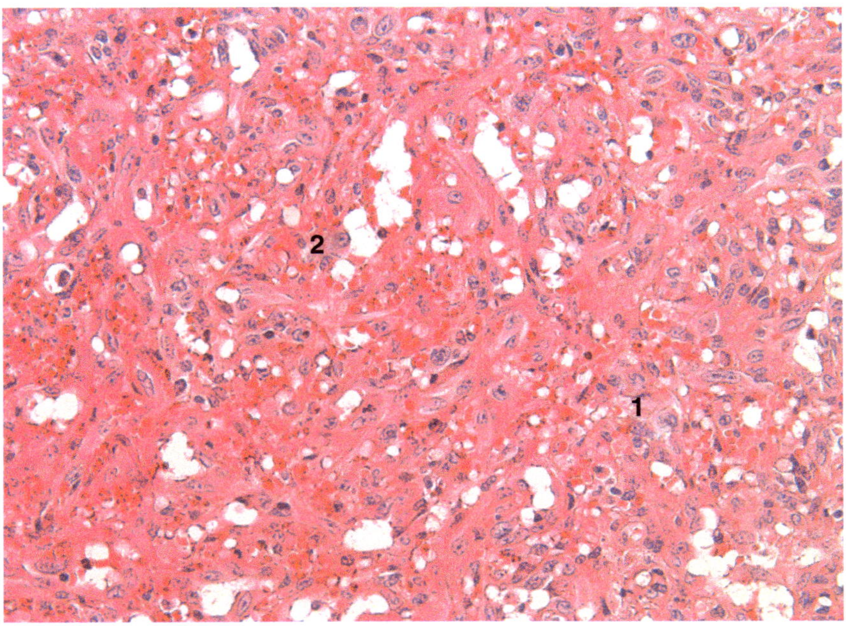


Hemangioendothelioma: Low-grade. The overall view shows tumoral cells with large, pleomorphic, and mildly hyperchromic nuclei with evident nucleoli. (*1*) Vascular cavities matted with neoplastic endothelial cells. (*2*) Neoplastic endothelial cells line the vascular cavities matted in one or more layers

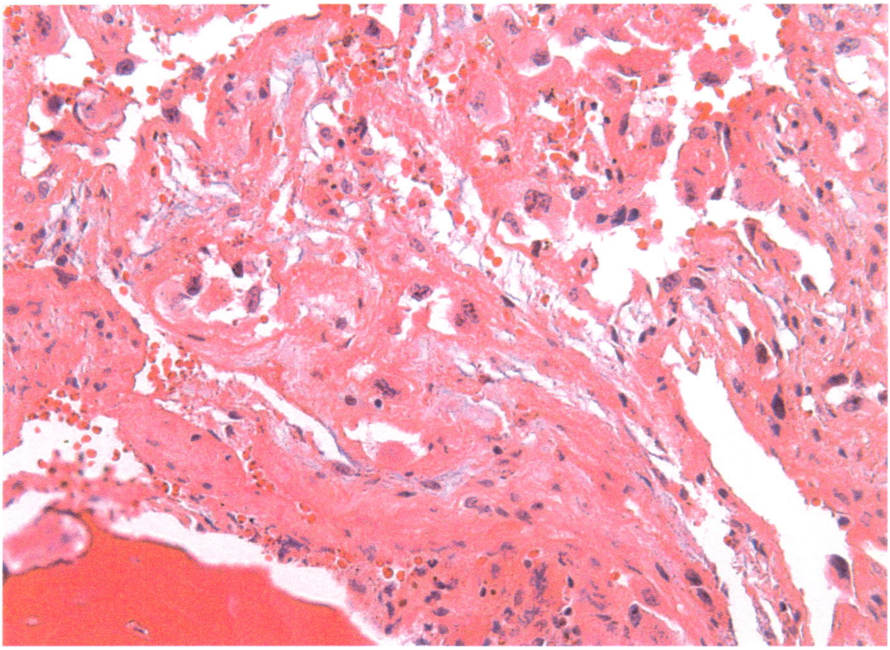

Angiosarcoma. Irregular and haphazarded blood-filled cavities rimmed by highly malignant atypical cells, diffusely permeating the host bone trabeculae

## Selected Bibliography

### *Hemangioendothelioma (Low Grade)*

Aflatoon K, Staals E, Bertoni F, Bacchini P, Donati D, Fabbri N, Boriani S, Frassica FJ (2004) Hemangioendothelioma of the spine. Clin Orthop Relat Res 418:191–197

Benassi MS, Gamberi G, Ragazzini P, Sollazzo MR, Ferrari C, Sangiorgi L, Treré D, Derenzini M, Picci P (1995) Bone hemangioendothelioma: an immunohistochemical study related to histological malignancy and proliferative activity (NORs). Tumori 81(3):179–184

Bruder E, Perez-Atayde AR, Jundt G, Alomari AI, Rischewski J, Fishman SJ, Mulliken JB, Kozakewich HP (2009) Vascular lesions of bone in children, adolescents, and young adults. A clinicopathologic reappraisal and application of the ISSVA classification. Virchows Arch 454(2):161–179

Errani C, Sung YS, Zhang L, Healey JH, Antonescu CR (2012) Monoclonality of multifocal epithelioid hemangioendothelioma of the liver by analysis of WWTR1-CAMTA1 breakpoints. Cancer Genet 205:12–17

Evans HL, Raymond AK, Ayala AG (2003) Vascular tumors of bone: a study of 17 cases other than ordinary hemangioma, with an evaluation of the relationship of hemangioendothelioma of bone to epithelioid hemangioma, epithelioid hemangioendothelioma, and high-grade angiosarcoma. Hum Pathol 34(7):680–689

Gill R, O'Donnell RJ, Horvai A (2009) Utility of immunohistochemistry for endothelial markers in distinguishing epithelioid hemangioendothelioma from carcinoma metastatic to bone. Arch Pathol Lab Med 133(6):967–972

## *Angiosarcoma (High Grade)*

Deshpande V, Rosenberg AE, O'Connell JX, Nielsen GP (2003) Epithelioid angiosarcoma of the bone: a series of 10 cases. Am J Surg Pathol 27(6):709–716
Dunlap JB, Magenis RE, Davis C, Himoe E, Mansoor A (2009) Cytogenetic analysis of a primary bone angiosarcoma. Cancer Genet Cytogenet 194(1):1–3
Hart J, Mandavilli S (2011) Epithelioid angiosarcoma: a brief diagnostic review and differential diagnosis. Arch Pathol Lab Med 135(2):268–272
Verbeke SL, Bertoni F, Bacchini P, Sciot R, Fletcher CD, Kroon HM, Hogendoorn PC, Bovée JV (2011) Distinct histological features characterize primary angiosarcoma of bone. Histopathology 58(2):254–264
Wenger DE, Wold LE (2000) Malignant vascular lesions of bone: radiologic and pathologic features. Skeletal Radiol 29(11):619–631. Review

# Chapter 52
# Chordoma

**Nicola Fabbri and Pietro Ruggieri**

**Definition**: Malignant tumor from vestigial remnants of the notochord in the cranial base and the vertebral column.
**Epidemiology**: Rare. Males. More than 50 years of age.

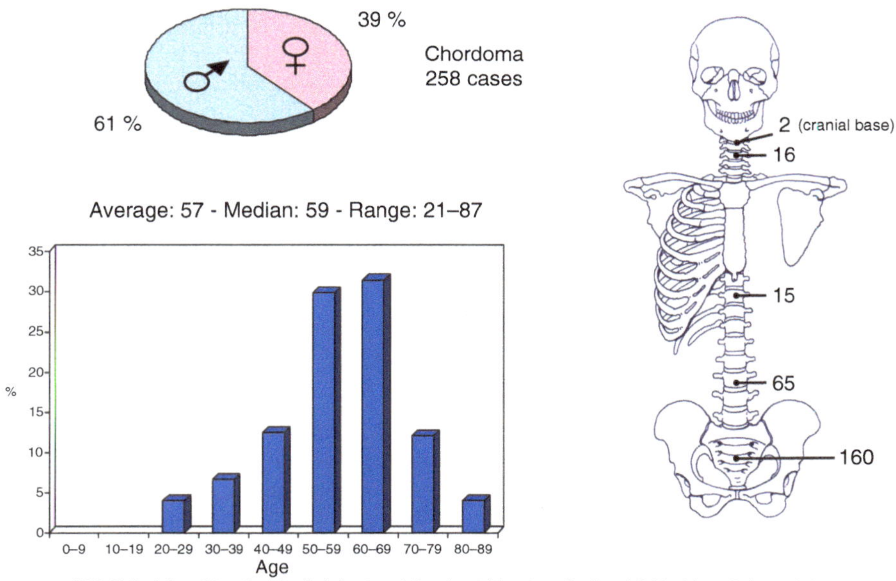


1900–2012 - Istituto Ortopedico Rizzoli - Laboratory of Experimental Oncology - Section of Epidemiology - Bologna - Italy

N. Fabbri, MD (✉)
Department of Surgery, Orthopaedic Service, Memorial Sloan-Kettering Cancer Center, 1275 York Avenue, New York, NY 10065, USA
e-mail: fabbrin@mskcc.org

P. Ruggieri, MD, PhD (✉)
2nd Orthopaedic and Traumatologic Clinic, Istituto Ortopedico Rizzoli, Bologna, Italy
e-mail: pietro.ruggieri@ior.it

P. Picci et al. (eds.), *Atlas of Musculoskeletal Tumors and Tumorlike Lesions*,
DOI 10.1007/978-3-319-01748-8_52, 

**Location**: 85 % in the sacrococcyx or in the spheno-occipital region. Fifteen percentage in the vertebral body.
**Clinical**: Early symptoms in C. of the base due to compression of the hypophysis, the optic, oculomotor nerves, obstruction of the nose-pharynx, expansion in the pontocerebellar angle, and increase of intracranial pressure. Late symptoms in C. of the sacrum: insidious pain, constipation, hemorrhoids, dysuria, limping, hypoesthesia, and sphincteric paresis. Rare palpable swelling. In vertebral locations: signs of myeloradicular compression.
**Imaging**: On X-ray—osteolytic lesion of the clivus and sella turcica, of the sacrum. The outlines of bone and lines of the sacral foramina have disappeared. Ill-defined margins of the lesion that is central. Faded radiopaque spots of intratumoral calcifications. In the vertebral body relatively well-defined edge, with sclerotic rim, calcifications. On bone scan—uptake. On CT—extraosseous mass protruding more anteriorly than posteriorly in the sacrum, well-defined borders, infiltration of the adjacent muscles, and dislocation of the viscera or of the dural sac. On MRI—relationship between tumor and rectum, level of rising in the sacral canal, hypointense, homogeneous on T1, inhomogeneous, hyperintense, bright on T2.
**Histopathology**: Small in the base of the skull, very large in the sacrum. Lobular, grayish, soft, gelatinous, mucoid aspect with frequent areas of hemorrhage, necrosis, and cystic changes. Lobular pattern with long ribbon or nests of large round cells embedded in a pinkish gelatinous appearance, with oval, round nuclei, and abundant cytoplasm with many fine, clear, mucous-filled vacuoles of different size which give a bubbly aspect, hence the term physaliferous cell. Cells arranged in strands separated by a faintly eosinophilic amorphous mucous, few cells in a large mass of mucoid to myxoid stroma. Nuclei mostly lie in the center of the cell giving a bull's-eye or target appearance. Relatively vascular, frequent blood vessels in fibrovascular shoots among the tumoral lobules. Frequent cartilaginous differentiation is observed in C. of the base of the skull.
**Course and Staging**: Slow growth, frequent local recurrences if surgery is not adequate, rare, and late metastases (10 % of C. of the sacrum). Usually, stage IB, very few cases as stage IA, rarely as stage IIB.
**Treatment**: Wide resection is mandatory, but often impossible. Little hope of avoiding recurrence if marginal or contaminated excision is performed. Radiotherapy is effective and useful. Many C. of the base of the skull cause death within 2–3 years. Cases with cartilaginous differentiation have a better prognosis. Although recurrence may be reoperated many times and metastases are late, survival after 10 years for C. of the sacrum is very low because surgery is late and inadequate.

**Key Points**

| | |
|---|---|
| Clinical | Old patients, pain and compression symptoms |
| Radiological | Pure lytic lesion |
| Histological | Lobular pattern with physaliferous cells |
| Differential diagnosis | Bone metastasis and all other primary purely osteolytic lesions of adults—benign notochordal cell tumor |

**Immunohistochemical Panel**

| | |
|---|---|
| VIM | + |
| CK | + |
| EMA | + |
| S-100 | ± |
| Brachyury | + (nuclear) |

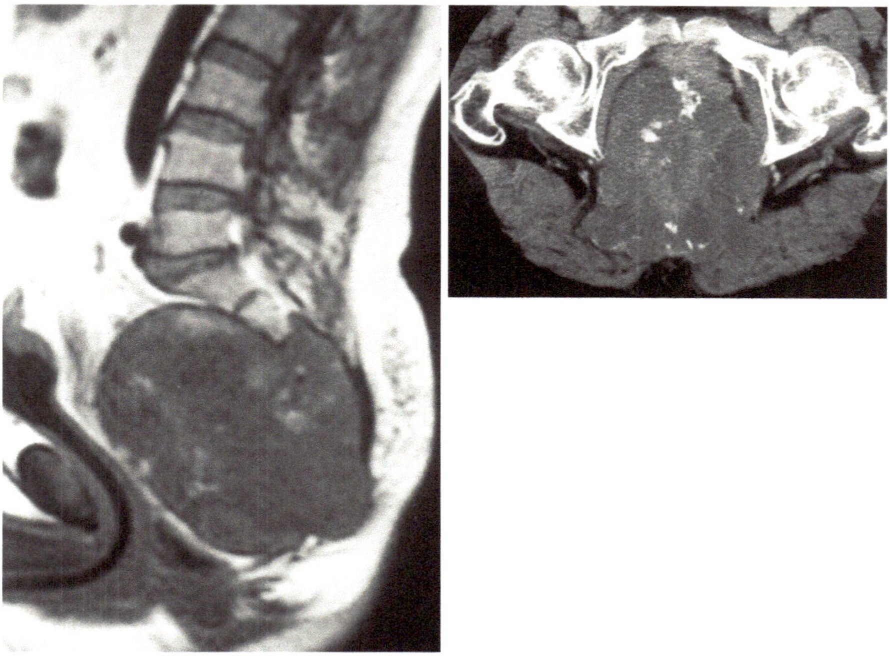

CT and sagittal T1 MR image. Huge tumor destroying the last sacral vertebrae, invading the anterior soft tissues, pushing the rectum, which remains free. On CT remaining pieces of bone are well visible

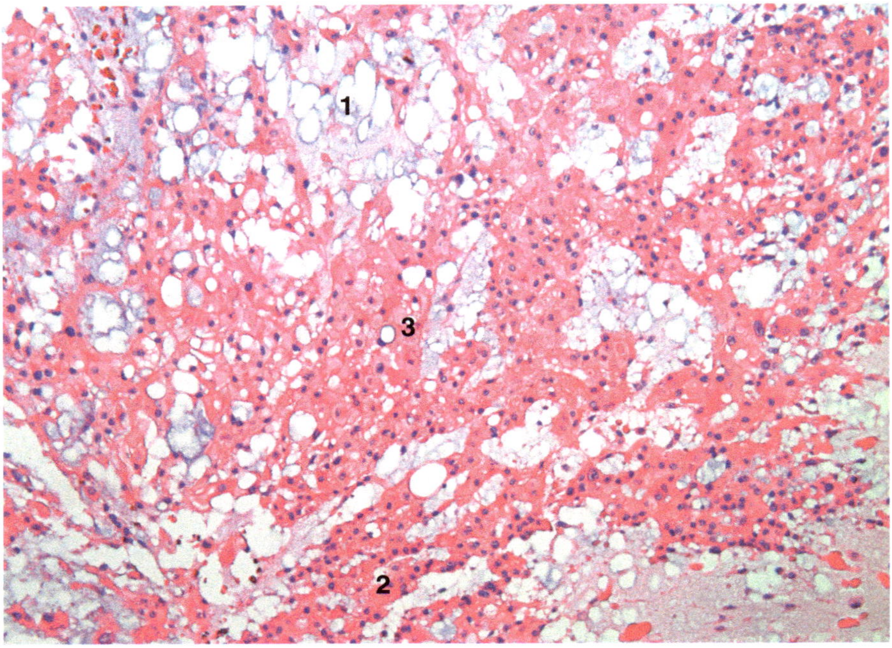


The histological picture of the tumor may vary from field to field. However, the architecture is usually characterized by cellular cords isolated and anastomized, producing mucin and contained in the mucoid substance accumulated on the outside. (*1*) Mucoid substance. (*2*) Large, deeply eosinophil cells organized in cords. (*3*) Signet ring cells and physaliphorous cells may be observed in more differentiated areas

## 52.1 Differential Diagnosis with Benign Notochordal Cell Tumor

Benign notochordal cell tumor (previously called giant notochordal rest or ecchordosis physaliphora) is a benign intravertebral lesion of notochord origin. It is very frequent on autopsies but rare on imaging exams (probably because a lot of cases are small). It involves the axial skeleton and is located within bone without soft tissue infiltration or cortex destruction. Sacrococcygeal area and the base of the skull are the most common sites involved. Conventional radiographs are often nondiagnostic while CT scan usually shows a sclerotic lesion within bone, without lysis. Magnetic resonance imaging is the gold standard: the homogenous intraosseous lesion has a low signal on T1-weighted images and high signal on T2-weighted images. The lesion does not take up contrast medium. Histologically BNCT is characterized by ovoid cells with eccentrically located round nuclei. Myxoid matrix is absent and mitotic figures are not usually found.

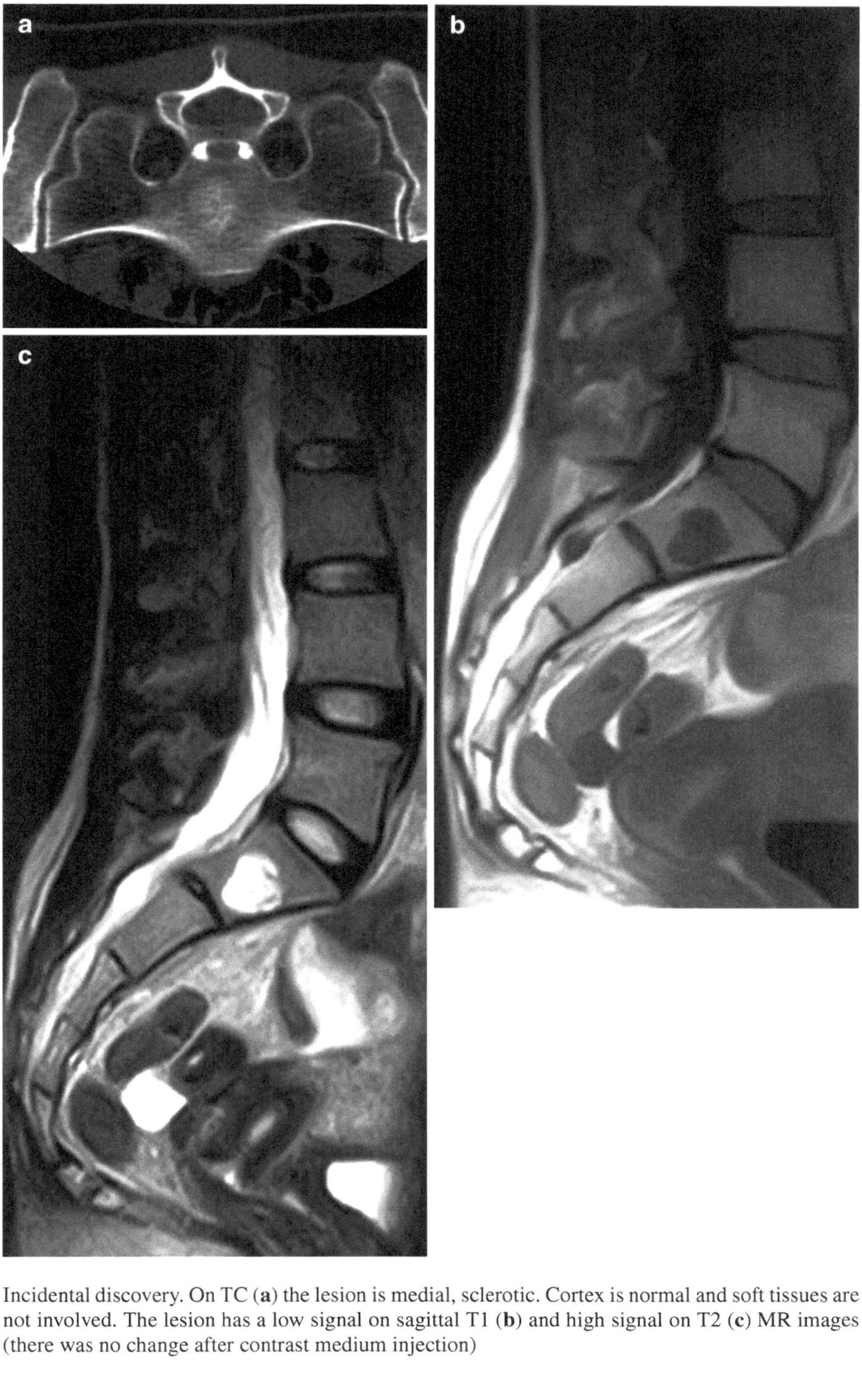


Incidental discovery. On TC (**a**) the lesion is medial, sclerotic. Cortex is normal and soft tissues are not involved. The lesion has a low signal on sagittal T1 (**b**) and high signal on T2 (**c**) MR images (there was no change after contrast medium injection)

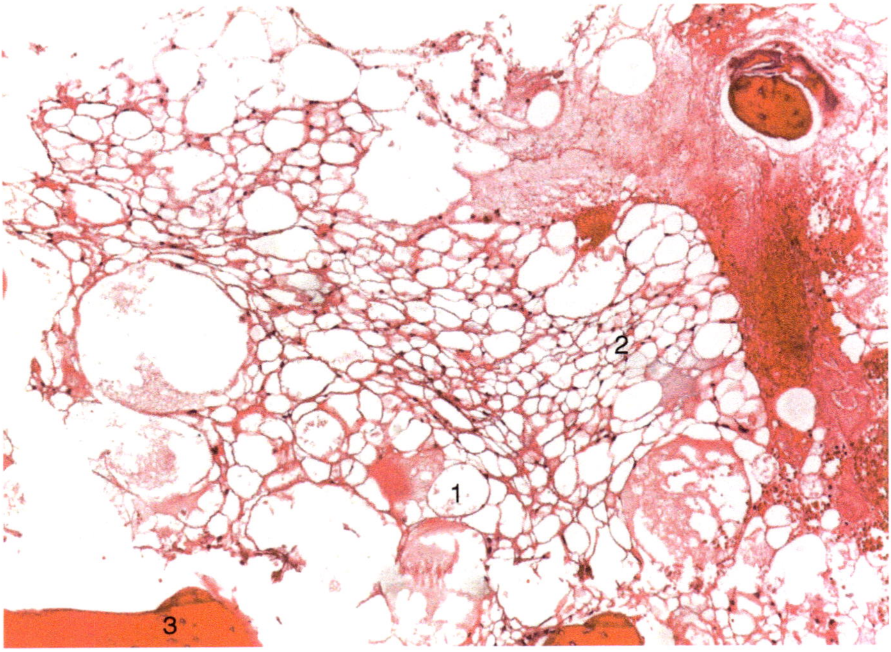


Benign notochordal cell tumor. Intraosseous tumor in the vertebral body composed of aggregates of large "adipocyte-like" cells. (*1*) Notochordal cells are large and show abundant clear cytoplasm. (*2*) The cells have well-defined cell membranes and centrally or peripherally placed nuclei with no atypia. (*3*) Affected bone trabeculae are often sclerotic

## Selected Bibliography

Ferraresi V, Nuzzo C, Zoccali C, Marandino F, Vidiri A, Salducca N, Zeuli M, Giannarelli D, Cognetti F, Biagini R (2010) Chordoma: clinical characteristics, management and prognosis of a case series of 25 patients. BMC Cancer 10:22

Ruggieri P, Angelini A, Ussia G, Montalti M, Mercuri M (2010) Surgical margins and local control in resection of sacral chordomas. Clin Orthop Relat Res 468(11):2939–2947

Sciubba DM, Cheng JJ, Petteys RJ, Weber KL, Frassica DA, Gokaslan ZL (2009) Chordoma of the sacrum and vertebral bodies. J Am Acad Orthop Surg 17(11):708–717. Review

Sen C, Triana AI, Berglind N, Godbold J, Shrivastava RK (2010) Clival chordomas: clinical management, results, and complications in 71 patients. J Neurosurg 113(5):1059–1071

Stacchiotti S, Casali PG, Lo Vullo S, Mariani L, Palassini E, Mercuri M, Alberghini M, Pilotti S, Zanella L, Gronchi A, Picci P (2010) Chordoma of the mobile spine and sacrum: a retrospective analysis of a series of patients surgically treated at two referral centers. Ann Surg Oncol 17(1): 211–219

Walter BA, Begnami M, Valera VA, Santi M, Rushing EJ, Quezado M (2011) Gain of chromosome 7 by chromogenic in situ hybridization (CISH) in chordomas is correlated to c-MET expression. J Neurooncol 101(2):199–206

# Chapter 53
# Primary Lymphoma of Bone

**Roberto Casadei**

**Definition**: Non-Hodgkin primary lymphoma of bone: intramedullary malignant lymphoma, without extraskeletal involvement within 6 months from onset of symptoms.

**Epidemiology**: Primary lymphoma of bone is infrequent (5 % of all extranodal lymphomas). The male gender is preferred. The mean age is around 45.

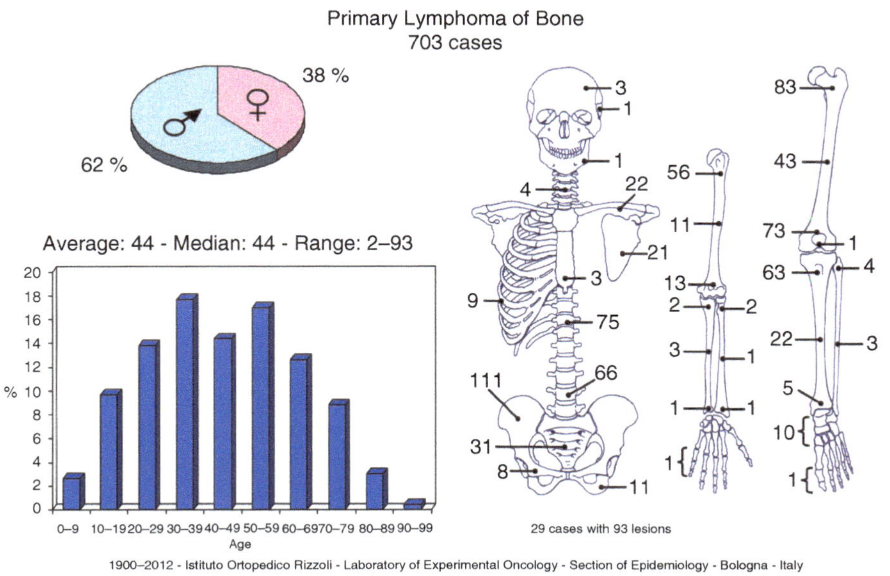


R. Casadei, MD
3rd Orthopaedic and Traumatologic Clinic Prevalently Oncologic,
Istituto Ortopedico Rizzoli, Bologna, Italy
e-mail: roberto.casadei@ior.it

P. Picci et al. (eds.), *Atlas of Musculoskeletal Tumors and Tumorlike Lesions*,
DOI 10.1007/978-3-319-01748-8_53, © Springer International Publishing Switzerland 2014

**Localization**: About half of the cases are localized in the skeleton of the trunk and skull (particularly jaws). The other half occurs in the long bones (metaphysis or diaphysis). Not rarely (around 20 %) involves two or more adjacent or distant bones.
**Clinical**: Essentially local: pain, swelling, pathological fracture, and neurological deficits in vertebral localization. Not uncommonly symptoms are mild and of long duration.
**Imaging**: On X-rays: small, diffuse, ill-defined lytic lesions that give a permeative pattern are the most frequent appearance of L. It is difficult to see the end of the lesion. Often a moth-eaten radiographic pattern is observed in L where lytic lesions are greater and spotted. The cortex may be broken or destroyed with a large mass in the soft tissue mimicking a bone sarcoma, and sometimes, the cortex can be thickened mimicking an osteomyelitis. Usually, a periosteal reaction is absent, but in rare cases a specular or lamellar reaction may be observed, and X-rays may mimic those of an osteosarcoma or an Ewing sarcoma. At times, L has an eccentric aggressive lytic lesion like a metastasis, a pure well-defined lytic lesion like a benign bone tumor, or a bubbly, diffused lytic lesions like a myeloma. More rarely, NHL can appear sclerotic, but a sclerotic appearance is more common in H lymphoma, and it may be the first sign of the disease when changes in the lymph nodes are still not very evident. In 30 % of cases, a pathologic fracture is the first appearance of L, and it may occur both in lytic and in sclerotic lesions. On bone scan an uptake shows the lesion, and it defines better than X-rays the real extension of the disease. CT scan shows the typical permeative or moth-eaten radiographic features of L, the mottled appearance due to reactive hyperostosis, and it is very useful to evaluate the relationship with the neurovascular bundles and the involvement of the joint and for the staging of the lesion showing lung metastasis or secondary localization to lymph nodes. MRI is not characteristic in L of long bones. It shows the intramedullary extension of the lesion and the relationship with the joint and the soft tissue, but MRI is really necessary in L of the spine because it can show exactly spinal cord compression and the extension of the disease.
**Histopathology**: Viable tumor is grayish, soft, and encephaloid. Hemorrhage, necrosis, and colliquation (sometimes with the impression of purulence) are frequent. Tumor manifests a prominent permeation and inflammatory response. Histologically the tumor is composed by sheets of dense roundish cells, clearly separated from each other. Cytoplasm is scant or abundant, pale to clear. Nuclei are large, pleomorphic, often with a vesicular appearance. Chromatin ranges from loose and granular to heavily clumped. Nucleoli are irregular in size, sometimes prominent. Mitoses are usually frequent, sometimes atypical. Reticulin fibers surround small groups of cells and even single cells. PAS does not reveal cytoplasmic glycogen. Immunostains (in undecalcified specimens) are positive for B and T cell markers, immunoglobulins, and common leukocyte antigens. Inflammatory cells are frequently spread in and around the tumor, possibly obscuring the malignant cell population. The majority of primary lymphomas of bone are of intermediate or high histological grade of malignancy. Regarding the phenotype, the majority are B cell lymphomas.

**Course and Staging**: Often it has a slow course. Pulmonary metastases occur rarely. When lymphoma is suspected or diagnosed, it is necessary to stage the lesion, by studying the skeleton, the regional and distant lymph nodes, the abdominal and thoracic organs and sites, and the central nervous system. Usually NHL remains localized in the bone and it may heal. At times, it may diffuse to regional lymph nodes and it also may heal. In other cases it is present a second or many bone localizations that may involve only regional lymph nodes. Finally, there are cases of systemic NHL that begins from the bone and then diffuse in other sites. At 10 years, the survival is increased from 30 to 60–80 % due to the introduction of polychemotherapy in the treatment protocols of L.

**Treatment and Prognosis**: Radiation therapy combined to chemotherapy is the treatment of choice. Surgery may be indicated in spine localizations to decompress marrow and roots or in pathologic or impending fractures in the long bones (internal fixation or segmental resection and endoprosthesis). Resection is also indicated in expendable bones. Surgery is always combined with radio- and chemotherapy. Disease-free survival at 10 years is highly significant. Local recurrence is rare. Unfavorable prognostic factors: bulk of disease, dissemination to nodes and/or other bones, and pelvic and spine localizations.

**Key Points**

| | |
|---|---|
| Clinical | Adults, pain and swelling, possible pathologic fracture |
| Radiological | Pure lytic lesion, no periosteal reaction |
| Histological | Pleomorphic blue round cells |
| Differential diagnosis | Bone metastasis and all other primary purely osteolytic lesions of adults |

**Immunohistochemical Panel**

| | |
|---|---|
| VIM | – |
| CD45 | + |
| CD3 | ± |
| CD20 | ± |
| TdT | ± |

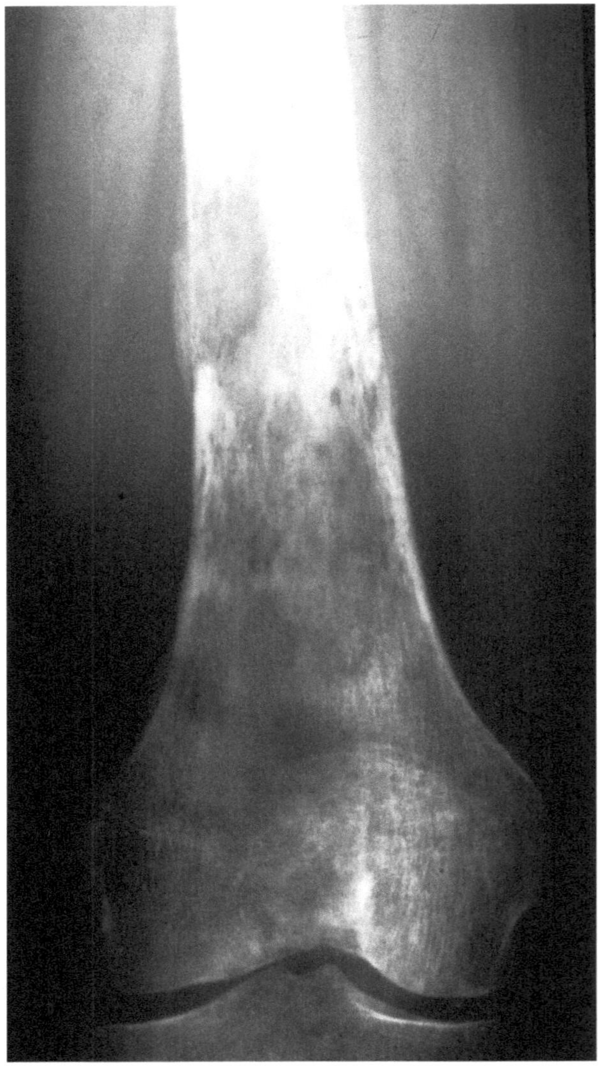

Radiograph. Mixed lytic and sclerotic femoral tumor, with limited cortical destruction, and moth-eaten limits

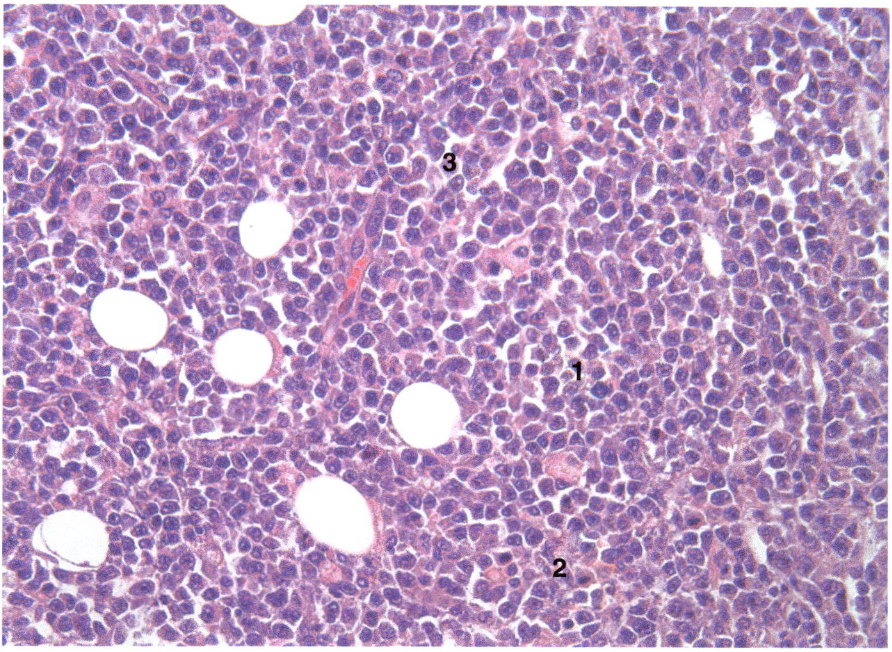


It is also a richly cellular tissue. There may be a delicate intercellular reticulum formed by the union between cytoplasmic processes. (*1*) Cells have pale and scarce cytoplasm, but in some, it is more abundant and clearly delimited. (*2*) Nuclei are generally larger than those of Ewing sarcoma, with considerable polydimensionality and pleomorphism. Some are roundish or oval; others are kidney shaped, indented, or lobulated. Some cells may have a rather large nucleus or double nucleus, but there are not giant cells. At times, the chromatin is pulverulent; other times it is in rougher blocks. The nucleus may tend to be vesicular, and often one or more nucleoli, even large ones, are evident. (*3*) Mitotic figures are frequent

## Selected Bibliography

Alencar A, Pitcher D, Byrne G, Lossos IS (2010) Primary bone lymphoma–the University of Miami experience. Leuk Lymphoma 51(1):39–49. Review

Bhagavathi S, Fu K (2009) Primary bone lymphoma. Arch Pathol Lab Med 133(11):1868–1871. Review

O'Connor AR, Birchall JD, O'Connor SR, Bessell E (2007) The value of 99mTc-MDP bone scintigraphy in staging primary lymphoma of bone. Nucl Med Commun 28(7):529–531

Ramadan KM, Shenkier T, Sehn LH, Gascoyne RD, Connors JM (2007) A clinicopathological retrospective study of 131 patients with primary bone lymphoma: a population-based study of successively treated cohorts from the British Columbia Cancer Agency. Ann Oncol 18(1):129–135

Reddy N, Greer JP (2010) Primary bone lymphoma: a set of unique problems in management. Leuk Lymphoma 51(1):1–2

# Chapter 54
# Multiple Myeloma

**Roberto Casadei**

**Definition**: Systemic malignant tumor of the bone marrow, originating from B-lymphoid cells and differentiating in plasma cells. Rarely solitary in bone or in the soft tissues (solitary plasmacytoma of bone, extramedullary plasmacytoma).
**Epidemiology**: 20 cases/million/year. The male sex is preferred. Mean age around 60.

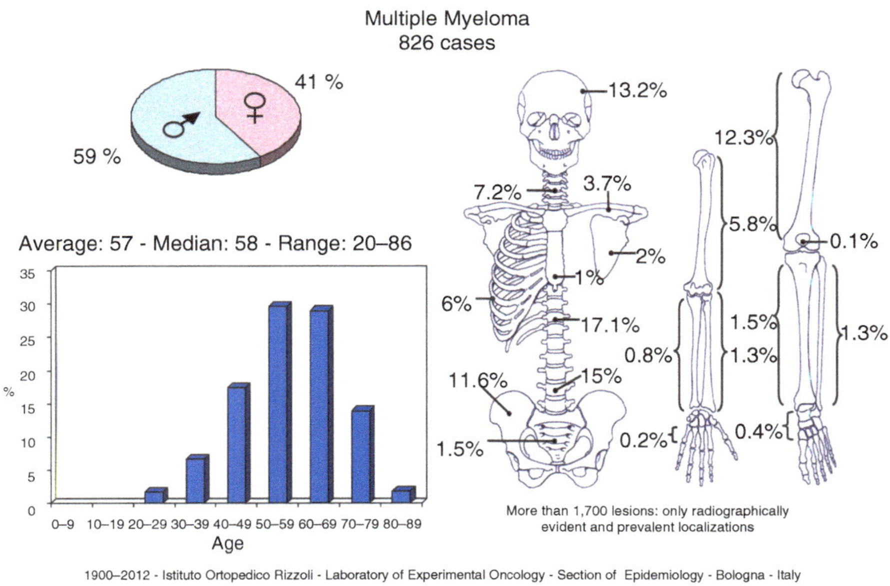


R. Casadei, MD
3rd Orthopaedic and Traumatologic Clinic Prevalently Oncologic,
Istituto Ortopedico Rizzoli, Bologna, Italy
e-mail: roberto.casadei@ior.it

P. Picci et al. (eds.), *Atlas of Musculoskeletal Tumors and Tumorlike Lesions*,
DOI 10.1007/978-3-319-01748-8_54, © Springer International Publishing Switzerland 2014

**Localization**: The axial skeleton is the preferred site. In the appendicular skeleton, the metaepiphyses are preferred. Very rare in bones of the hand and foot.

**Clinical**: Pain (usually relieved by rest), occasionally pathologic fracture (more often at a later stage). In advanced stages, there may be swelling, weight loss, anemia, bleeding diathesis, propensity to infections, hypercalcemic and hyperuricemic syndromes, subcutaneous myeloma masses, uremic renal insufficiency, jaundice, and ascites. Congestive heart failure, nephrotic syndrome, peripheral neuropathy, and carpal tunnel syndrome may result from amyloidosis. In 90 % of cases, serum and/or urinary immunoelectrophoresis shows a spike due to the excess of a monoclonal protein. Bence Jones proteinuria. Erythrocyte sedimentation rate is usually elevated. Elevated serum creatinine, hyperazotemia, renal clearances. Hypercalcemia (in advanced disease), due to massive osteolysis. Hyperuricemia is rather frequent, due to the release of purines and pyrimidines from dying neoplastic plasma cells. Anemia, leukopenia, and thrombocytopenia are due to substitution of bone marrow by the tumor.

**Imaging**: Radiographic skeletal changes remain absent, until the proliferative medullary disease has caused enough bone resorption.

(1) Diffused osteoporosis (osteopenia), particularly seen in the spine; (2) "punched-out" round lytic areas without associated osteoblastic changes, with sharp limits; no rim of sclerosis around the osteolyses; (3) large or massive osteolytic tumors (trabeculated, honeycombed, or even bubbly).

The cortices are thinned or canceled, and extension of the tumor in the soft tissues is frequent, particularly in the ribs and spine. Pathologic fractures are frequent. Periosteal reaction is absent. Rarely (1–2 %), multiple myeloma is osteosclerotic. This variant affects a slightly younger age and has fewer tumor cells, lower levels of monoclonal proteins, and more elevated serum alkaline phosphatase. It reflects a less aggressive infiltrative process, which permits time for osteoblastic reaction. MRI is the most sensitive to discover diffused and nodular disease in cancellous bone. Isotope bone scan is usually negative or scarcely positive.

Bone marrow aspiration. If necessary, it should be repeated from multiple sites. It is diagnostic in about 90 % of cases of multiple myeloma. If the smear contains over 3–5 % of plasma cells, diagnosis may be suspected. If it contains 10 %, the disease is very likely. If the plasma cell content is over 10–20 % and particularly if immature or anaplastic elements are seen, the diagnosis is certain.

**Histopathology**: The myelomatous tissue is soft to gelatinous, grayish to dark red. Paramyloid substance may form grossly appreciable osseous or soft tissue masses, with a homogeneous and gelatinous appearance, yellowish gray or pink. It is composed by cells differentiating in plasma cells and producing a specific protein, but no matrix nor eliciting any fibrous or bone reactive tissue around them.

Histologically myelomatous tissue is composed by regular sheets of round to polygonal cells, separated from each other, without any matrix around them.

- Differentiated myeloma: Plasma cells are mature, small, with abundant basophilic cytoplasm, a clear perinuclear halo (Golgi apparatus), and a round eccentric nucleus with chromatin clumping giving the "leopard skin" appearance. There is a moderate pleomorphism, some cells have two or more nuclei, and mitoses are rare.
- Intermediate myeloma: Pleomorphism is more prominent; cells have a larger, frequently central hyperchromatic nucleus, often without the perinuclear halo. The nucleolus is more evident. Mitoses are not rare, both typical and atypical.
- Anaplastic myeloma: Many cells have large, multiple, and bizarre nuclei; pleomorphism and hyperchromia are intense, nucleoli are prominent, and plasma cell features are not clearly evident.

Immunohistology shows a monoclonal protein marker in all types of myeloma.

**Treatment and Prognosis**: Treatment is based on chemotherapy, particularly using alkylating agents (cyclophosphamide, melphalan), nitrosourea (BCNU), prednisone, thalidomide, and biphosphonates. In younger patients, allogenic marrow transplantation has been employed. Radiation treatment is used, to relieve pain, decrease spinal cord compression, and prevent pathologic fractures. Surgery may be indicated to decompress the spinal cord or stabilize the spine with internal fixation, treat or prevent pathologic fractures (internal fixation), and occasionally manage destructive lesions or fractures in the long bones (resection with endoprosthesis). With rare exceptions, prognosis is fatal, with only around 30 % survivors at 5 years. Median survival is around 4 years. S. myeloma has a slower course, but a systemic diffusion is very frequent even if it may occur also 8–12 years after the diagnosis. Real S. myeloma is very rare, and 45 % of patients remain healed 10–15 years after treatment. Death occurs due to complications: pneumonia, cystopyelitis, cardiocirculatory and kidney insufficiency, and cachexia.

Adverse prognostic factors are the extent of bone involvement and the severity or values of anemia, hypercalcemia, impaired renal function, circulating monoclonal proteins, hyperuricemia, and alkaline phosphatase.

**Key Points**

| | |
|---|---|
| Clinical | Adults, pain and swelling |
| Radiological | Pure lytic lesion frequently multifocal. Bone scan usually negative |
| Histological | Pleomorphic blue round cells with chromatin arranged in a "cartwheel aspect" |
| Differential diagnosis | Bone metastasis, vascular tumors, and all other primary purely osteolytic lesions of adults |

**Immunohistochemical Panel**

| | |
|---|---|
| VIM | – |
| CD138 | + |
| Kappa | ± |
| Lambda | ± |

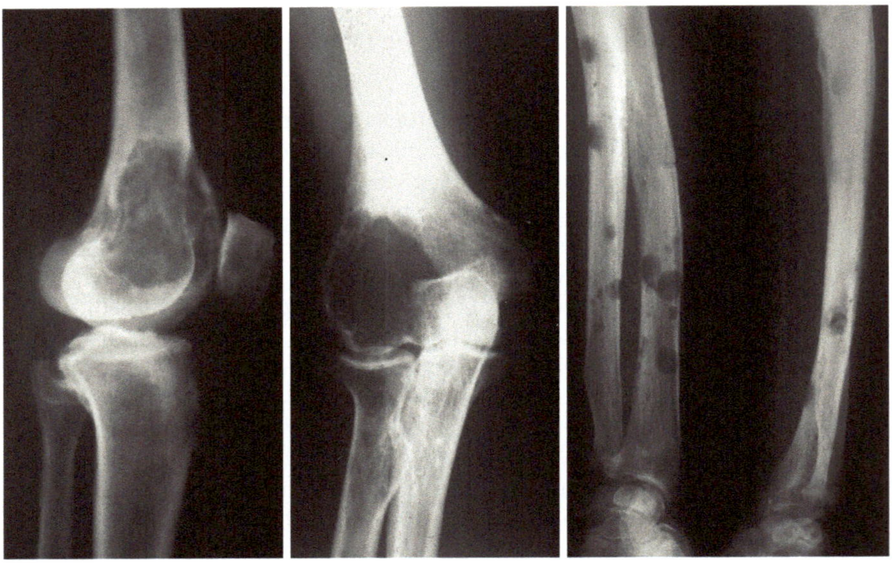

Radiographs. Multiple purely lytic well-limited lesions

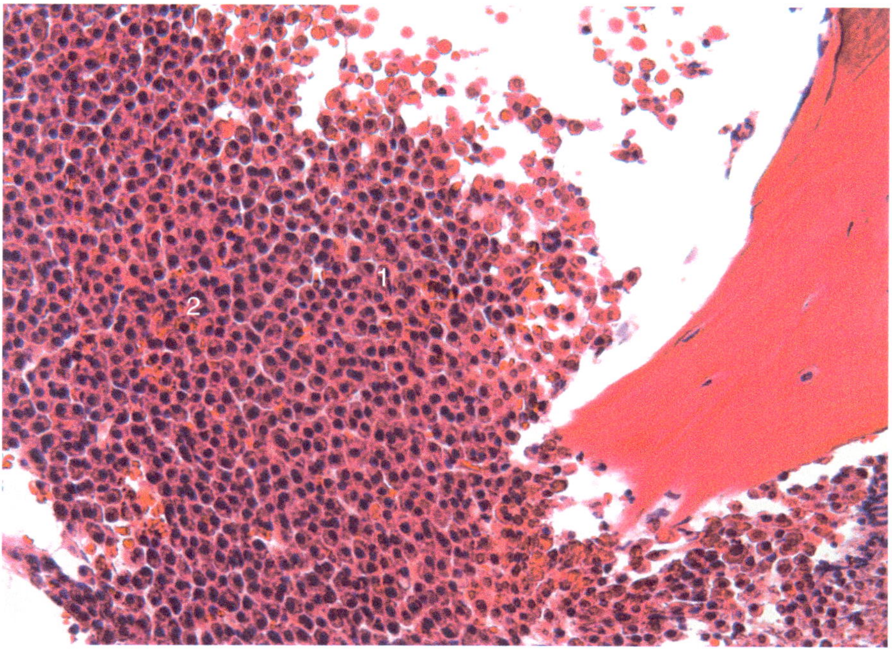


Tissue is entirely constituted of a very thick mat of cells with no intercellular matrix. (*1*) Tumoral cells with plasma cellular features: round eccentric nucleus with chromatin in distinct blocks. (*2*) Intercellular stroma is absent

## 54.1 Solitary Plasmacytoma of Bone

**Definition**: It is rare, more commonly localized in the axial skeleton. Monoclonal proteins in the serum and urine may be detected, but not as frequently as in multiple myeloma.

**Imaging**: Shows a localized and often massive osteolysis, homogeneous, trabeculated, or bubbling, with frequent destruction or inflation of the cortex.

**Histology**: Plasmacytoma is identical to myeloma, of the differentiated or intermediate type.

**Treatment and Prognosis**: It is based on radiation therapy and/or surgery, sometimes chemotherapy. Local recurrence is uncommon. Major percentages of solitary plasmacytomas of bone eventually give metastases (to regional lymph nodes, to distant bones) or become disseminated. This unfavorable evolution is unpredictable and can occur even more than 10 or 15 years after diagnosis. Prognosis is uncertain for many years. Reported overall survival after 10 years is around 50 %.

## Selected Bibliography

Dimopoulos M, Terpos E, Comenzo RL, Tosi P, Beksac M, Sezer O, Siegel D, Lokhorst H, Kumar S, Rajkumar SV, Niesvizky R, Moulopoulos LA, Durie BG, IMWG (2009) International myeloma working group consensus statement and guidelines regarding the current role of imaging techniques in the diagnosis and monitoring of multiple Myeloma. Leukemia 23(9): 1545–1556

Harousseau JL (2010) Multiple myeloma in the elderly: when to treat, when to go to transplant. Oncology (Williston Park) 24(11):992–998. Review

Kumar A, Galeb S, Djulbegovic B (2011) Treatment of patients with multiple myeloma: an overview of systematic reviews. Acta Haematol 125(1–2):8–22. Review

Nahi H, Sutlu T, Jansson M, Alici E, Gahrton G (2011) Clinical impact of chromosomal aberrations in multiple myeloma. J Intern Med 269(2):137–147

Rajkumar SV (2011) Multiple myeloma: 2011 update on diagnosis, risk-stratification, and management. Am J Hematol 86(1):57–65. Review

# Chapter 55
# Metastatic Carcinoma

**Roberto Casadei and Marco Gambarotti**

**Definition**: Secondary localization to the bone of any primary carcinoma.
**Importance**: Metastatic carcinoma represents the most common malignancy to the bone.
**Epidemiology**: At autopsy at least 70 % of cancer patients present metastases to the bone: breast 84 %, prostate 84 %, thyroid 50 %, lung 44 %, and kidney 37 %. They are obviously characteristic of adult or old age, but thyroid or breast cancers may also be observed in the fourth or even third decade of life.

R. Casadei (✉)
3rd Orthopaedic and Traumatologic Clinic Prevalently Oncologic,
Istituto Ortopedico Rizzoli, Bologna, Italy
e-mail: roberto.casadei@ior.it

M. Gambarotti (✉)
Department of Anatomy and Pathological Histology, Istituto Ortopedico Rizzoli,
Via del Barbiano 1/10, Bologna, 40136, Italy
e-mail: marco.gambarotti@ior.it

P. Picci et al. (eds.), *Atlas of Musculoskeletal Tumors and Tumorlike Lesions*,
DOI 10.1007/978-3-319-01748-8_55, 

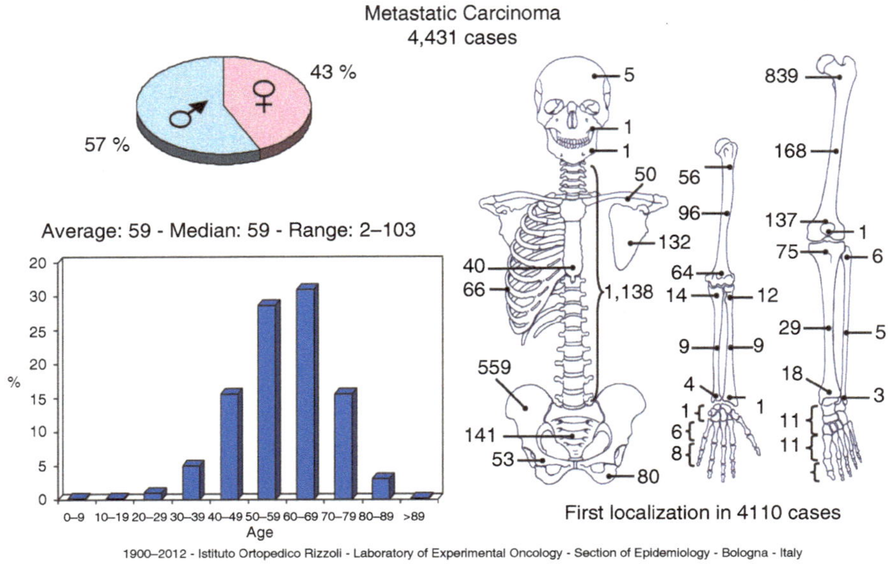


**Localization**: The most common locations of metastatic carcinoma are the spine, pelvis, femur, ribs, and skull. Metastases may affect any bone but are rare distal to elbow and knee.

**Clinical**: Pain, mass, pathological fracture, and hypercalcemia. At times (about 20 %) the metastasis to the bone is diagnosed first, without knowledge of the primary lesion even if at the end of staging, primary tumor is often observed in the lung or in the bowel.

**Imaging**: Standard radiography may be negative (40 % of cases). Frequently, metastases to the bone present as purely osteolytic lesions (kidney), often subperiosteal (lung) or blastic (prostate, breast). Isotope scan and MRI are very sensitive in detecting bone mets (spine!) before radiological changes become apparent on plain x-rays. Today, PET-CT is very useful to detect bone lesions not identified with other techniques and to evaluate relapse of the disease.

**Histopathology**: Different features related to the original primary lesion, usually squamous or gland patterns, are present. Origin of metastatic disease is histologically clear only in a few well-differentiated carcinomas (thyroid, kidney, prostate). However, immunohistochemical patterns can help to identify primary tumor.

**Treatment**: In metastatic patients to maintain or to restore a good quality of life is the most important object. Palliative treatment is necessary when multiple and diffused disease is observed and the patient has a poor prognosis. An adjuvant treatment using chemotherapy, low-dose radiotherapy, and simple surgery is useful in patients with a fair prognosis. In those with a good prognosis, chemotherapy, high-dose radiotherapy, and wide resection with stable reconstruction can be combined together to obtain a long survival. Today, new drugs and new target therapies are more and more spread out in clinical practice.

## Key Points

| | |
|---|---|
| Clinical | More frequent bone lesion. Adults, pain. Pathologic fracture |
| Radiological | Permeative or pure lytic, mixed, blastic features |
| Histological | Epithelial aspects, depending on primary lesion |
| Differential diagnosis | All other primary lesions of adults |

## Immunohistochemical Panel

| | Breast | Prostate | Lung | Thyroid | Kidney | Colon | Urothelial | Melanoma |
|---|---|---|---|---|---|---|---|---|
| CK MNF-116 | + | + | + | + | + | + | + | – |
| CK 7 | + | – | + | + | – | – | + | – |
| CK 20 | – | – | + | – | – | + | + | – |
| TTF1 | – | – | ± | + | – | – | – | – |
| Thyroglobulin | – | – | – | + | – | – | – | – |
| CDX-2 | – | – | – | – | – | + | – | – |
| EST | ± | – | – | – | – | – | – | – |
| PROG | ± | – | – | – | – | – | – | – |
| PSA | – | + | – | – | – | – | – | – |
| PAP | – | + | – | – | – | – | – | – |
| AMACR | – | + | – | – | – | – | – | – |
| CD10 | – | – | – | – | + | – | – | – |
| S-100 | – | – | – | – | – | – | – | + |
| HMB-45 | – | – | – | – | – | – | – | + |
| MART-1 | – | – | – | – | – | – | – | + |
| MYTF | – | – | – | – | – | – | – | + |

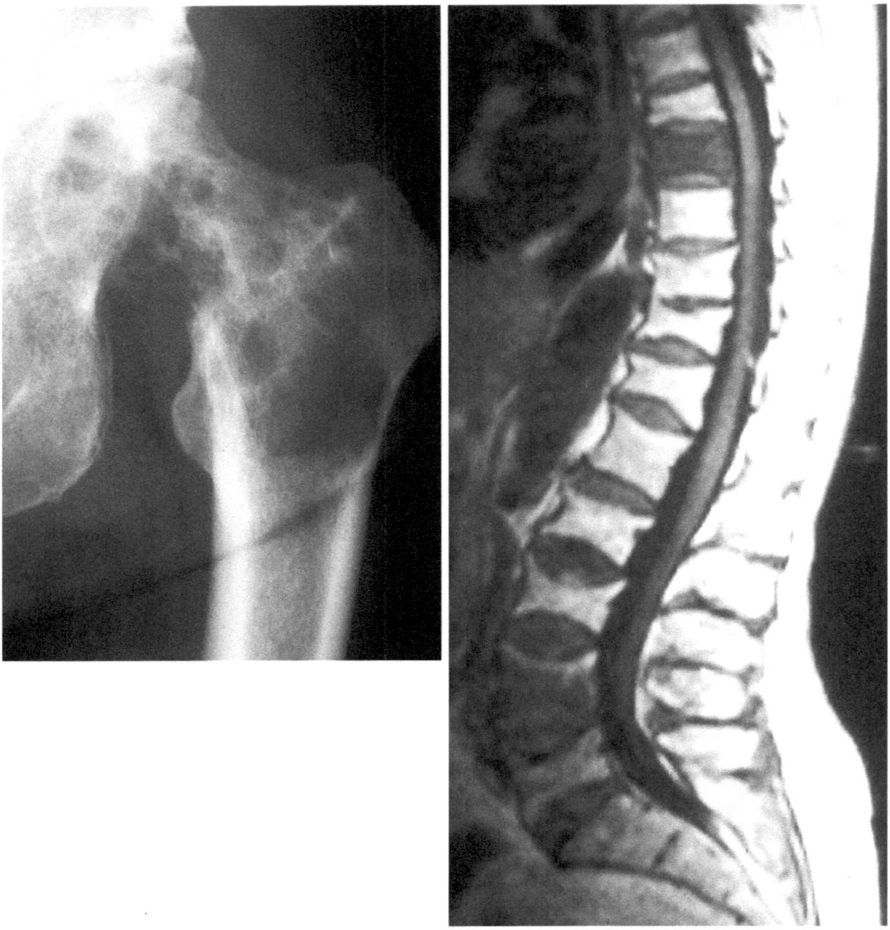

Radiograph and sagittal spine T1 MR image. Heterogeneous mainly lytic lesion, destroying the femoral neck cortex. Multiple metastases of the spine

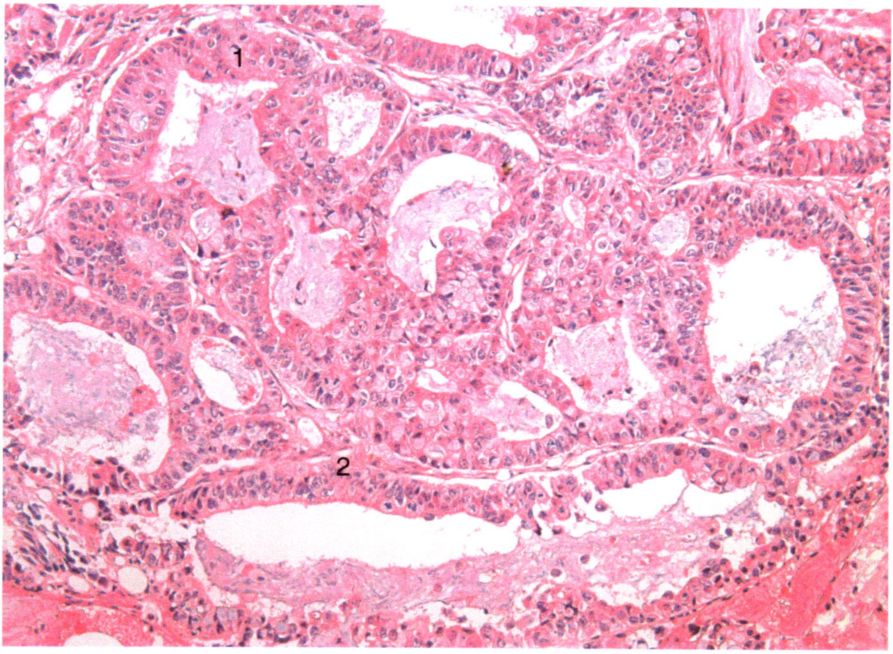


Nearly always diagnosis is easy: finding neoplastic epithelial tissue inside the bone cannot but signify metastasis. (*1*) Malignant epithelial cells forming glandular structures in a metastasis from colon mucinous adenocarcinoma. (*2*) Basement membranes of carcinomatous cells

## Selected Bibliography

Buijs JT, van der Pluijm G (2009) Osteotropic cancers: from primary tumor to bone. Cancer Lett 273(2):177–193. Review

Elgazzar AH, Kazem N (2009) Metastatic bone disease: evaluation by functional imaging in correlation with morphologic modalities. Gulf J Oncolog 5:9–21. Review

Kaijzel EL, Snoeks TJ, Buijs JT, van der Pluijm G, Löwik CW (2009) Multimodal imaging and treatment of bone metastasis. Clin Exp Metastasis 26(4):371–379. Review

Roberts CC, Daffner RH, Weissman BN, Bancroft L, Bennett DL, Blebea JS, Bruno MA, Fries IB, Germano IM, Holly L, Jacobson JA, Luchs JS, Morrison WB, Olson JJ, Payne WK, Resnik CS, Schweitzer ME, Seeger LL, Taljanovic M, Wise JN, Lutz ST (2010) ACR appropriateness criteria on metastatic bone disease. J Am Coll Radiol 7(6):400–409. Review. Erratum in: J Am Coll Radiol. 2010 Sep;7(9):e1 Review

Santini D, Galluzzo S, Zoccoli A, Pantano F, Fratto ME, Vincenzi B, Lombardi L, Gucciardino C, Silvestris N, Riva E, Rizzo S, Russo A, Maiello E, Colucci G, Tonini G (2010) New molecular targets in bone metastases. Cancer Treat Rev 36(Suppl 3):S6–S10. Review

# Part II
# Soft Tissue Tumors

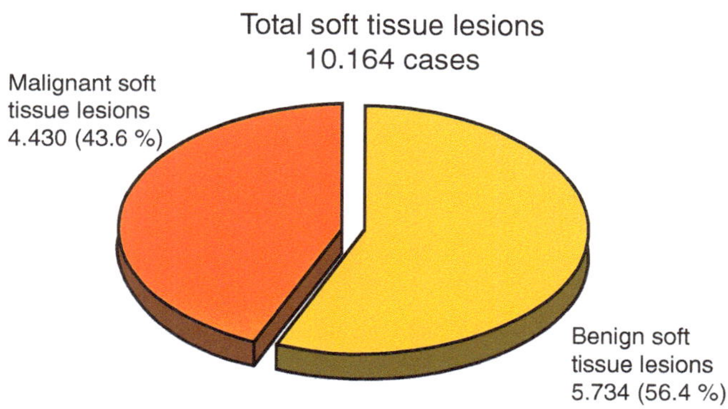


1900–2012 - Istituto Ortopedico Rizzoli - Laboratory of Experimental Oncology - Section of Epidemiology - Bologna - Italy

## Epidemiology

**(Updated by Piero Picci)**

Soft tissue tumors (STT) are a heterogeneous group of benign and malignant processes. Some are assumed reactive, others are clearly neoplastic. These lesions account for less than 4 % of all tumors in adult patients, and for 7–10 % of all tumors in pediatric patients. More than 99 % of STT are benign, and the incidence of malignant STT is about 4–5 times that of malignant tumors arising from bone.

The rarity of STT causes problems with respect to diagnosis, grading, or optimal therapeutic approach. Over the last few decades, there have been significant changes in diagnostics and treatment of STT. Several developments in the field of radiology, pathology and surgery have significantly changed the way STT are currently diagnosed and treated, improving prognosis and quality of life for patients with these rare diseases.

# Chapter 56
# Epidemiology

**Piero Picci, Daniel Vanel, Marco Gambarotti, Pietro Ruggieri, and Stefano Ferrari**

Soft tissue tumors (STT) are a heterogeneous group of benign and malignant processes. Some are assumed reactive, others are clearly neoplastic. These lesions account for less than 4 % of all tumors in adult patients and for 7–10 % of all tumors in pediatric patients. More than 99 % of STT are benign, and the incidence of malignant STT is about 4–5 times that of malignant tumors arising from the bone.

The rarity of STT causes problems with respect to diagnosis, grading, or optimal therapeutic approach. Over the last few decades, there have been significant changes in diagnostics and treatment of STT. Several developments in the field of radiology, pathology, and surgery have significantly changed the way STT are currently diagnosed and treated, improving prognosis and quality of life for patients with these rare diseases.

P. Picci, MD (✉)
Laboratory of Experimental Oncology, Musculoskeletal Oncology,
Istituto Ortopedico Rizzoli, Bologna, Italy
e-mail: piero.picci@ior.it

D. Vanel, MD (✉)
Department of Anatomy and Pathology, Istituto Ortopedico Rizzoli,
Via del Barbiano 1, Bologna 40136, Italy
e-mail: daniel.vanel@ior.it

M. Gambarotti, MD (✉)
Department of Anatomy and Pathological Histology, Istituto Ortopedico Rizzoli,
Via del Barbiano 1/10, Bologna 40136, Italy
e-mail: marco.gambarotti@ior.it

P. Ruggieri, MD, PhD (✉)
2nd Orthopaedic and Traumatologic Clinic, Istituto Ortopedico Rizzoli, Bologna, Italy
e-mail: pietro.ruggieri@ior.it

S. Ferrari, MD (✉)
Chemotherapy Ward of Musculoskeletal Tumors, Istituto Ortopedico Rizzoli, Bologna, Italy
e-mail: stefano.ferrari@ior.it

P. Picci et al. (eds.), *Atlas of Musculoskeletal Tumors and Tumorlike Lesions*,
DOI 10.1007/978-3-319-01748-8_56, 

## 56.1 Developments in the Field of Radiology

Daniel Vanel

Until the introduction of ultrasound and computed tomography (CT), STT were usually detected at a late stage of disease due to their low incidence and nonspecific clinical findings. The only imaging technique available was conventional radiography with a very poor sensitivity for soft tissue lesions. Although these new methods had an important impact on detection of STT, they also suffered from inherent drawbacks, such as the poor specificity of ultrasound and the poor contrast resolution of CT. Many of these problems were resolved by the introduction of magnetic resonance imaging (MRI).

Thanks to its high contrast resolution and its multiplanar imaging capability, new horizons were opened for imaging of STT. It is now possible to provide an anatomic and topographic depiction of the lesion, defining with great precision the relationship to muscle compartments, fascial planes, bones, and neurovascular structures. Modern imaging techniques are not only of great importance for differential diagnosis of STT, they also play an important role in monitoring effects of nonsurgical treatments and are essential in early detection of local and distant recurrence during postoperative follow-up.

Most recent developments are the introduction of positron-emission tomography (PET) and the application of dynamic MRI modalities. Although experience with these new techniques is still rather limited for STT, it is likely that they will have an increasing role in diagnosis, staging, monitoring, and follow-up of these lesions in the near future.

## 56.2 Developments in the Field of Pathology

Marco Gambarotti

STT are classified according to their histogenesis and biological behavior (see table). Because of the rarity and vast histological varieties of STT (there are at least 50 histological subtypes of malignant STT and an equal number of benign entities), they frequently pose diagnostic problems for surgical pathologists. Where in the past clinical and morphologic features were the only diagnostic parameters, nowadays pathologists can increasingly rely on ancillary diagnostic methods to aid with differential diagnosis. Introduction of immunohistochemistry has had a significant impact on the classification of STT. Up until the 1980s, diagnosis of "malignant fibrous histiocytoma" represented about 75 % of all soft tissue sarcomas. With the application of immunohistochemistry, it became possible to reclassify many of these lesions. Therefore, diagnosis of "malignant fibrous histiocytoma," which according to WHO Classification 2013 is now called "undifferentiated high-grade pleomorphic sarcoma," has become increasingly rare and accounts for about 20 % of malignant STT.

Introduction of cytogenetic and molecular genetic analysis in the 1990s has not only increased our knowledge of the biology of STT, it also resulted in a further subclassification of these lesions. Selective cytogenetic and molecular genetic alterations nowadays play an important role in diagnostics of many STT, such as Ewing sarcoma/PNET, synovial sarcoma, alveolar rhabdomyosarcoma, clear-cell sarcoma, liposarcoma, dermatofibrosarcoma protuberans, and extraskeletal myxoid chondrosarcoma.

The grading system for soft tissue sarcomas used at the Rizzoli Institute is the FNCLCC system.

The classification of STT, based on immunohistochemical, molecular, and cytogenetic expression, allows for diagnostic and prognostic stratification of these lesions, which could lead to the development of more tumor-specific, targeted treatment strategies in the near future.

## 56.3 Developments in the Field of Surgery

Pietro Ruggieri

Surgery is the main treatment modality for most STT. The extent of surgery depends on histological grade and biological behavior of the tumor. In designing a treatment plan, the surgeon should balance the goal of minimizing local and distant recurrence with the aim of preserving function and quality of life.

Up to the 1970s, amputation was the most frequently performed surgery for malignant STT. Through the years there has been a trend toward limb-salvage surgery, which nowadays is the mainstay of treatment (> 95 % of cases). Several developments have allowed for this more conservative treatment. As already mentioned, better understanding of biological behavior of these lesions and improved imaging techniques have had an important contribution. In addition, the introduction of a multimodality treatment, with adjuvant radiotherapy and/or chemotherapy, has had a significant impact on surgical planning. Finally, a multidisciplinary surgical approach, involving, besides orthopedic surgeons, general, vascular, neuro-, and plastic surgeons, has allowed for the excision of lesions in complex anatomic areas and an appropriate coverage of extensive tissue defects after tumor excisions.

## 56.4 Developments in the Field of Chemotherapy

Stefano Ferrari

Adjuvant chemotherapy might improve disease-free survival in high-risk patients as reported in two meta-analyses. However, studies are conflicting, and the use of adjuvant chemotherapy in a clinical setting is still controversial.

Data from clinical trials suggest that chemotherapy can be proposed to high-risk patients having a G3, deep, >5 cm tumor.

In patients with G2–3, deep, >5 cm soft tissue sarcomas, regional hyperthermia in addition to systemic chemotherapy may give local and disease-free survival advantage.

In non-resectable tumors or those amenable only to mutilating surgery, chemotherapy and/or radiotherapy or isolated hyperthermic limb perfusion with tumor necrosis factor alpha (TNFa) + melphalan, if the tumor is confined to an extremity, or regional hyperthermia combined with chemotherapy are options.

Standard chemotherapy is based on anthracyclines. However, a higher response rate may be expected by multiagent treatments as reported in several clinical trials.

Chemotherapy with doxorubicin plus ifosfamide may be of choice, especially when tumor response is felt to be able to give an advantage and the performance status is good.

The histological type may be considered in decision-making, since some types are felt to be more chemosensitive, while others less.

In myxoid liposarcoma a peculiar pattern of tumor response to trabectedin has been reported, with an early phase of tissue changes preceding tumor shrinkage. Responses have been obtained in synovial sarcoma with high-dose ifosfamide. Randomized evidence provided that gemcitabine + docetaxel is more effective than gemcitabine alone as second-line chemotherapy, and gemcitabine was shown to have antitumor activity in leiomyosarcoma also as a single agent. Dacarbazine has some activity as second-line therapy (possibly mostly in leiomyosarcoma). In angiosarcoma, taxanes are an alternative option, given their high antitumor activity in this specific histological type. Imatinib is the standard medical therapy for those rare patients with dermatofibrosarcoma protuberans who are not amenable to surgery or with metastases deserving medical therapy.

Classification of soft tissue tumors

| Histogenesis | Benign | Uncertain behavior (locally aggressive and/or rarely metastasizing) | Malignant |
|---|---|---|---|
| Adipocytic | Lipoma | ALT/WDL | Dedifferentiated liposarcoma |
| | Lipomatosis | | Myxoid liposarcoma |
| | Lipomatosis of nerve | | Pleomorphic liposarcoma |
| | Lipoblastoma/lipoblastomatosis | | Liposarcoma, NOS |
| | Angiolipoma | | |
| | Myolipoma | | |
| | Chondroid lipoma | | |
| | Extrarenal angiomyolipoma | | |
| | Extra-adrenal myelolipoma | | |
| | Spindle cell/pleomorphic lipoma | | |
| | Hibernoma | | |
| Fibroblastic/ myofibroblastic | Nodular fasciitis | Palmar/plantar fibromatosis | Adult fibrosarcoma |
| | Proliferative fasciitis | Desmoid-type fibromatoses | Myxofibrosarcoma |
| | Proliferative myositis | Lipofibromatosis | Low-grade fibromyxoid sarcoma |
| | Myositis ossificans | Giant-cell fibroblastoma | Sclerosing epithelioid fibrosarcoma |
| | Fibro-osseous pseudotumor of digits | DFSP | Myofibroblastic sarcoma |
| | Ischemic fasciitis | Solitary fibrous tumor | Fibrosarcomatous DFSP |
| | Elastofibroma | Inflammatory myofibroblastic tumor | Malignant solitary fibrous tumor |
| | Fibrous hamartoma of infancy | Myxoinflammatory fibroblastic sarcoma | |
| | Fibromatosis colli | Infantile fibrosarcoma | |
| | Juvenile hyaline fibromatosis | | |
| | Inclusion body fibromatosis | | |
| | Fibroma of tendon sheath | | |
| | Desmoplastic fibroblastoma | | |
| | Mammary-type myofibroblastoma | | |

(continued)

| Histogenesis | Benign | Uncertain behavior (locally aggressive and/or rarely metastasizing) | Malignant |
|---|---|---|---|
| | Calcifying aponeurotic fibroma | | |
| | Angiomyofibroblastoma | | |
| | Cellular angiofibroma | | |
| | Nuchal-type fibroma | | |
| | Gardner fibroma | | |
| | Calcifying fibrous tumor | | |
| Fibrohistiocytic | GCT of tendon sheath | PVNS | Malignant PVNS |
| | Benign fibrous histiocytoma | Plexiform fibrohistiocytic tumor | |
| | | Giant-cell tumor of soft tissues | |
| Smooth muscle | Leiomyoma | | Leiomyosarcoma |
| Skeletal muscle | Rhabdomyoma | | Rhabdomyosarcoma |
| Pericitic (Perivascular) | Glomus tumor | | Malignant glomus tumor |
| | Myopericytoma/myofibroma | | |
| | Angioleiomyoma | | |
| Vascular | Hemangiomas | Kaposiform hemangioendothelioma | Kaposi sarcoma |
| | Epithelioid hemangioma | Retiform hemangioendothelioma | Epithelioid hemangioendothelioma |
| | Angiomatosis | Papillary intralymphatic angioendothelioma | Angiosarcoma |
| | Lymphangioma | Composite hemangioendothelioma | |
| | | Pseudomyogenic hemangioendothelioma | |

| | | | |
|---|---|---|---|
| Nerve Sheath | Schwannoma | Melanotic schwannoma | MPNST |
| | Neurofibroma | | Epithelioid MPNST |
| | Perineurioma | | Malignant triton tumor |
| | Granular cell tumor | | Malignant perineurioma |
| | Dermal nerve sheath myxoma | | Malignant granular cell tumor |
| | Solitary circumscribed neuroma | | Ectomesenchymoma |
| | Ectopic meningioma | | |
| | Nasal glial heterotopia | | |
| | Benign triton tumor | | |
| | Hybrid nerve sheath tumors | | |
| Cartilaginous and osseous | Soft tissue chondroma | Hemosiderotic fibrolipomatous tumor | Mesenchymal chondrosarcoma |
| | | | Extraskeletal osteosarcoma |
| Tumors of uncertain differentiation | Acral fibromyxoma | Atypical fibroxanthoma | Synovial sarcoma |
| | Intramuscular myxoma | Angiomatoid fibrous histiocytoma | Epithelioid sarcoma |
| | Juxta-articular myxoma | Pleomorphic hyalinizing angiectatic tumor | Alveolar soft part sarcoma |
| | Deep ("aggressive") angiomyxoma | Myoepithelioma | Clear-cell sarcoma |
| | Ectopic hamartomatous thymoma | Ossifying fibromyxoid tumor | Extraskeletal myxoid chondrosarcoma |
| | | Phosphaturic mesenchymal tumor | Extraskeletal Ewing sarcoma |
| | | PEComa | Desmoplastic small round cell tumor |
| | | | Extrarenal rhabdoid tumor |
| | | | Intimal sarcoma |
| | | | Malignant myoepithelioma |
| | | | Malignant ossifying fibromyxoid tumor |
| | | | Malignant phosphaturic m. t. |
| | | | Malignant PEComa |
| | | | Undifferentiated/unclassified sarcoma |

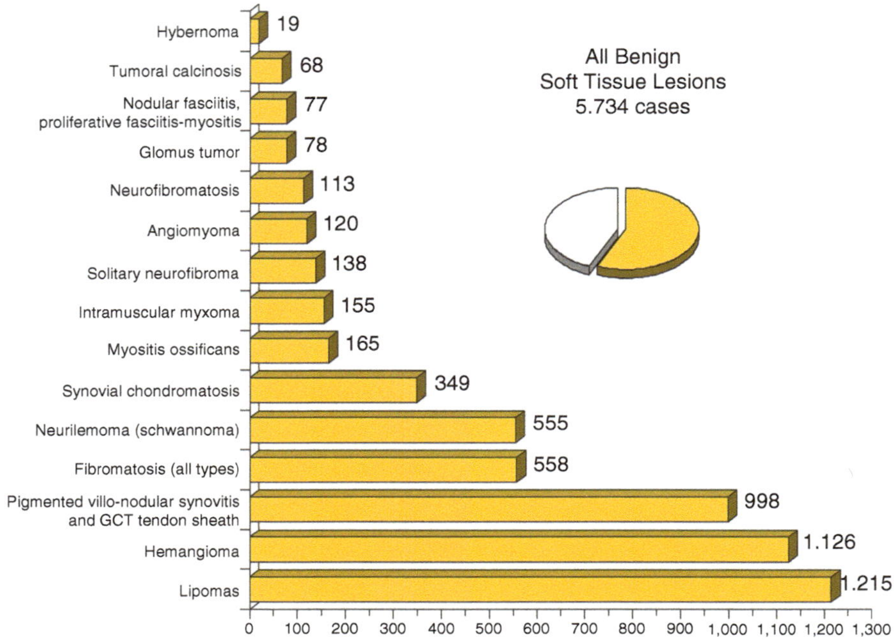


1900–2012 - Istituto Ortopedico Rizzoli - Laboratory of Experimental Oncology - Section of Epidemiology - Bologna - Italy

**ALL BENIGN SOFT TISSUE LESIONS**
**5,734 cases**

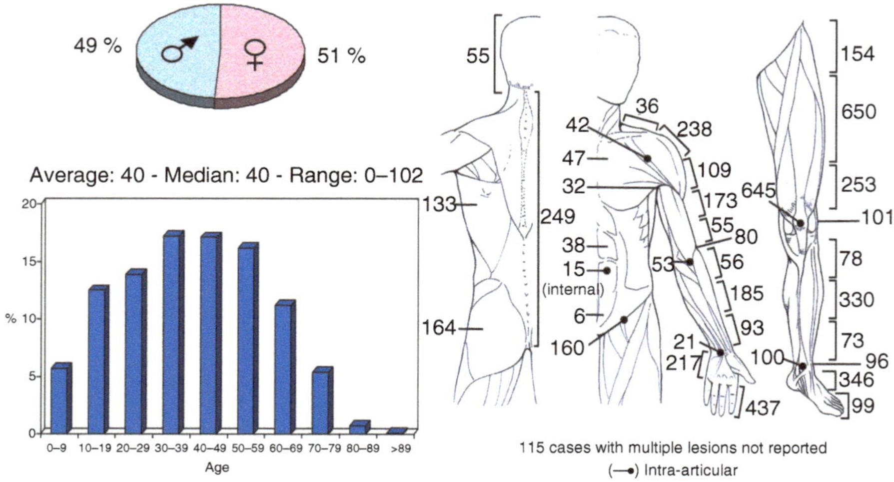


1900–2012 - Istituto Ortopedico Rizzoli - Laboratory of Experimental Oncology - Section of Epidemiology - Bologna - Italy

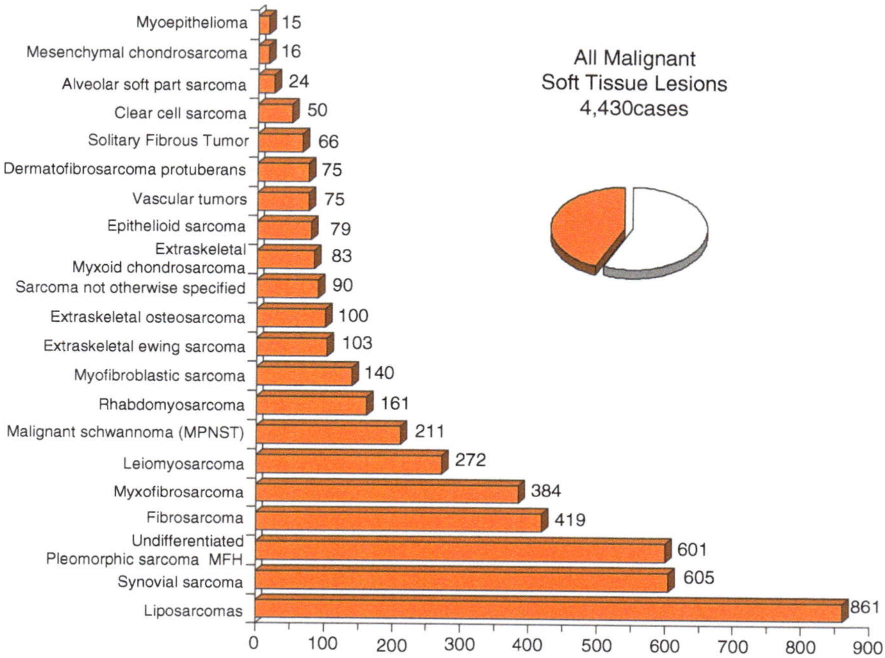


1900-2012 - Istituto Ortopedico Rizzoli - Laboratory of Experimental Oncology - Section of Epidemiology - Bologna - Italy

ALL MALIGNANT SOFT TISSUE LESIONS
4,430 cases

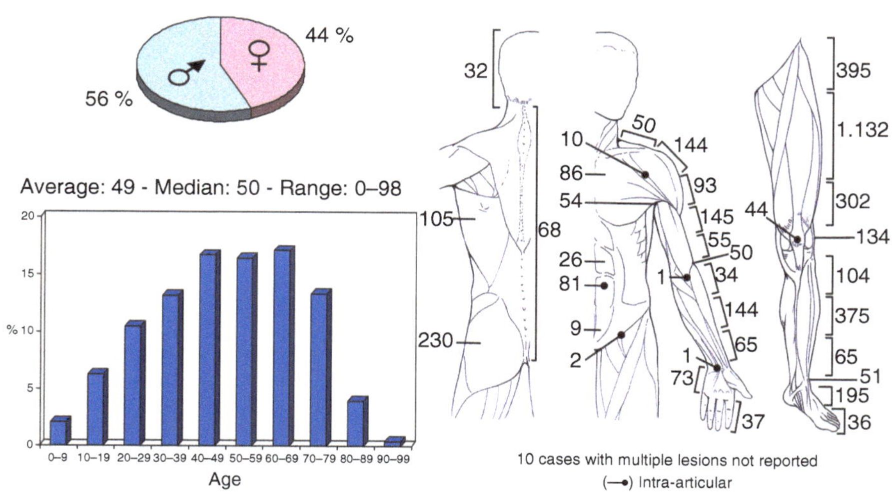


1900–2012 - Istituto Ortopedico Rizzoli - Laboratory of Experimental Oncology - Section of Epidemiology - Bologna - Italy

# Selected Bibliography

## *Imaging*

Frassica FJ, Khanna JA, McCarthy EF (2000) The role of MR imaging in soft tissue tumor evaluation: perspective of the orthopedic oncologist and musculoskeletal pathologist. Magn Reson Imaging Clin N Am 8(4):915–927. Review

Kind M, Stock N, Coindre JM (2009) Histology and imaging of soft tissue sarcomas. Eur J Radiol 72(1):6–15. Review

Murphey MD (2007) World Health Organization classification of bone and soft tissue tumors: modifications and implications for radiologists. Semin Musculoskelet Radiol 11(3):201–214. Review

van Vliet M, Kliffen M, Krestin GP, van Dijke CF (2009) Soft tissue sarcomas at a glance: clinical, histological, and MR imaging features of malignant extremity soft tissue tumors. Eur Radiol 19(6):1499–1511. Review

Wu JS, Hochman MG (2009) Soft-tissue tumors and tumor like lesions: a systematic imaging approach. Radiology 253(2):297–316. Review

## *Classification*

Bovée JV, Hogendoorn PC (2004) Pitfalls in pathology of soft tissue sarcomas. Cancer Treat Res 120:81–97. Review

Coindre JM (2003) Immunohistochemistry in the diagnosis of soft tissue tumours. Histopathology 43(1):1–16. Review

Coindre JM (2010) Molecular biology of soft-tissue sarcomas. Bull Cancer 97(11):1337–1345. Review

Davies CE, Davies AM, Kindblom LG, James SL (2010) Soft tissue tumors with muscle differentiation. Semin Musculoskelet Radiol 14(2):245–256. Review

Deyrup AT, Weiss SW (2006) Grading of soft tissue sarcomas: the challenge of providing precise information in an imprecise world. Histopathology 48(1):42–50. Review

Fletcher CD (2006) The evolving classification of soft tissue tumours: an update based on the new WHO classification. Histopathology 48(1):3–12. Review

Fletcher CDM, Bridge JA, Hogendoorn PCW, Mertens F (eds) (2013) WHO classification of tumors of soft tissue and bone, 4th edn. International Agency for Research on Cancer, Lyon

Jain S, Xu R, Prieto VG, Lee P (2010) Molecular classification of soft tissue sarcomas and its clinical applications. Int J Clin Exp Pathol 3(4):416–428. Review

Mankin HJ, Hornicek FJ (2005) Diagnosis, classification, and management of soft tissue sarcomas. Cancer Control 12(1):5–21. Review

Miettinen M (2006) From morphological to molecular diagnosis of soft tissue tumors. Adv Exp Med Biol 587:99–113. Review

Ordóñez JL, Osuna D, García-Domínguez DJ, Amaral AT, Otero-Motta AP, Mackintosh C, Sevillano MV, Barbado MV, Hernández T, de Alava E (2010) The clinical relevance of molecular genetics in soft tissue sarcomas. Adv Anat Pathol 17(3):162–181. Review

Thway K (2009) Pathology of soft tissue sarcomas. Clin Oncol (R Coll Radiol) 21(9):695–705. Review

Wu JM, Montgomery E (2008) Classification and pathology. Surg Clin North Am 88(3):483–520, v-vi. Review

## Chemotherapy

Blay JY, Cassier PA, Ray-Coquard I (2011) Soft tissue sarcomas: are all soft tissue sarcomas treated with the same drugs? Eur J Cancer 47(Suppl 3):S385–S388

Eggermont AMM, de Wilt JHW, ten Hagen TLM (2003) Current uses of isolated limb perfusion in the clinic and a model system for new strategies. Lancet Oncol 4:429–437

ESMO, European Sarcoma Network Working Group (2012) Soft tissue and visceral sarcomas: ESMO Clinical Practice Guidelines for diagnosis, treatment and follow-up. Ann Oncol 23(Suppl 7):vii92–vii99

Frustaci S, Gherlinzoni F, De Paoli A et al (2001) Adjuvant chemotherapy for adult soft tissue sarcomas of the extremities and girdles: results of the Italian randomized cooperative trial. J Clin Oncol 29:1238–1247

Gronchi A, Frustaci S, Mercuri M et al (2012) Short full dose adjuvant chemotherapy in high risk adult soft tissue sarcomas: a randomized clinical trial from the Italian Sarcoma Group and the Spanish Sarcoma Group. J Clin Oncol 30:850–856

Issels RD, Lindner LH, Verweij J et al (2010) Neo-adjuvant chemotherapy alone or with regional hyperthermia for localised high-risk soft-tissue sarcoma: a randomised phase 3 multicentre study. Lancet Oncol 11(6):561–570

Italiano A, Delva F, Mathoulin-Pelissier S, Le Cesne A, Bonvalot S, Terrier P, Trassard M, Michels JJ, Blay JY, Coindre JM, Bui B (2010) Effect of adjuvant chemotherapy on survival in FNCLCC grade 3 soft tissue sarcomas: a multivariate analysis of the French Sarcoma Group Database. Ann Oncol 21(12):243

Pervaiz N, Colterjohn N, Farrokhyar F et al (2008) A systematic meta-analysis of randomized controlled trials of adjuvant chemotherapy for localized resectable soft-tissue sarcoma. Cancer 113:573–581

Sarcoma Meta-analysis Collaboration (1997) Adjuvant chemotherapy for localised resectable soft-tissue sarcoma in adult: meta-analysis of individual data. Lancet 350:1647–1654

Woll PJ, Reichardt P, Le Cesne A et al (2012) Adjuvant chemotherapy with doxorubicin, ifosfamide, and lenograstim for resected soft-tissue sarcoma (EORTC 62931): a multicentre randomised controlled trial. Lancet Oncol 13(10):1045–1054

## Pathology

Antonescu CR (2008) Molecular profiling in the diagnosis and treatment of high grade sarcomas. Ultrastruct Pathol 32(2):37–42. Review

Bovée JV, Hogendoorn PC (2010) Molecular pathology of sarcomas: concepts and clinical implications. Virchows Arch 456(2):193–199. Review

Fletcher CDM, Bridge JA, Hogendoorn PCW, Mertens F (eds) (2013) WHO classification of tumors of soft tissue and bone, 4th edn. International Agency for Research on Cancer, Lyon

Trojani M, Contesso G, Coindre JM et al (1984) Soft-tissue sarcomas of adults; study of pathological prognostic variables and definition of a histopathological grading system. Int J Cancer 33:37–42

Unni K et al (2005) Afip atlas of tumor pathology – series 4 – tumors of the bones and joints. ARP Press Silver Spring, Washington, DC

# Chapter 57
# Myositis Ossificans

**Nicola Fabbri**

**Definition**: Hyperplastic and reactive process involving muscle fibers, fascia, and periosteum and leading to soft tissue ossifications.
**Importance**: Two types are recognized: posttraumatic type (localized, self-limited, and cured by rest) and genetic type (fibrodysplasia ossificans progressiva: diffuse, sometimes generalized and fatal).
**Epidemiology**: Posttraumatic type occurs rarely, more frequently in young males. The genetic type is exceptionally rare, probably due to dominant mutations. It is usually discovered before the age of 4.

N. Fabbri, MD
Department of Surgery, Orthopaedic Service, Memorial Sloan-Kettering Cancer Center, 1275 York Avenue, New York, 10065, NY, USA
e-mail: fabbrin@mskcc.org

P. Picci et al. (eds.), *Atlas of Musculoskeletal Tumors and Tumorlike Lesions*,
DOI 10.1007/978-3-319-01748-8_57, 

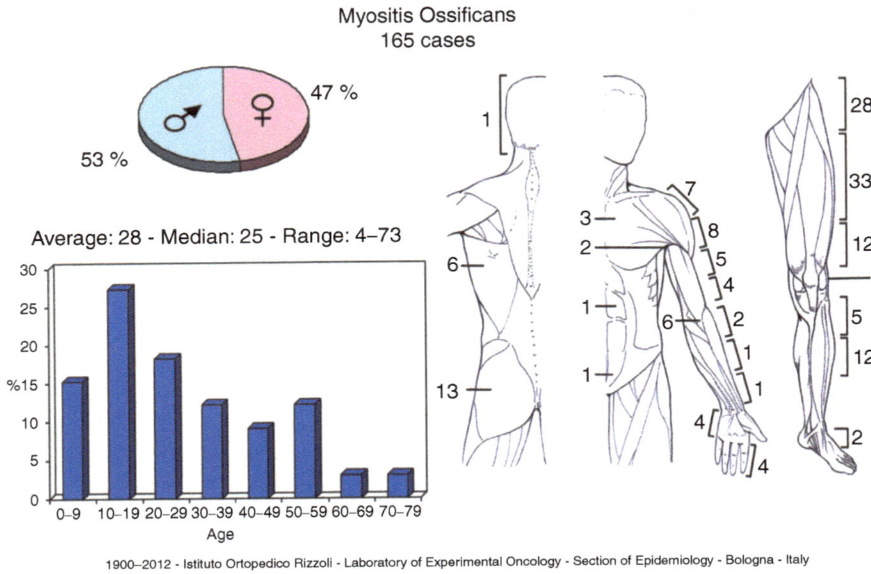


**Clinical**: Posttraumatic type: painful enlarging mass with intense inflammation of the surrounding tissues. If the involved anatomic compartment is immobilized, inflammation resolves, and the mass slowly regresses. Complete maturation of the mass occurs in 6–12 months. Genetic type: progressive and diffused ossification of tendons, ligaments, and connective tissue around muscles, micro-clinodactyly.

**Localization**: The posttraumatic type may develop in any site, but it is more frequent in the anterior thigh, upper arm (brachialis muscle), and adductor or gluteal muscles. The genetic type usually starts in the neck, shoulder, axilla, or paraspinal muscles.

**Imaging**: Radiological appearance is related to the phase of the process: in the early period only a fading round radiodensity is shown; a peripheral rim becomes visible in few weeks, then the osseous radiodensity enlarges its borders, but a radiolucent central zone remains for a long period. Remodeling of the area occurs for years.

**Histopathology**: Early phase: spindle cell proliferation, high mitotic activity with hyperchromatic nuclei (d. d. with sarcomas). Then, the peripheral mature and the central immature zone are easily distinguished (zonation phenomenon): loosely arranged spindle cells in center with bone formation at periphery. Mitotically active but without true cellular atypia.

**Course**: Posttraumatic type: self-limited after rest and anti-inflammatory drugs. Genetic type: progressive severe disability. If respiratory muscles are involved, disease may be fatal.

**Treatment**: No surgical treatment is usually required. In genetic diseases, good general care and avoidance of trauma (particularly the iatrogenic ones from im injections, biopsies, surgery) are emphasized.

Radiograph of the arm. The lesion is more ossified at the periphery, with a partially clear center. This type of ossification, starting 3 weeks after clinical onset, is very typical of the diagnosis

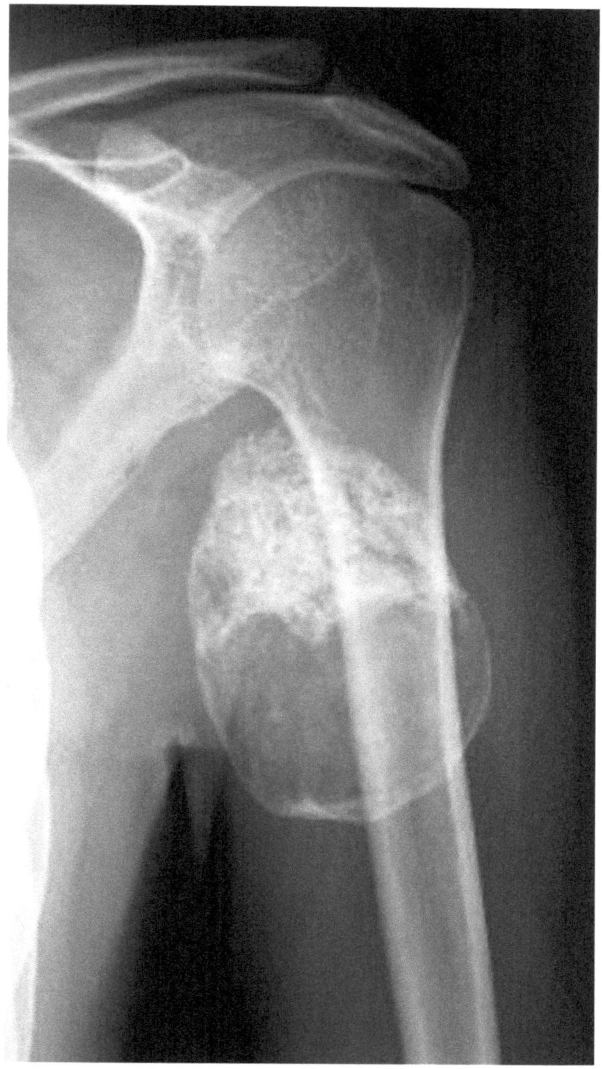

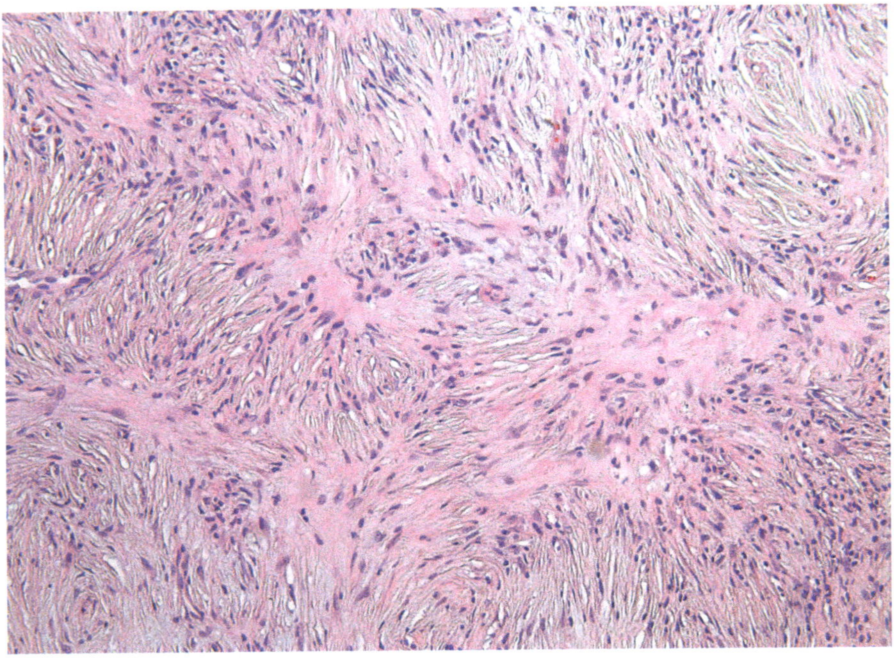

Storiform pattern of bland-appearing spindle cells and scattered inflammatory cells in the center of the lesion

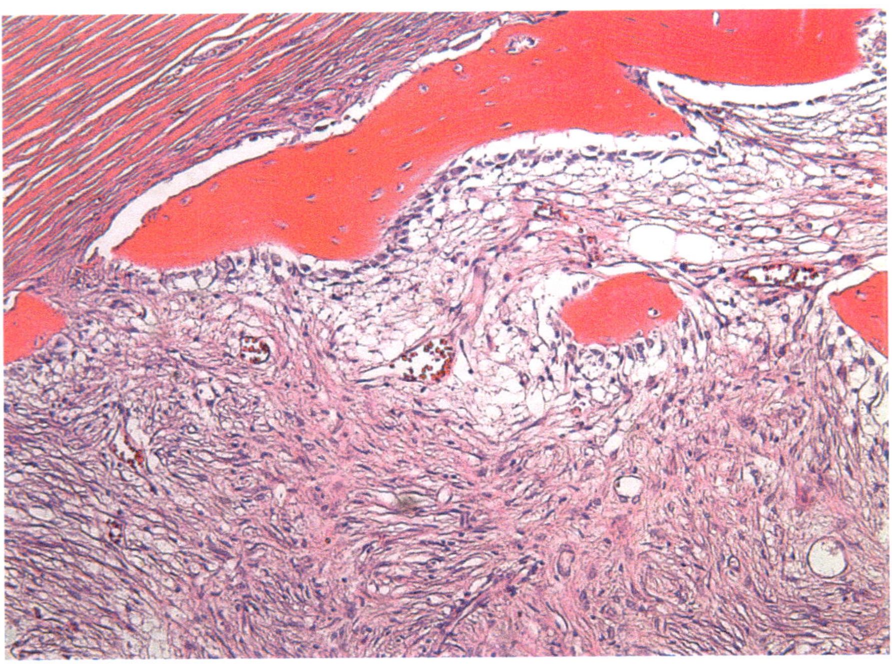

Ossified rim at the periphery (zoning phenomenon)

## Selected Bibliography

Abate M, Salini V, Rimondi E, Errani C, Alberghini M, Mercuri M, Pelotti P (2011) Post traumatic myositis ossificans: sonographic findings. J Clin Ultrasound 39:135–140

Martin DA, Senanayake S (2011) Images in clinical medicine. Myositis ossificans. N Engl J Med 364(8):758

McCarthy EF, Sundaram M (2005) Heterotopic ossification: a review. Skeletal Radiol 34(10): 609–619. Review

Parikh J, Hyare H, Saifuddin A (2002) The imaging features of post-traumatic myositis ossificans, with emphasis on MRI. Clin Radiol 57(12):1058–1066. Review

Rosenberg AE (2008) Pseudosarcomas of soft tissue. Arch Pathol Lab Med 132(4):579–586. Review

# Chapter 58
# Pigmented Villonodular Synovitis (PVNS) and Giant-Cell Tumor of Tendon Sheath

**Nicola Fabbri**

**Definition**: Benign, composed of synovial-like mononuclear cells of the joint, the tendon sheaths, and the mucous bursae.
**Epidemiology**: No sex predilection. 20–40 years of age.

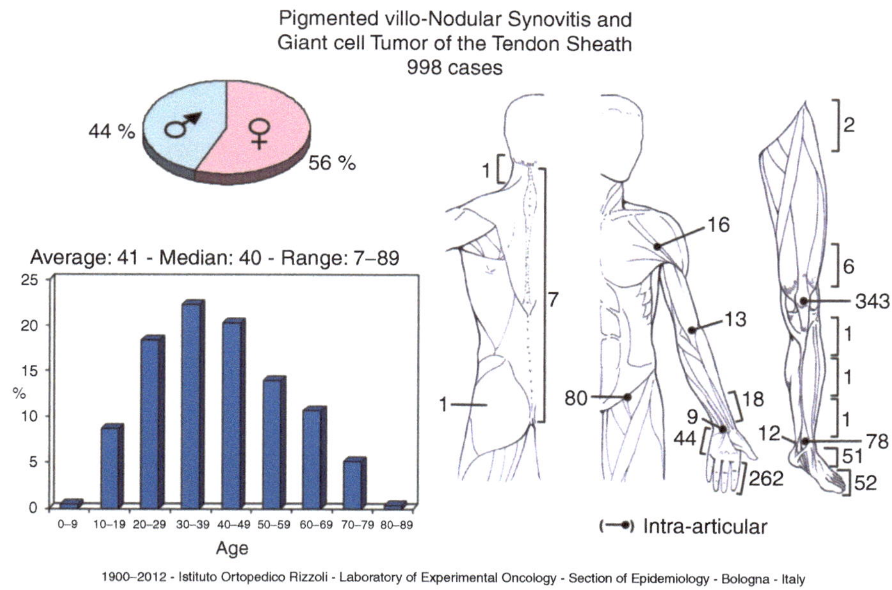


N. Fabbri, MD
Department of Surgery, Orthopaedic Service, Memorial Sloan-Kettering Cancer Center, 1275 York Avenue, New York 10065, NY, USA
e-mail: fabbrin@mskcc.org

P. Picci et al. (eds.), *Atlas of Musculoskeletal Tumors and Tumorlike Lesions*,
DOI 10.1007/978-3-319-01748-8_58, © Springer International Publishing Switzerland 2014

**Location**: (a) Paratendinous (frequent): in the sheath of a flexor tendon of the fingers, in the palm of the hand close to a metacarpophalangeal joint, in the wrist, on the dorsum of a finger adjacent to an extensor tendon, and rare in the foot; (b) in the joint (rare): >75 % in the knee, then in the hip, wrist, ankle, and shoulder; and (c) in the bursae (exceptional).
**Clinical**: Pain, swelling, effusion, or completely asymptomatic.
**Diagnosis**: On x-ray thickening of the synovial membrane with the same density of the soft tissue and without calcification. Skeletal erosions due to compression are frequent as rounded or multilobulated osteolytic lesions with well-defined margins, on the perimeter of the joint. Erosions are multiple and superficial, with sclerotic edges. On CT—lobulated newly formed tissue in the joint with considerable uptake of contrast dye. On bone scan—bone uptake and increased flow and blood pool in the mass. On MRI—heterogeneous, mostly low signal on T1 and T2 is characteristic. Intra- and peritumoral enhancing curvilinear regions on contrast T1.
**Histopathology**: Roughly lobulated, single, soft, yellow-white to pale-brown nodules with smooth surface. In advanced stages it matures in a fibrous scar. It becomes hard, compact, and white with some yellow or brown bands and adheres to the surrounding tissues, to bone, and to tendon. The synovial membrane appears thickened, leathery yellow, and matted by long large villi like a "ruffled beard," with multiple, soft, yellow-brown, lobulated nodules of varying size. Fibrin membranes cover the villi surface. Soft, pasty, friable, and yellow-brown tissue fills joint space in more advanced lesions. Pathologic tissue may be easily enucleated from the bone lesions that have a smooth bony wall. PVNS may invade the joint capsule and expand into the muscles, between the tendons, but it never infiltrates them.
**Course and Staging**: Slowly growing, it may be an aggressive lesion. Usually, stage 2, but also stage 3 in late cases of enormous size.
**Treatment**: Marginal excision is easy and adequate. In PVNS of the joint, surgery is difficult and sometimes it is impossible to eradicate the whole lesion. A wide synovectomy is indicated, but often local recurrence occurs. Radiotherapy is useful as adjuvant treatment when complete excision is not feasible. When the disease destroys the joint cartilage, an arthrodesis or prosthesis may be required.

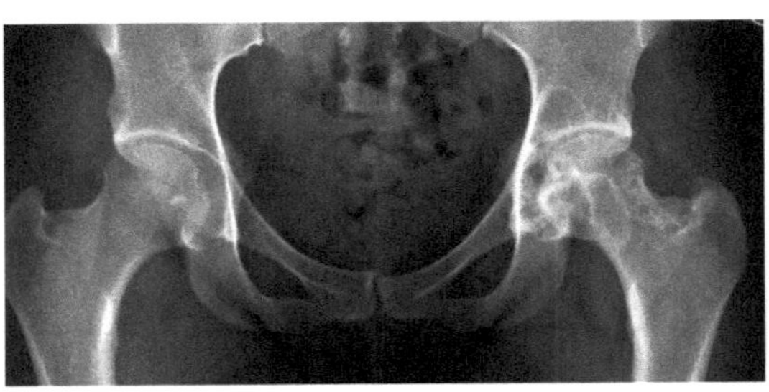

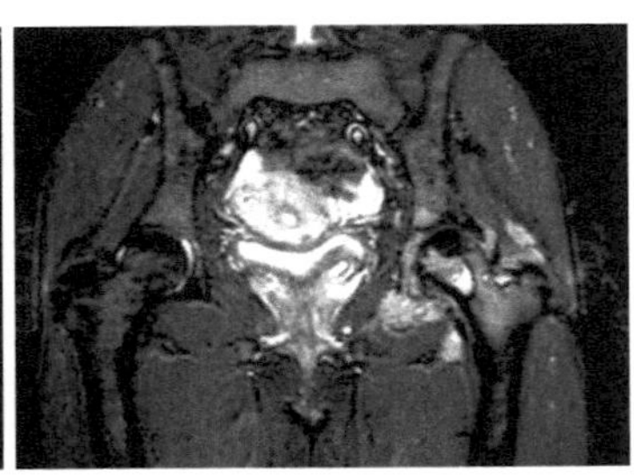

Radiograph and T2 coronal MR image. Hip lesion with multiple erosions of the bone of both parts of the joint. Lesions have a low signal on T2 image due to iron deposits

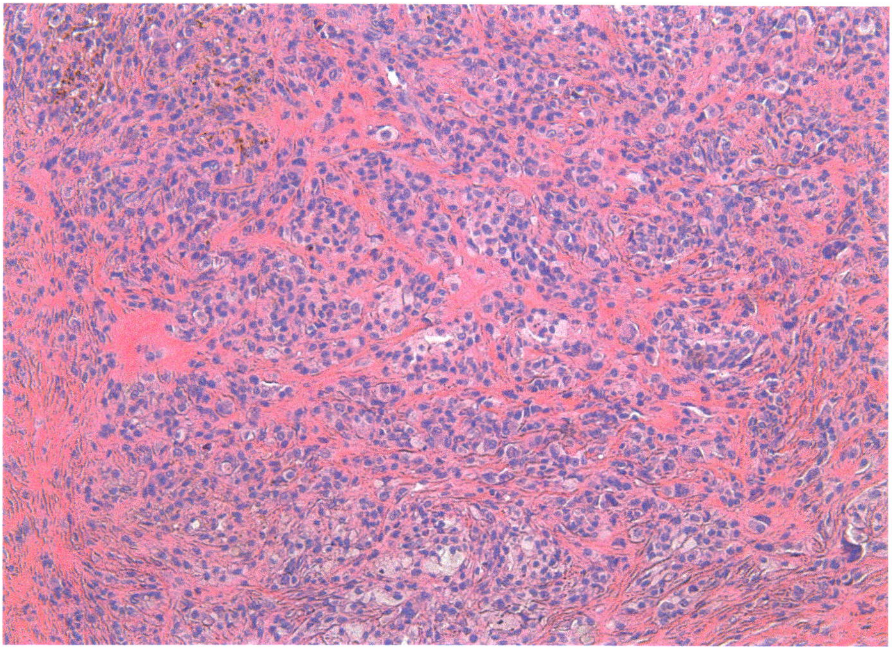

Sheets of synovial mononuclear cells with abundant hemosiderin pigment and scattered groups of foam cells. Vaguely nodular pattern on panoramic view

## Selected Bibliography

Adelani MA, Wupperman RM, Holt GE (2008) Benign synovial disorders. J Am Acad Orthop Surg 16(5):268–275. Review

Frick MA, Wenger DE, Adkins M (2007) MR imaging of synovial disorders of the knee: an update. Radiol Clin North Am 45(6):1017–1031. Review

Mendenhall WM, Mendenhall CM, Reith JD, Scarborough MT, Gibbs CP, Mendenhall NP (2006) Pigmented villonodular synovitis. Am J Clin Oncol 29(6):548–550. Review

Tyler WK, Vidal AF, Williams RJ, Healey JH (2006) Pigmented villonodular synovitis. J Am Acad Orthop Surg 14(6):376–385. Review

Wan JM, Magarelli N, Peh WC, Guglielmi G, Shek TW (2010) Imaging of giant cell tumour of the tendon sheath. Radiol Med 115(1):141–151. Review

# Chapter 59
# Synovial Chondromatosis

**Davide Donati**

**Definition**: A benign neoplasm of well-differentiated hyaline cartilage within the joint synovial membrane or the tendinous sheath or the bursae mucosae.
**Epidemiology**: Rare. Males. 30–50 years old.

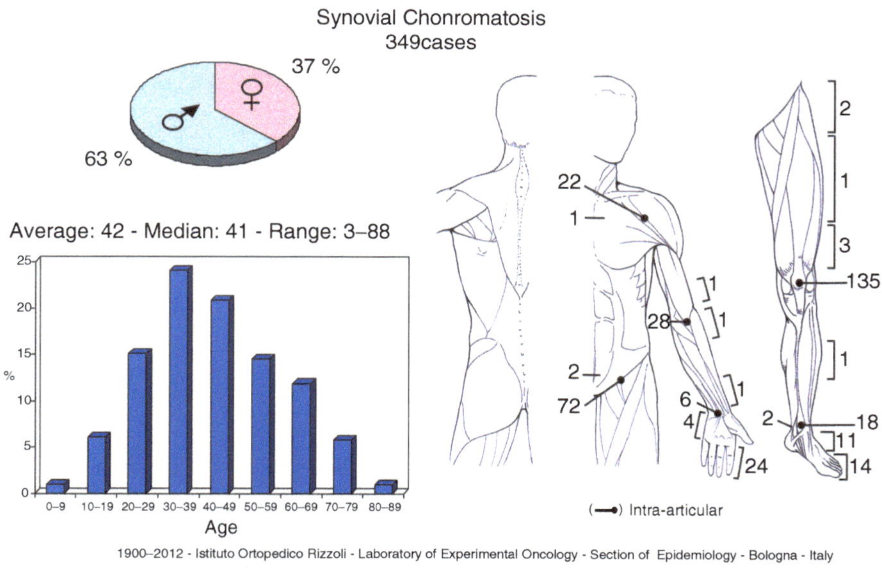


**Location**: >50 % knee, then elbow, shoulder, wrist, hip, and ankle. Extra-articular form in the fingers.

D. Donati, MD
3rd Orthopaedic and Traumatologic Clinic Prevalently Oncologic,
Istituto Ortopedico Rizzoli, Bologna, Italy
e-mail: davide.donati@ior.it

P. Picci et al. (eds.), *Atlas of Musculoskeletal Tumors and Tumorlike Lesions*, 281
DOI 10.1007/978-3-319-01748-8_59, © Springer International Publishing Switzerland 2014

**Clinical**: Pain, limited joint motion, and sensation of joint crackling. Rare blocking and joint effusion. Palpated loose bodies. Recurring and slowly progressive symptoms. Plurilobulated hard-elastic mass expanding around a joint in aggressive type.
**Imaging**: On X-ray—negative if not calcified or small-sized calcified nodules into the joint, typical rice grain-like. The joint is normal. Arthrosis is frequent in inveterate cases and bone erosion or cortical scalloping in the aggressive ones. On CT—more evident pseudotumoral mass, calcifications, bone lesions, and invasion of the surrounding tissues. On MRI—increased joint fluid, lobular intra-articular mass, intermediate intensity if uncalcified or with white punctuated appearance if ossified on T1, round, ringlike, dark signal voids in strong enhancement of the synovial tissue on contrast T1 and in bright signal of the joint fluid on T2.
**Histopathology**: White, smooth, lucent, translucid soft or firm, loose bodies of different size. Thickened and seeded with cartilaginous nodule synovial membrane; hard, granular, yellowish ossified areas in the nodules. Minute nests of cartilage enmeshed in primitive mesenchymal tissue in all stages of maturation. Well-differentiated rather cellular hyaline cartilage. Cells in small bunches, divided by bands of ground substance. Well-defined boundaries of the lobuli; large, numerous cells with a plump, double nucleus, with nuclear pleomorphism and abnormalities like in chondrosarcoma grade 1 or 2. Endochondral ossification areas.
**Course and Staging**: Slow growth and rare recurrence even if surgery is inadequate. Usually, stage 2 or 3. Sometimes it may spontaneously regress. Very rare cases can transform into chondrosarcoma.
**Treatment**: Intralesional excision with wide synovectomy.

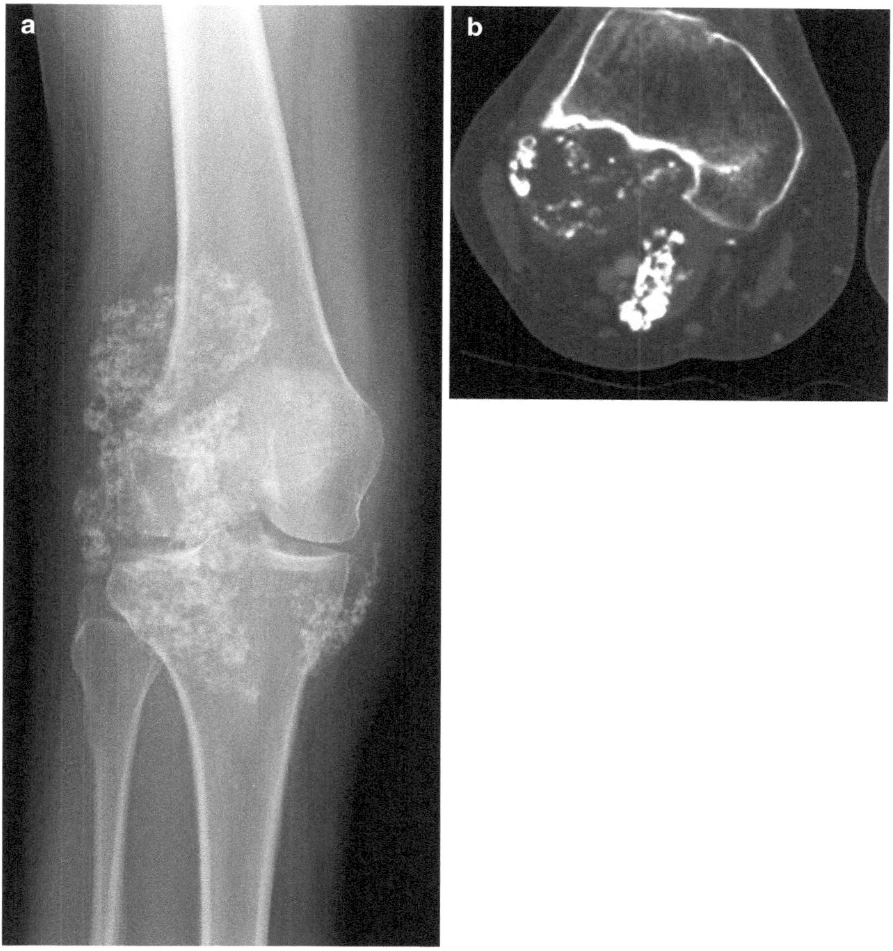


Radiograph and CT. Mass in the joint, eroding bone in a well-limited way and containing cartilaginous calcifications

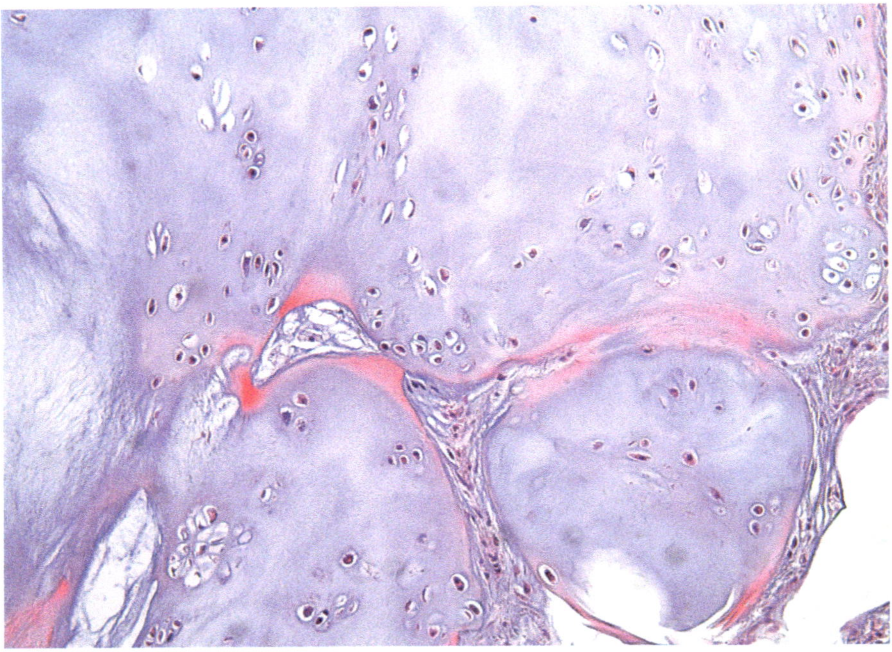

Lobular pattern, clustering of chondrocytes, hyaline matrix

## Selected Bibliography

Adelani MA, Wupperman RM, Holt GE (2008) Benign synovial disorders. J Am Acad Orthop Surg 16(5):268–275. Review

Frick MA, Wenger DE, Adkins M (2007) MR imaging of synovial disorders of the knee: an update. Radiol Clin North Am 45(6):1017–1031. Review

McKenzie G, Raby N, Ritchie D (2008) A pictorial review of primary synovial osteochondromatosis. Eur Radiol 18(11):2662–2669

Murphey MD, Vidal JA, Fanburg-Smith JC, Gajewski DA (2007) Imaging of synovial chondromatosis with radiologic-pathologic correlation. Radiographics 27(5):1465–1488. Review

Zamora EE, Mansor A, Vanel D, Errani C, Mercuri M, Picci P, Alberghini M (2009) Synovial chondrosarcoma: report of two cases and literature review. Eur J Radiol 72(1):38–43

# Chapter 60
# Fibromatosis

**Nicola Fabbri**

It covers a broad group of benign fibroblastic tissue proliferation of similar microscopic appearance.

A. Superficial: (1) palmar f. (Dupuytren's contracture), (2) plantar f. (Ledderhose's disease), (3) penile f. (Peyronie's disease) and (4) knuckle pads.
B. Deep (desmoid-type fibromatosis): (1) extra-abdominal f. and (2) abdominal f.

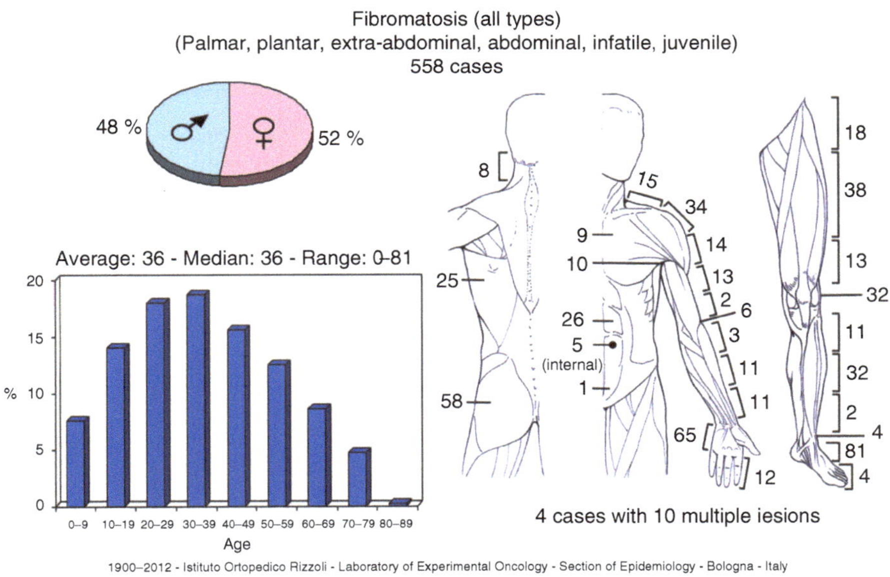


N. Fabbri, MD
Department of Surgery, Orthopaedic Service, Memorial Sloan-Kettering Cancer Center, 1275 York Avenue, New York, 10065, NY, USA
e-mail: fabbrin@mskcc.org

P. Picci et al. (eds.), *Atlas of Musculoskeletal Tumors and Tumorlike Lesions*,
DOI 10.1007/978-3-319-01748-8_60, © Springer International Publishing Switzerland 2014

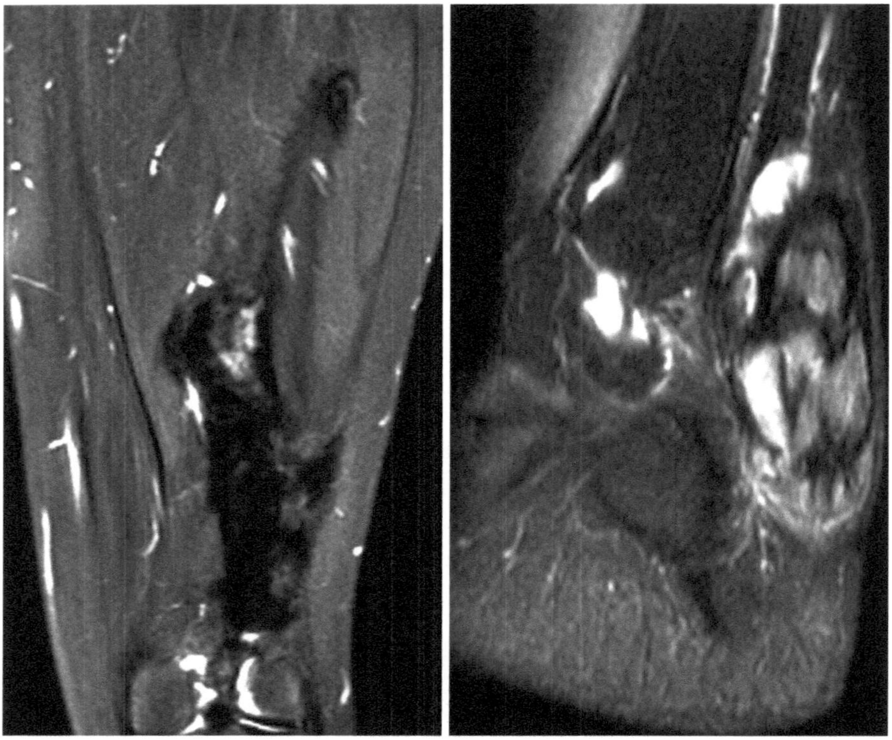

T2 MR images, coronal of the thigh and sagittal of the foot. The thigh lesion has a low signal and infiltrates the muscles, whereas one in the foot is nodular with a high signal. Both behaviors are possible on different masses in the same patient

# Chapter 61
# Extra-abdominal Fibromatosis

**Nicola Fabbri**

**Definition**: An invasive, permeative, and progressive tumor uniformly composed of fibroblasts and abundant collagen.
**Epidemiology**: Children and young males. Peak incidence 25–35 years.
**Localization**: Proximal part of the limbs (shoulder and buttock), then trunk, thigh, popliteal fossa, sura, arm, and forearm. Usually in the deep soft tissues. Skin invasion when the tumor expands to the superficial fascia.

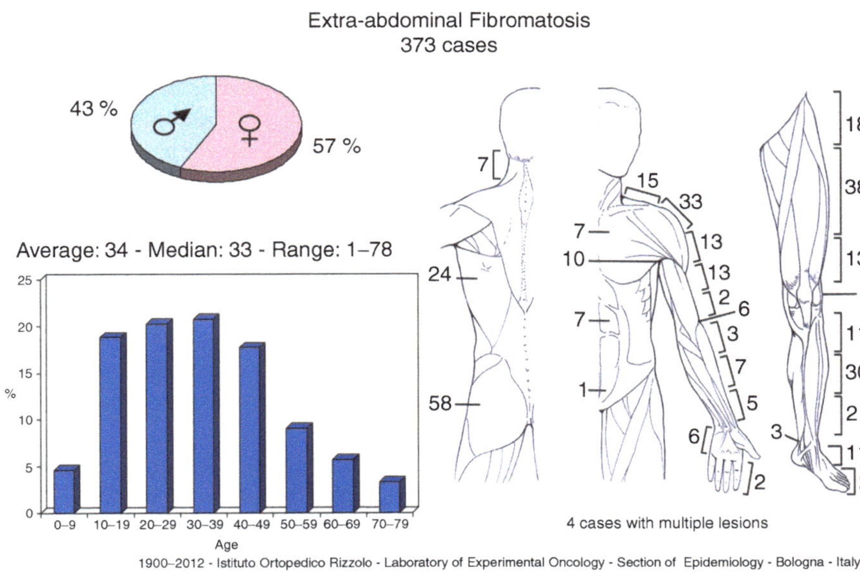


N. Fabbri, MD
Department of Surgery, Orthopaedic Service, Memorial Sloan-Kettering Cancer Center,
1275 York Avenue, New York 10065, NY, USA
e-mail: fabbrin@mskcc.org

P. Picci et al. (eds.), *Atlas of Musculoskeletal Tumors and Tumorlike Lesions*,
DOI 10.1007/978-3-319-01748-8_61, © Springer International Publishing Switzerland 2014

**Clinical**: Deep, painless, firm or hard, and poorly circumscribed mass that may be adherent to the skin or bone and that grows insidiously and slowly for several months. Tumor mass is elongated along the muscle, or it is made up of large masses disseminated along the limb and joined together. The tumor grows by infiltrative ramifications along the muscle, the aponeuroses, and the tendons. When large, muscular retraction phenomena and limited joint motion occur. Pain, numbness, hypoesthesia, and motor weakness when it infiltrates a large nerve.
**Imaging**: On X-ray a dense mass interrupts the adjacent intermuscular and soft tissue planes. Adherent to bone (37 %) with saucerized cortex without periosteal reaction. On angiography a thick intratumoral capillary network, with few vessels of large caliber. On CT intensely and diffusely takes up contrast medium, easily distinguished from surrounding muscles, irregular shape of the tumor, many nodules joined together, and faded boundaries and digitations along the fascial and interstitial planes. On MRI infiltrative pattern (63 %) in youngsters and nodular pattern (81 %) in adults; hypo-/isointense to muscle on T1 and intermediate on T2; inhomogeneous (75 %), bright in the center (cellular area) and dark in the periphery (collagen area) on T2; on T2 white zones enhance after gadolinium administration, whereas dark zones remain hypointense; bright mass on T2 are aggressive lesions; natural evolution shows proximal migration; and gradual decrease of enhancing cellular areas and increasing of nonenhancing collagen zones.
**Histopathology**: Dense, white, firm lesion that may be of large size (up to 20 cm). Ill-defined margins, digitations, and satellite nodules near the tumoral limits. Firmly adherent to the aponeuroses, tendons, joint capsule, vessels, and nerves. These are compressed but not invaded. Rarely, compression on the bone causes a superficial scalloping of the cortex. A densely collagenized mature fibrous connective tissue with a relatively sparse number of fibroblasts. Mitoses are rare or absent. Variably prominent ectatic blood vessels. Spindle cells in intertwined bundles that are wavy and have no particular order. Infiltration of the muscle and the surrounding tissues. On immunohistochemistry, desmoid tumor is diffusely positive for vimentin and shows nuclear expression of beta-catenin in 85 % of cases. It may be positive for pan-muscle actin and smooth muscle actin.
**Course and Staging**: Locally aggressive tumor not associated with metastases. Stage 3. An involution to stage 2 or even stage 1 lesion is possible. Local recurrence is frequent even if removed with adequate margins. Malignant transformation has never been observed.
**Treatment**: Wide or better radical excision. When adequate excision is not possible, radiotherapy may give favorable results if it is used in addition to surgery. When the tumor compresses and surrounds main vessels of the limb, recurs several times, and pushes toward the pelvis and the axilla, wide amputation is necessary. Prognosis quoad vitam is good, but 2 % of patients die from vascular and/or interference from vital internal organs if the tumor enters the neck, chest, and abdomen. Low-dose, long-term chemotherapy with methotrexate shows good results and is used for recurrent disease or tumors that are difficult to excise.

**Immunohistochemical Panel**

| | |
|---|---|
| VIM | + |
| MS Act | ± |
| Smooth M Act | ± |
| Beta-catenin | + (nuclear staining, 70–75 %) |

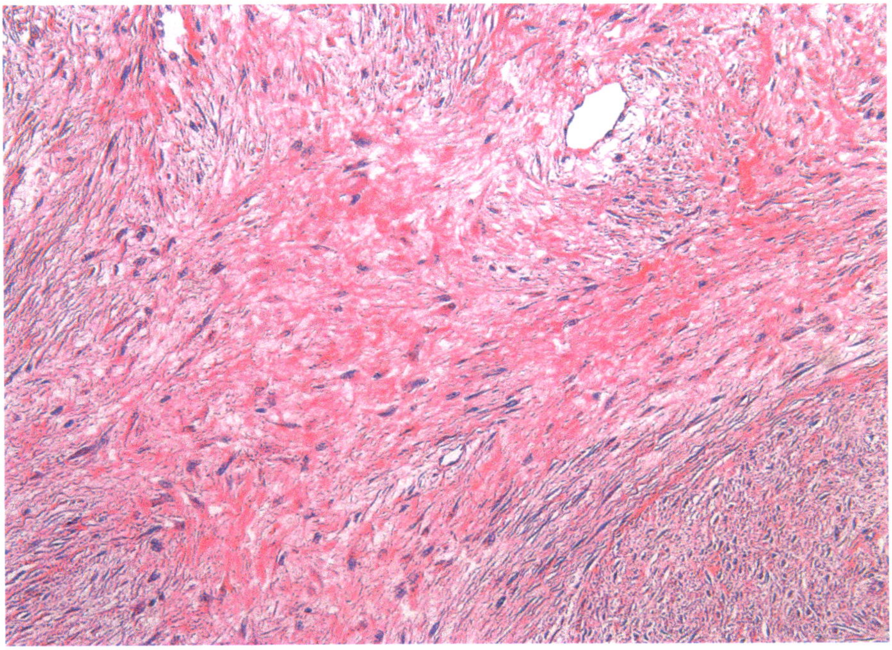

Heavily collagenized tissue with low cellularity, spindle cells with bland appearance, intersected fascicles, and ectatic vascular clefts

## Selected Bibliography

Desmoid Lazar AJ, Hajibashi S, Lev D (2009) Tumor: from surgical extirpation to molecular dissection. Curr Opin Oncol 21(4):352–359. Review

Guglielmi G, Cifaratti A, Scalzo G, Magarelli N (2009) Imaging of superficial and deep fibromatosis. Radiol Med 114(8):1292–1307. Review

Lee JC, Thomas JM, Phillips S, Fisher C, Moskovic E (2006) Aggressive fibromatosis: MRI features with pathologic correlation. AJR Am J Roentgenol 186(1):247–254

McCarville MB, Hoffer FA, Adelman CS, Khoury JD, Li C, Skapek SX (2007) MRI and biologic behavior of desmoid tumors in children. AJR Am J Roentgenol 189(3):633–640

Murphey MD, Ruble CM, Tyszko SM, Zbojniewicz AM, Potter BK, Miettinen M (2009) From the archives of the AFIP: musculoskeletal fibromatoses: radiologic-pathologic correlation. Radiographics 29(7):2143–2173

# Chapter 62
# Lipomas

**Nicola Fabbri**

**Definition**: A benign tumor constituted by well-differentiated adipocytes.
**Epidemiology**: The most common among soft tissue tumors. It is more frequently observed between 40 and 60 years of age and prevails in females when it is superficial, whereas in males when it is deep and multiple.

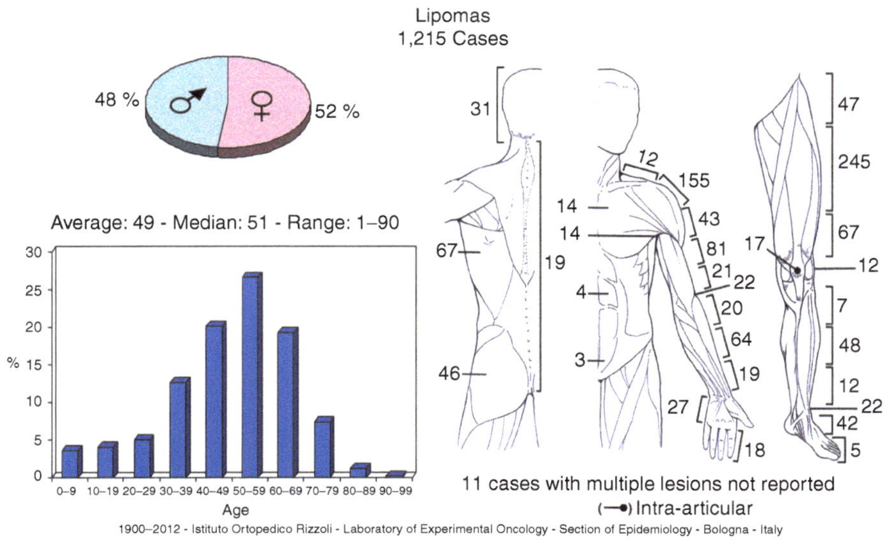


**Localization**: (a) Superficial (frequent): in the subcutaneous tissue of the back, shoulder, neck, and proximal extremities. (b) Deep (rare): within or between

N. Fabbri, MD
Department of Surgery, Orthopaedic Service, Memorial Sloan-Kettering Cancer Center, 1275 York Avenue, New York 10065, NY, USA
e-mail: fabbrin@mskcc.org

P. Picci et al. (eds.), *Atlas of Musculoskeletal Tumors and Tumorlike Lesions*,
DOI 10.1007/978-3-319-01748-8_62, 

muscles or adherent to bone, tendons, joints or nerves. In 5 % of cases lipomas are multiple with symmetrical distribution in the dorsum and proximal upper limb.
**Clinical**: Solitary lump, slow growing, and painless unless there is nerve compression. Superficial lipoma never grows large size (average 4 cm.) and it is movable. Deep lipoma tends to be larger (average 10 cm) and with a spherical, fixed, and firm mass.
**Imaging**: On X-ray, a radiolucent mass with or without calcification or ossification; mild cortical thickening when parosteal. On CT scan, a lobulated, sharply marginated radiolucency with homogeneous density. On MRI, an encapsulated, bright mass without enhancement after contrast administration; signal intensity equal to that of fat; regular thin septation. On angiography, avascular. On bone scan, there is no uptake.
**Histopathology**: It is often lobulated with a very thin true capsule and pseudocapsule fibrous, thick, and adherent to the surrounding tissues. Soft on palpation, pale yellow in color, lipoma is constituted by lipocytes, mature "signet ring" cells. Vessels are not very apparent, because they are thin and compressed by lipocytes. On immunohistochemistry, all lipomas are diffusely positive for S-100 and vimentin. Spindle cell and pleomorphic lipoma are also positive for CD34.
**Course and Staging**: a) Superficial lipoma: easily diagnosed, asymptomatic, generally stage 1, but it may behave as active stage 2 lesion. According to AJC classification, lipoma is more frequently stage Ia. b) Deep lipoma: an extensive anatomo-pathological study with multiple specimens is necessary to exclude liposarcoma lipoma-like. Usually, stage 2 or stage Ib according to AJC classification. Malignant changes are exceptional.
**Treatment**: Marginal excision is curative. Recurrence is rare (<5 %).

| Variants | Age | Sex | Clinical | Gross | Histology |
|---|---|---|---|---|---|
| Angiolipoma | 20 | Male | <2 cm/forearm<br>Subcutaneous<br>Pain | Firm<br>Yellow/reddish | Lipocytes +<br>Network of capillaries with fibrin thrombi |
| Spindle cell lipoma | Adult | Male | 4 cm/back<br>Subcutaneous<br>Painless | Soft<br>Yellow/whitish | Lipocytes +<br>Vessels + spindle cells +<br>Myxoid matrix +<br>Collagenous bands |
| Pleomorphic lipoma | Adult | Male | 4 cm/back<br>Subcutaneous<br>Painless | Firm<br>Yellow/whitish | Lipocytes +<br>Bizarre floret-like multinucleated cells |
| Lipoblastoma | <2 | Male | 3 cm/limbs<br>Subcutaneous<br>Painless | Lobulated translucid | Like myxoid liposarcoma |
| Lipomatosis | 10 | – | Large/diffused<br>Pain | Dense<br>Tissue infiltrating | Mature adipose |
| Intranervous L | <30 | Male | Hand/wrist<br>Pain + neuropathy | Hard | Surrounds and infiltrates the nerve |
| Hibernoma | Adult | Male | 4 cm/scapular<br>Subcutaneous<br>Painless | Firm | Central nucleus + Foam cytoplasm = Brown fat |

**Immunohistochemical Panel**

| | |
|---|---|
| Usual Lipoma | |
| VIM | + |
| S100 | + |

**Immunohistochemical Panel**

| | |
|---|---|
| Spindle Cell Lipoma | |
| VIM | + |
| S100 | + |
| CD34 | + |

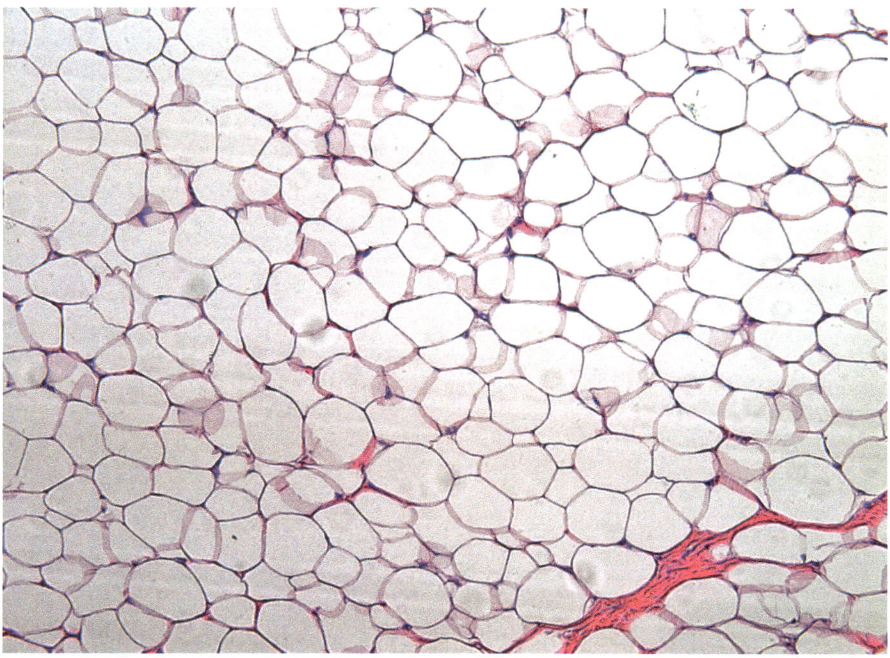

Mature adipocytic cells organized in lobules with flat nuclei at the periphery and optically empty cytoplasms. No atypia of the cells

## Selected Bibliography

Drevelegas A, Pilavaki M, Chourmouzi D (2004) Lipomatous tumors of soft tissue: MR appearance with histological correlation. Eur J Radiol 50(3):257–267. Review

Goodwin RW, O'Donnell P, Saifuddin A (2007) MRI appearances of common benign soft-tissue tumours. Clin Radiol 62(9):843–853. Review

Lee JC, Gupta A, Saifuddin A, Flanagan A, Skinner JA, Briggs TW, Cannon SR (2006) Hibernoma: MRI features in eight consecutive cases. Clin Radiol 61(12):1029–1034

Rubin BP, Dal CP (2001) The genetics of lipomatous tumors. Semin Diagn Pathol 18(4):286–293. Review

Zamecnik M, Michal M (2007) Angiomatous spindle cell lipoma: Report of three cases with immunohistochemical and ultrastructural study and reappraisal of former 'pseudoangiomatous' variant. Pathol Int 57(1):26–31. Review

# Chapter 63
# Liposarcomas

**Marco Gambarotti**

**Definition**: Malignant tumor in which cells differentiate in lipoblasts and lipocytes.
**Epidemiology**: The most frequent sarcoma of the soft tissues. Prevails in males when it is localized in the limbs, but not when in retroperitoneum. Very rare prior to 20 years of age, whereas it is typical of adult and advanced age.

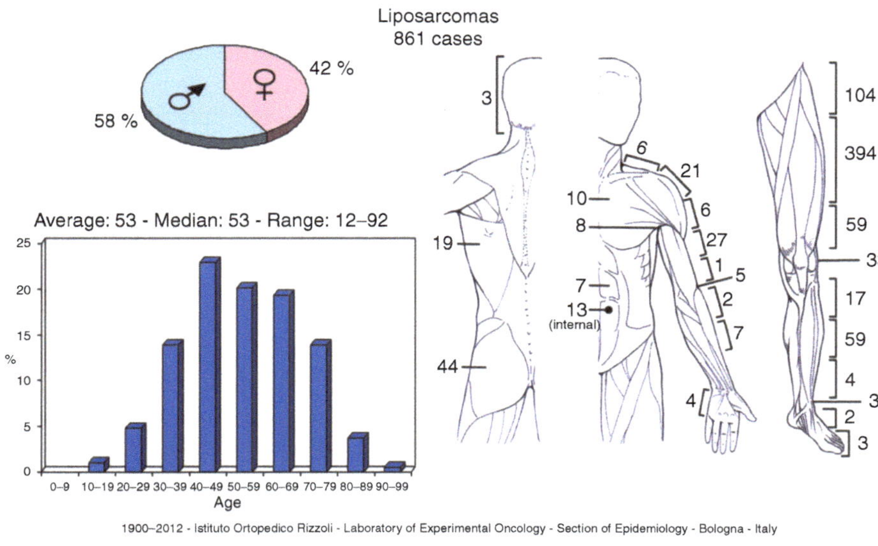


M. Gambarotti, MD
Department of Anatomy and Pathological Histology, Istituto Ortopedico Rizzoli,
Via del Barbiano 1/10, Bologna 40136, Italy
e-mail: marco.gambarotti@ior.it

P. Picci et al. (eds.), *Atlas of Musculoskeletal Tumors and Tumorlike Lesions*,
DOI 10.1007/978-3-319-01748-8_63, © Springer International Publishing Switzerland 2014

**Localization**: Anywhere there is fat. Most common in thigh, retroperitoneum, inguinal region, and popliteal fossa. Never in the hands, feet, and neck.
**Clinical**: Insidiously growing, deep-seated, ill-defined mass that usually reaches a large size. Pain, tenderness, and functional disturbances occur in 10–15 % of cases, but these are usually late complaints. Tumor may be painful due to compression on the nerve or cause edema of the limb due to venous occlusion. Retroperitoneal tumor may cause hydronephrosis, intestinal compression, inguinal hernia, and edema of the lower limbs.
**Imaging**: On X-ray: in more differentiate varieties, well-delineated radiolucency clearly distinguished from the surrounding muscles. Occasionally calcification, little vascularity. In more undifferentiated cases, less well-defined masses with greater density. On CT scan: in lipoma-like type, well-marginated mass mimicking lipoma. Frequently, thickened linear, nodular septa that enhance after contrast. In myxoid type, encapsulated, septated, slightly heterogeneous mass mimicking a cyst. Density of water. On MRI the myxoid type shows a homogeneous low signal intensity (dark) on T1 and high (white) on T2, strong enhancement, and minute foci of fatty tissue; small bright streaks on T1 in an overall dark lesion are typical. Nonspecific appearances are displayed by the other varieties: heterogeneous signal intensity on T1, white and dark areas on T2, and enhancement of all components. In dedifferentiated liposarcoma, a homogeneous mass of low signal intensity on T1 converts to a lesion with bright and intermediate areas on T2. Often the tumor is adherent or erodes adjacent cortical bone that responds with modest periosteal reaction. On bone scan liposarcomas are inordinately hot in contrast to other malignant soft tissue tumors. On angiography: richly vascular with intense intra- and peritumoral neoangiogenesis.
**Histopathology**: Usually very large, multilobulated, and delimited by a thin reactive tissue (pseudocapsule) interrupted by some tumor nodules. Satellites are present.

**Varieties:**

(a) *Well-Differentiated Liposarcoma/Atypical Lipomatous Tumor (WDL/ALT)*: 40 % of all cases.

1. Liposarcoma lipoma-like: The most common. Yellowish, soft, oily, and friable. Mature fat mixed with collagen. Lipoblasts and lipocytes with hyperchromic nuclei and hyperchromic stromal cells within fibrous septa. Easy to confuse with lipoma.
2. Sclerosing liposarcoma: More frequent in retroperitoneum. Whiter and firmer. Scattered bizarre stomal cells associated with multivacuolated lipoblasts. Abundant collagen. Easily confused with a scar.
3. Inflammatory: More frequent in retroperitoneum. Like 1 and 2 admixed with extensive chronic inflammation. Easily mistaken for inflammatory process involving fat.

(b) *Myxoid (ML)*: Most frequent in the limbs (30–35 %). Soft, pale yellow, mucoid, and translucid or bright cherry red. Proliferating lipoblasts in different stages. A myxomatous background between cellular areas. Myxoid substance rich in mucopolysaccharides with a large quantity of hyaluronic acid. A network of thin capillaries of uniform caliber with an arborizing pattern. Oval to stellate cells scattered in the myxoid substance. Mitotic activity is rare. Hypercellular areas can be present showing ovoid-round, hyperchromic cells with scanty lipoblasts isolated or collected in small clusters. These are the features of a neoplasm that was formerly known as round cell liposarcoma. Myxoid and round cell liposarcoma represent a morphologic continuum of myxoid adipocytic neoplasia. This is supported by genetic analysis demonstrating the exact same characteristic chromosome translocations. The two main karyotypic aberrations are t(12;16) that fuses the DDIT3 gene on 12q13 with the FUS gene on 16p11 and t(12;22) that fuses DDIT3 with EWS on 22q12.

(c) *Pleomorphic (PL)*: Frequent (20 %). Softer, encephaloid, very cellular tissue with a nondescriptive pattern recognizable as sarcoma but with no unique features. On high power pleomorphism is prominent. Pleomorphic lipoblasts with vacuolated cytoplasm are scattered among undifferentiated hyperchromatic neoplastic cells. Large atypical, multinucleated giant cells are scattered through the tissue. Many atypical mitosis. S-100 immunoreactivity may help highlight the presence of multivacuolated lipoblasts in those cases in whom adipocytic differentiation tends to be focal and, therefore, easily overlooked. Similar to most pleomorphic high-grade sarcomas, pleomorphic liposarcomas tend to exhibit complex karyotypes.

(d) *Dedifferentiated (DL)*: Rare (<5 %). Most cases from retroperitoneal WDL. WDL/ALT with transformation to a sarcoma of variable histological grade. On immunohistochemistry, overexpression of MDM2 and CDK4 is consistently observed in both components.

**Course and Staging**: Type (a) are low-grade malignant tumors and present as stage IA. Type (b) purely myxoid liposarcomas are low-grade malignant tumors; they usually present as stage IA and seldom progress to higher stages. The presence of hypercellularity is associated with worsening of prognosis, and these lesions should be considered stage II. Type (c) are high-grade malignant tumors and present as stage IIB. Type (d) are high-grade malignant tumors and present as stage IIB. This is the proof of progression of malignancy. Type (a–b) recur also many years after excision, whereas in type (c–d) local recurrence occurs rapidly. Metastases are exceptional in type (a), rare in type (b), and frequent and early in type (c–d).

**Treatment**: In type (a–b) wide excision, while in type (c–d) radical removal. Radiotherapy is useful in type (c and d), and particularly effective in type (b). Conventional ifosfamide-based chemotherapy is preferred in type (c). Trabectedin (ET743) has shown to be effective in metastatic myxoid liposarcoma.

**Immunohistochemical Panel**

| | |
|---|---|
| VIM | + |
| S100 | + |

## Immunohistochemical Panel

| | |
|---|---|
| WDLPS/DDLPS (well-differentiated liposarcoma/dedifferentiated liposarcoma) | |
| VIM | + |
| S100 | + |
| MDM2 | + |
| CDK4 | + |

## Genetic Alterations

| | |
|---|---|
| Atypical lipomatous tumor/well-differentiated LPS/dedifferentiated LPS | |
| MDM2 | Amplification |
| CDK4 | Amplification |

## Chromosomal Translocations

| | | |
|---|---|---|
| Myxoid/round cell liposarcoma | | |
| t(12;16) (q13;p11) | TLS(FUS)-CHOP (type 1, type2) | 95 % |
| | TLS(FUS)-CHOP (type 3) | |
| t(12;22) (q13;q12) | EWS-CHOP (type 1) | 1–5 % |
| | EWS-CHOP (type 2, type3) | |

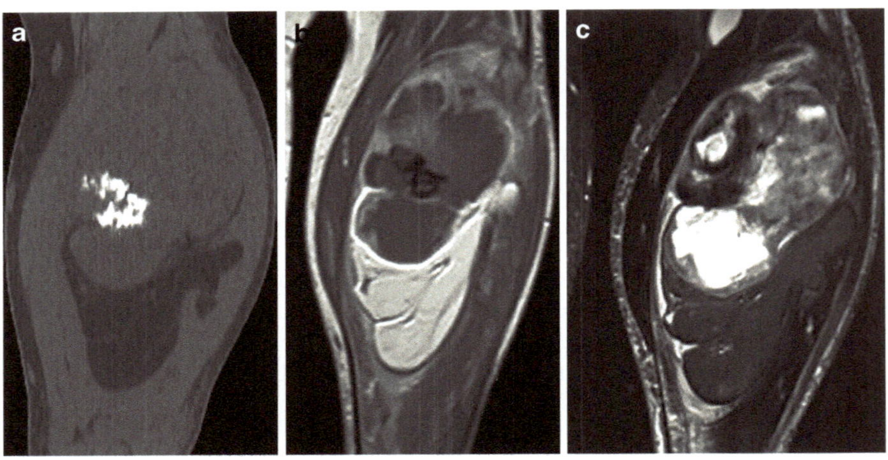


Well-limited mass of the thigh. On CT the mass is ossified. On T1 and T2 W FS coronal images, the lesion is made of two components. One is fat, corresponding to a well-differentiated liposarcoma; the second mainly fluid is partially ossified and corresponds to a myxoid component

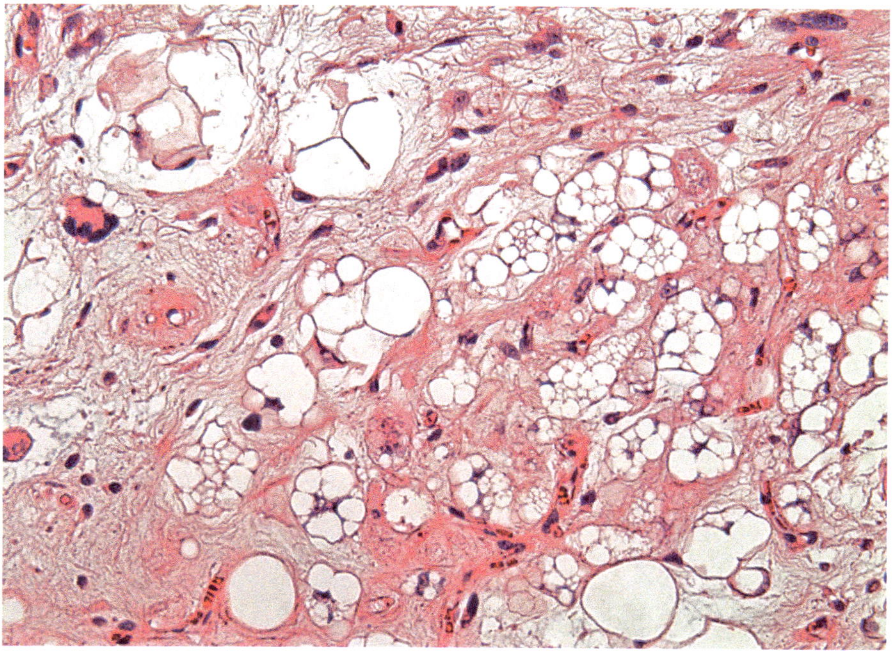

Well-differentiated liposarcoma/atypical lipomatous tumor. Mature adipocytic cells and scattered atypical adipocytes with hyperchromic nuclei and rare lipoblasts (unnecessary for diagnosis)

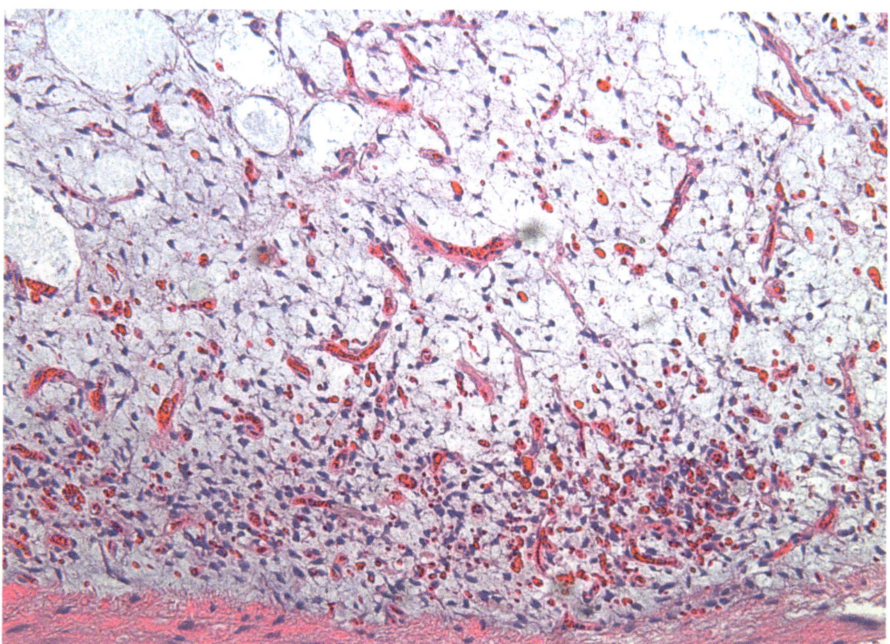

Myxoid liposarcoma. Oval to stellate cells embedded in a myxoid matrix with typical arborizing vascular pattern. Variable presence of round cell component

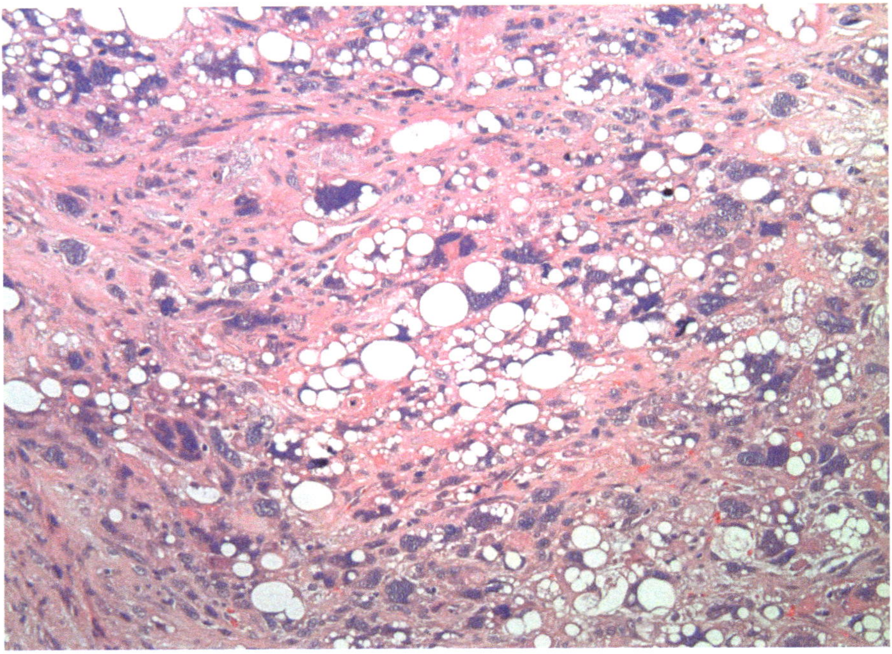

Pleomorphic liposarcoma. UPS-like high-grade sarcoma with scattered pleomorphic lipoblasts

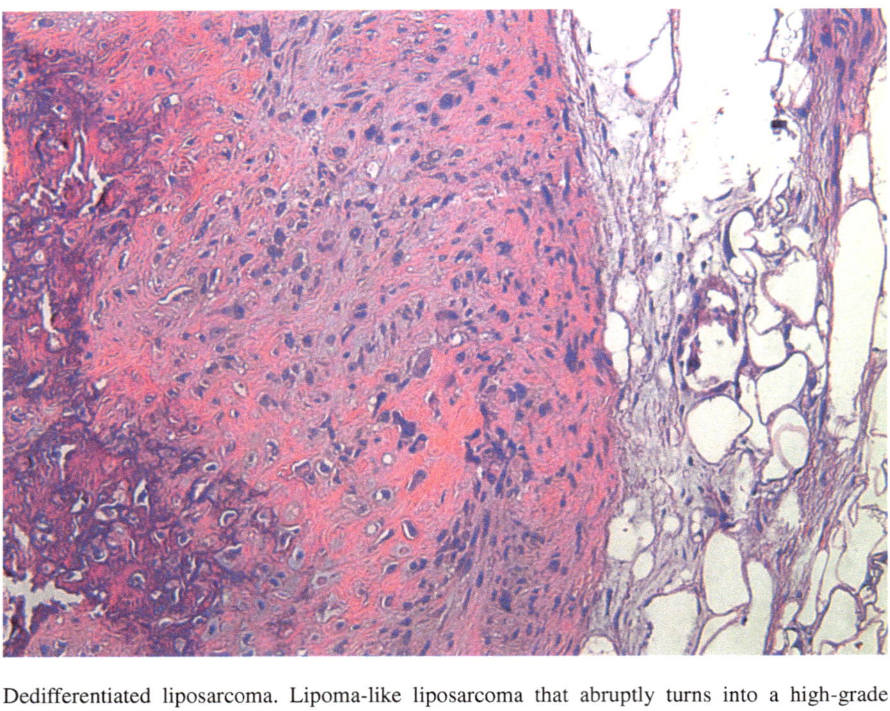

Dedifferentiated liposarcoma. Lipoma-like liposarcoma that abruptly turns into a high-grade spindle cell and/or pleomorphic sarcoma

## Selected Bibliography

Coindre JM, Pédeutour F, Aurias A (2010) Well-differentiated and dedifferentiated liposarcomas. Virchows Arch 456(2):167–179. Review

Dahlén A, Debiec-Rychter M, Pedeutour F, Domanski HA, Höglund M, Bauer HC, Rydholm A, Sciot R, Mandahl N, Mertens F (2003) Clustering of deletions on chromosome 13 in benign and low-malignant lipomatous tumors. Int J Cancer 103(5):616–623. Review

Dalal KM, Antonescu CR, Singer S (2008) Diagnosis and management of lipomatous tumors. J Surg Oncol 97(4):298–313. Review

Dei Tos AP (2000) Liposarcoma: new entities and evolving concepts. Ann Diagn Pathol 4(4): 252–266. Review

Peterson JJ, Kransdorf MJ, Bancroft LW, O'Connor MI (2003) Malignant fatty tumors: classification, clinical course, imaging appearance and treatment. Skeletal Radiol 32(9):493–503. Review

# Chapter 64
# Dermatofibrosarcoma Protuberans

**Marco Gambarotti**

**Definition**: Low-grade fibroblastic tumor of the skin in adult patients. There is a juvenile form of DFSP, called giant-cell fibroblastoma.

**Epidemiology**: Infrequent, males. DFSP: young to middle-aged adults, rare in children. Giant-cell fibroblastoma: infants and children younger than 5 years, exceptional in adults.

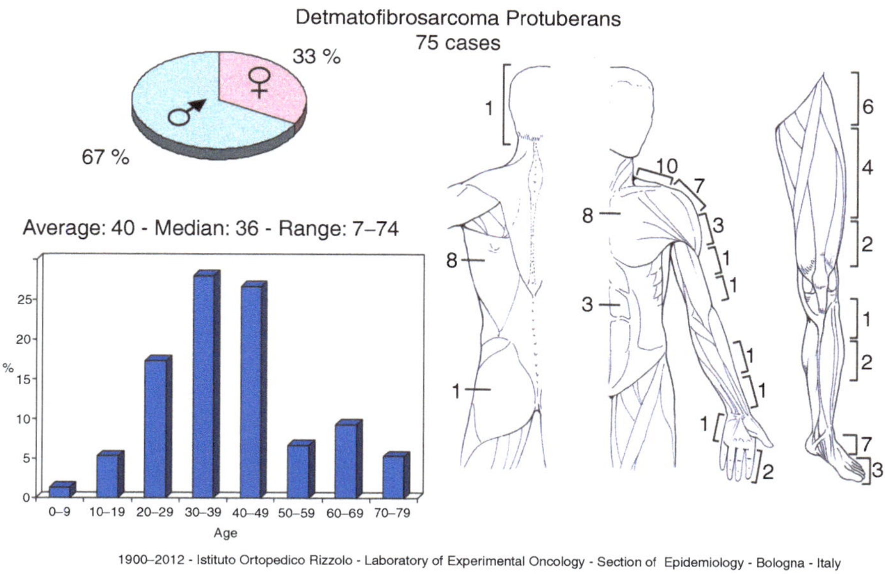


M. Gambarotti, MD
Department of Anatomy and Pathological Histology, Istituto Ortopedico Rizzoli,
Via del Barbiano 1/10, Bologna 40136, Italy
e-mail: marco.gambarotti@ior.it

P. Picci et al. (eds.), *Atlas of Musculoskeletal Tumors and Tumorlike Lesions*,
DOI 10.1007/978-3-319-01748-8_64, © Springer International Publishing Switzerland 2014

**Location**: Trunk, groin, and proximal extremities.
**Clinical**: Nodular or plaque-like painless cutaneous tumor, slowly growing.
**Diagnosis**: A well-defined superficial tumor. On MRI, in T1 usually iso- or hypointense to skeletal muscle. In T2 intermediate or high signal intensity compared to fat. On STIR high signal, similar to water or blood vessels. Uniform enhancement of gadolinium.
**Histopathology**: DFSP: Monotonous spindle cell proliferation, diffusely infiltrating the dermis, subcutis, or both, spindle cells with elongated wavy nuclei. Cells are organized in a storiform pattern, there is minimal cellular atypia, and mitoses are rare. Characteristically there is infiltration of the subcutaneous fat in a honeycomb pattern. Giant-cell fibroblastoma: spindle to pleomorphic cells, variably collagenized matrix, multinucleated-appearing giant cells bordering pseudovascular spaces. Mitoses are rare and necrosis is absent. Fibrosarcomatous DFSP is defined by herringbone pattern and increased atypia and mitoses. On immunohistochemistry DFSP are positive for CD34. Apolipoprotein A1 has also been reported as a sensitive marker of DFSP. Giant-cell fibroblastoma: positive for CD34 and negative for S-100, CD31, and epithelial markers. DFSP and giant-cell fibroblastoma share similar molecular abnormalities: the presence of supernumerary ring chromosomes consisting of amplified sequences from chromosomes 17 and 22 and/or the presence of t(17;22), a balanced reciprocal translocation that results in the fusion of COL1A1, a gene of collagen, and PDGFβ, a gene that encodes a growth factor. Ring chromosomes are predominantly observed in DFSP of adult patients, whereas the t(17;22) translocation is mostly seen in DFSP of children and in giant-cell fibroblastoma.
**Course and Staging**: Conventional DFSP recurs locally in 10–50 % of cases, often after incomplete excisions. Distant metastases are observed in less than 5 % of cases, almost all of which are associated with high-grade fibrosarcomatous changes. Giant-cell fibroblastoma recurs in up to 50 % but does not metastasize. Recurrence rates are closely related to surgical margins.
**Treatment**: Wide excision with tumor-free margins is curative. Radiotherapy has been proposed for unresectable tumors or after margin-positive resections. Imatinib mesylate (Gleevec), a tyrosine kinase inhibitor, may have potential value in the treatment of recurrent or metastatic DFSP.

**Chromosomal Translocations**

| | | |
|---|---|---|
| t(17;22) (q22;q13) | COL1A1-PDGFβ | >90 % |
| ring 17q, ring 22q, der(22) | COL1A1-PDGFβ | 75 % |

**Immunohistochemical Panel**

| | |
|---|---|
| Vimentin | + |
| CD34 | + |

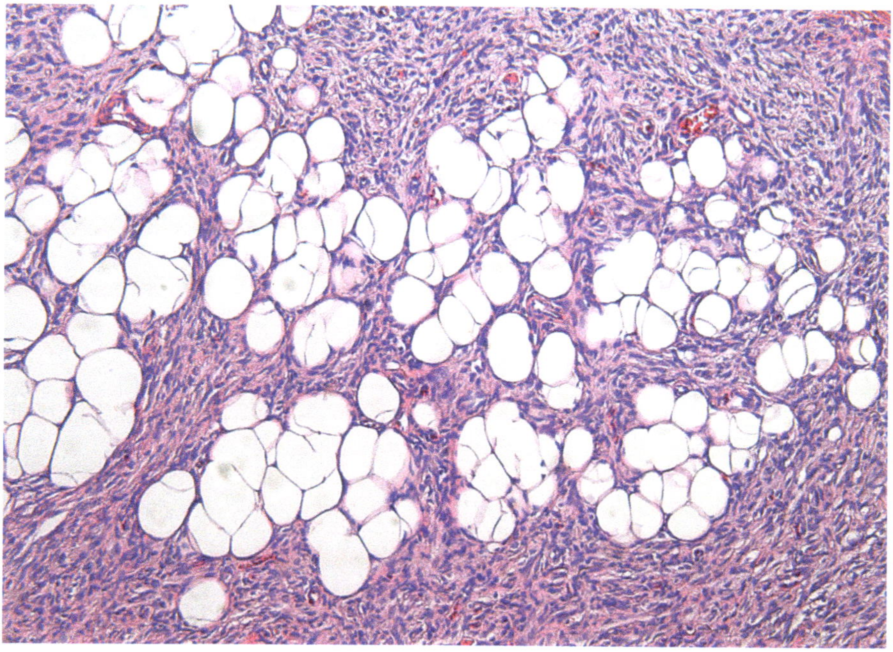

Bland-appearing spindle cells organized in a monotonous storiform pattern, with infiltrative margins that frequently surround lobules of fat

## Selected Bibliography


Archontaki M, Korkolis DP, Arnogiannaki N, Konstantinidou C, Georgopoulos S, Dendrinos P, Zarkadas G, Kokkalis G (2010) Dermatofibrosarcoma protuberans: a case series of 16 patients treated in a single institution with literature review. Anticancer Res 30(9):3775–3779. Review

Gooskens SL, Oranje AP, van Adrichem LN, de Waard-van der Spek FB, den Hollander JC, van de Ven CP, van den Heuvel-Eibrink MM (2010) Imatinib mesylate for children with dermatofibrosarcoma protuberans (DFSP). Pediatr Blood Cancer 55(2):369–373. Review

McArthur G (2007) Dermatofibrosarcoma protuberans: recent clinical progress. Ann Surg Oncol 14(10):2876–2886. Review

Palmerini E, Gambarotti M, Staals EL, Zanella L, Sieberova G, Longhi A, Cesari M, Bonarelli S, Picci P, Ruggieri P, Alberghini M, Ferrari S (2012) Fibrosarcomatous changes and expression of CD34+ and apolipoprotein-D in dermatofibrosarcoma protuberans. Clin Sarcoma Res 2(1):4. doi:10.1186/2045-3329-2-4

Stadler FJ, Scott GA, Brown MD (1998) Malignant fibrous tumors. Semin Cutan Med Surg 17(2):141–152. Review

Sundram UN (2009) Review: Dermatofibrosarcoma protuberans: histologic approach and updated treatment recommendations. Clin Adv Hematol Oncol 7(6):406–408. Review

# Chapter 65 Fibrosarcoma

**Marco Gambarotti**

**Definition**: Malignant tumor composed of fibroblasts. It is a diagnosis of exclusion.

**Epidemiology**: <1 % of adult soft tissues sarcomas. Medium age 45 years. No sex predilection.

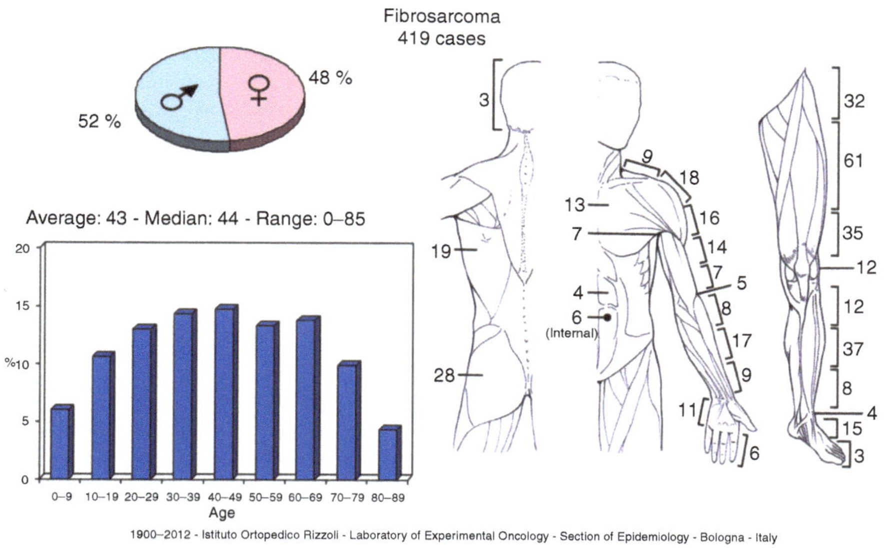


M. Gambarotti, MD
Department of Anatomy and Pathological Histology, Istituto Ortopedico Rizzoli,
Via del Barbiano 1/10, Bologna 40136, Italy
e-mail: marco.gambarotti@ior.it

P. Picci et al. (eds.), *Atlas of Musculoskeletal Tumors and Tumorlike Lesions*,
DOI 10.1007/978-3-319-01748-8_65, © Springer International Publishing Switzerland 2014

**Location**: Most are deep tumors. Anywhere where fibrous tissue is found. Thigh, knee, trunk, arm, forearm. Rare in the head and neck. Hand and foot are most frequent sites of fibrosarcoma in the child.
**Clinical**: Palpable, globose, well-defined mass rarely >10 cm. It is slow growing (few weeks to 20 years). It is nearly or totally painless except when it compresses a nerve. In advanced forms it may adhere to the bone and ulcerate the skin.
**Imaging**: On X-rays: a mass denser than muscles. Calcifications are exceptional. Bone may be eroded or saucerized with minimal periosteal reaction. On CT: homogeneous density, compact soft tissue mass with ill-defined margins and poor enhancement after contrast administration. On MRI: inhomogeneous, lower or isointense signal as those of muscle on T1, 90 % marked, peripheral enhancement on gadolinium, and dark areas on a background of intermediate or high intensity on T2.
**Histopathology**: Low-grade well-differentiated tumor: firm, scar-like consistency, white yellow, rounded limits, growing in an expansile fashion with a pseudocapsule of reactive tissue that sharply delimits it from surrounding normal tissues. High-grade poorly differentiated tumor: soft, fish-flesh consistency, gray-white, irregular margins, growing in an invasive fashion without a capsule and with multiple processes that infiltrate the surrounding tissues, and satellites nodules isolated from the main tumor mass. On low-power view a uniform aspect of interlacing bundles of monomorphic spindle cells with interwoven parallel arranged eosinophilic collagen fibers. The cells are characteristically arranged in long, sweeping fascicles in a herringbone pattern. The cells have tapered darkly stained nuclei. High-grade lesions with marked anaplasia and pleomorphism are better classified as undifferentiated pleomorphic sarcoma. Immunohistochemistry is consistently positive for vimentin and may be positive for pan-muscle actin and smooth muscle actin.
**Course and Staging**: Fibrosarcoma can come from transformation of preexisting benign processes (neurofibromatosis, burn scars) and from radiation exposure. Any stage: more frequently stage IA and stage IIB. Low-grade and age <10 years are favorable prognostic factors. In children metastases are rare (<10 %). Lymph node metastases are rare (<5 %).
**Treatment**: Wide excision. In the adult and in a high-grade lesion, radical margins may be indicated. Radiotherapy is less effective, and it is used as adjuvant in high-grade tumors. Recent studies have demonstrated that adjuvant chemotherapy is useful for a better prognosis. When surgical margins are inadequate, local recurrence occurs in 50 % of cases. Lung metastases occur in 60 % of cases. Ten-year survival rate is 60 % for low-grade and 30 % for high-grade tumors.

**Immunohistochemical Panel**

| | |
|---|---|
| VIM | + |
| MS Act | ± |
| Smooth M Act | ± |

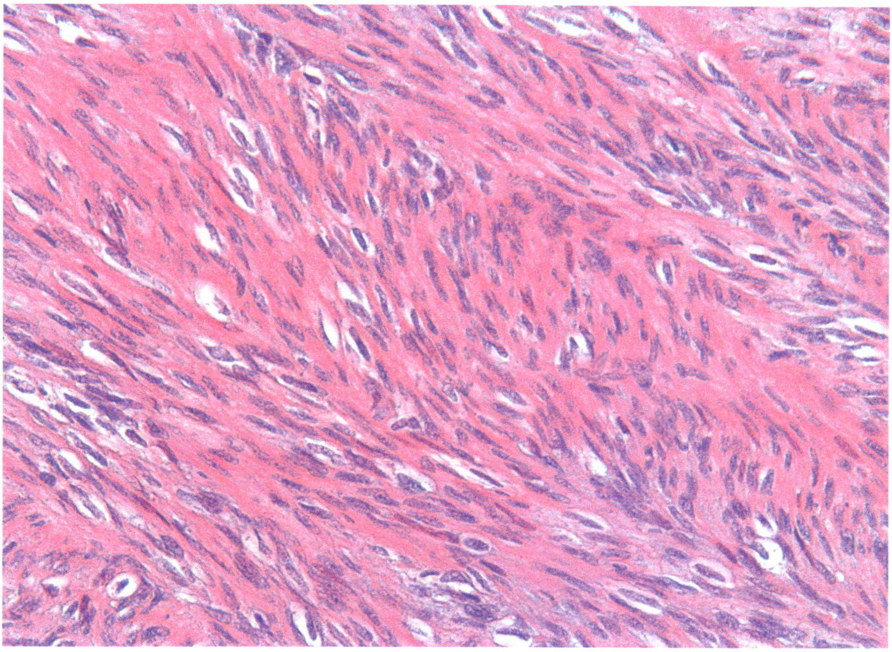

Spindle cell neoplasm with variable grade of atypia arranged in a herringbone pattern with intercellular fibrillary collagen

## Selected Bibliography


Alaggio R, Bisogno G, Rosato A, Ninfo V, Coffin CM (2009) Undifferentiated sarcoma: does it exist? A clinicopathologic study of 7 pediatric cases and review of literature. Hum Pathol 40(11):1600–1610. Review

Bahrami A, Folpe AL (2010) Adult-type fibrosarcoma: A reevaluation of 163 putative cases diagnosed at a single institution over a 48-year period. Am J Surg Pathol 34(10):1504–1513

Collini P, Sorensen PH, Patel S, Blay JY, Issels RD, Maki RG, Eriksson M, del Muro XG (2009) Sarcomas with spindle cell morphology. Semin Oncol 36(4):324–337. Review

Fletcher CD, Dal Cin P, de Wever I, Mandahl N, Mertens F, Mitelman F, Rosai J, Rydholm A, Sciot R, Tallini G, van den Berghe H, Vanni R, Willén H (1999) Correlation between clinicopathological features and karyotype in spindle cell sarcomas. A report of 130 cases from the CHAMP study group. Am J Pathol 154(6):1841–1847

Wong SL (2008) Diagnosis and management of desmoid tumors and fibrosarcoma. J Surg Oncol 97(6):554–558. Review

# Chapter 66
# Leiomyosarcoma

**Marco Gambarotti**

**Definition**: Malignant tumor showing smooth muscle differentiation. Type (a) retroperitoneal; (b) cutaneous; and c) vascular.
**Epidemiology**: 7 % of soft tissue sarcomas. Type (a): females. Median age 60 years. Type (b): 2–3 % of superficial sarcomas. No sex predilection. 40–70 years old. Type (c): rare, female in L. of the inferior vena cava, and no sex predilection in the other cases. Median age 50 years.

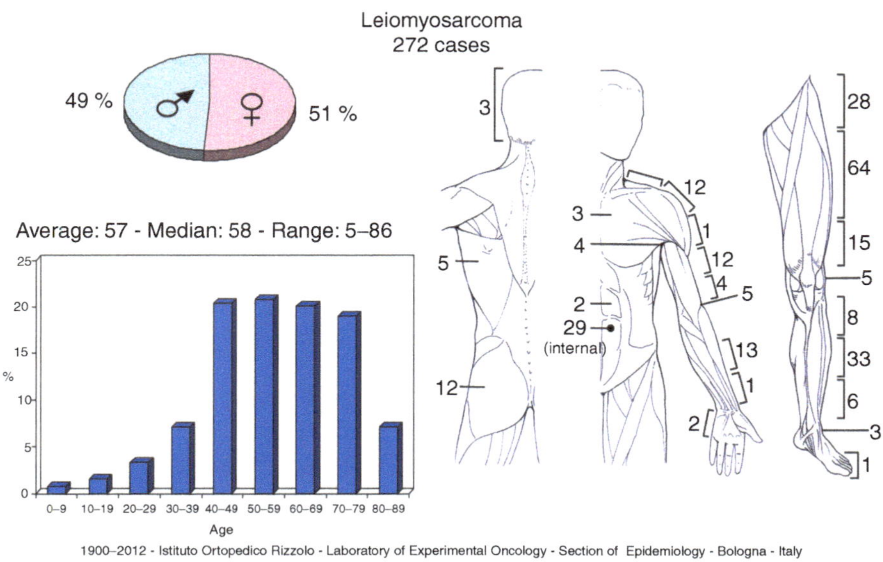


M. Gambarotti, MD
Department of Anatomy and Pathological Histology, Istituto Ortopedico Rizzoli,
Via del Barbiano 1/10, Bologna 40136, Italy
e-mail: marco.gambarotti@ior.it

P. Picci et al. (eds.), *Atlas of Musculoskeletal Tumors and Tumorlike Lesions*,
DOI 10.1007/978-3-319-01748-8_66, © Springer International Publishing Switzerland 2014

**Location**: Type (a): retroperitoneum. Commonly involves kidney, pancreas, and vertebral body for direct extension. Type (b): in the limbs, from piliferous areas. Quite small, less than 2 cm. subcutaneous tumors grow faster and reach a larger size. Type (c): in the lower limbs, more frequently involving the veins, exceptionally the arteries.
**Clinical**: Type (a): abdominal mass, pain, weight loss, nausea, or vomiting. Type (b): pain, surface changes in the epidermis. Type (c): pain for pressure on nerve close to the affected vessel, edema due to venous compression.
**Imaging**: On CT hypo- or moderately vascular lesions, type (a): found by CAT scan or angiography that does not show typical aspect. Dislocation of the most important vessels is usual. Type (c): angiographically highly vascularized. On MRI inhomogeneous, iso-/hypointense on T1, with a thick, irregular rim enhancement on contrast T1, and marked inhomogeneity with mixed but mainly high signal intensity on T2.
**Histopathology**: White-gray whorled appearance, more often fleshy masses with foci of necrosis and hemorrhage, and frequent cyst formation. Proliferation of spindle cells with an elongated nucleus with abundant cytoplasm. The nucleus is central and blunt ended or "cigar shaped." A vacuole close to one end of the nucleus produces a slight indentation so that it becomes concave rather than convex in the contour. Sometimes, cytoplasm may have a "clotted" appearance because of clumping of the myofilaments. Cells are arranged in bundles that form wide waves intersecting with different orientation. A right angle is frequent. Stroma is a delicate mesh between cells. At times, a palisade-like aspect of the nuclei similar to neurinoma is observed. Pleomorphic appearance with highly anaplastic to epithelioid cells is common. Mitotic figures are frequent. Type b is whitish-gray or pinkish, fasciculated aspect, has ill-defined margins by virtue of the intricate blending of tumor fascicles with surrounding collagen and pilar arrector muscle in dermis tumors and well-circumscribed lesions by a pseudocapsule in subcutis tumors. Mitotic figures are frequent. Pure cutaneous (dermal without subcutaneous involvement) leiomyosarcomas have been recently designed as "atypical smooth muscle tumors" because they show a 30 % rate of locale recurrence and no metastasis. Immunohistochemistry is positive for vimentin, pan-muscle actin, smooth muscle actin, desmin, and caldesmon.
**Course and Staging**: Type (a) highly aggressive, so they may cause death also by local extension. Survival from 0 to 29 % at 5 years. Usually, stage IIB. Type (b): good prognosis, metastases infrequent, local recurrence frequent (50 %), usually stage IA–B. Type (c): poor prognosis, 50 % have metastases; tumors of the small veins seem to have a better prognosis, usually stage I–IIB.

**Treatment**: Type (a): wide excision, but it is often so large that total resection is impossible. Local recurrence is frequent. Type (b): wide excision.

Type (c): wide excision is often impossible in large veins (hepatic, inferior, or middle cava). Necessary in tumors of the small veins.

**Immunohistochemical Panel**

| | |
|---|---|
| VIM | + |
| MS Act | + |
| Smooth M Act | + |
| Desmin | + |
| Caldesmon | + |

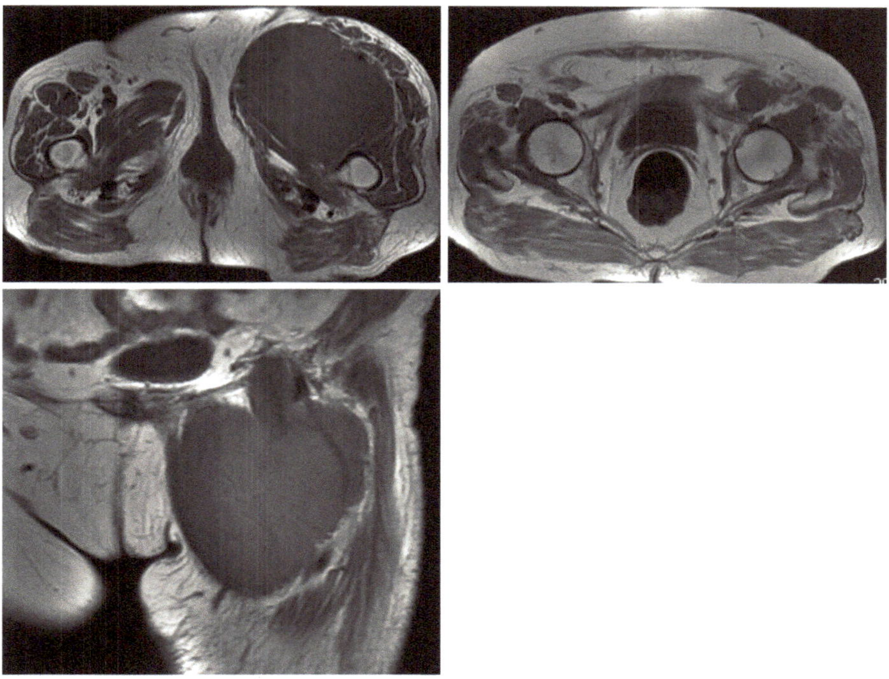

Axial and coronal T1 MR images. Mass of the root of the thigh, involving the femoral vein

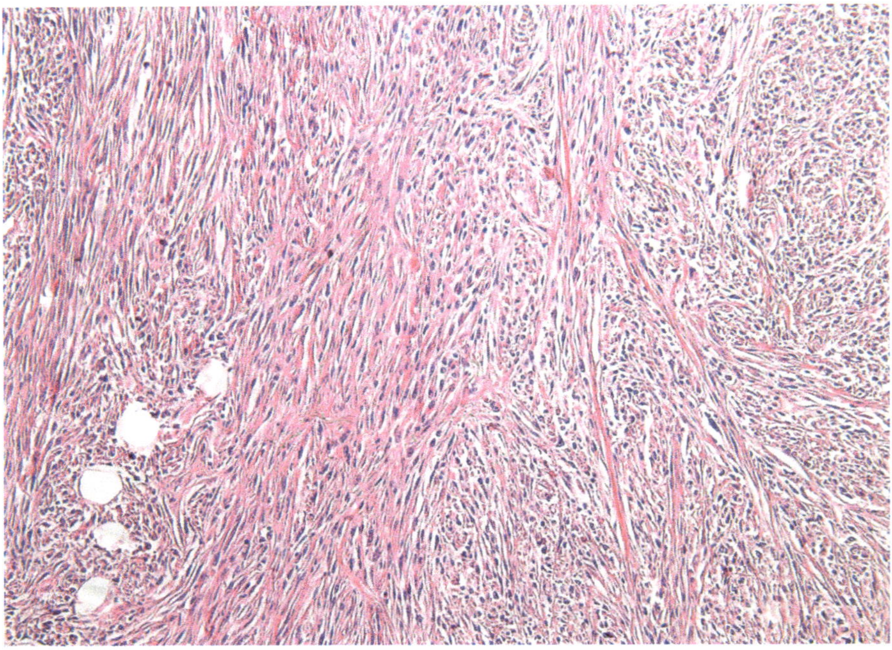

Spindle cell neoplasm with smooth muscle differentiation with variable pattern

## Selected Bibliography

Katz SC, DeMatteo RP (2008) Gastrointestinal stromal tumors and leiomyosarcomas. J Surg Oncol 97(4):350–359. Review

Miettinen M, Fetsch JF (2006) Evaluation of biological potential of smooth muscle tumours. Histopathology 48(1):97–105. Review

O'Sullivan PJ, Harris AC, Munk PL (2008) Radiological imaging features of non-uterine leiomyosarcoma. Br J Radiol 81(961):73–81. Review

West RB (2010) Expression profiling in soft tissue sarcomas with emphasis on synovial sarcoma, gastrointestinal stromal tumor, and leiomyosarcoma. Adv Anat Pathol 17(5):366–373. Review

Yang J, Du X, Chen K, Ylipää A, Lazar AJ, Trent J, Lev D, Pollock R, Hao X, Hunt K, Zhang W (2009) Genetic aberrations in soft tissue leiomyosarcoma. Cancer Lett 275(1):1–8. Review

# Chapter 67
# Rhabdomyosarcoma

**Marco Gambarotti**

**Definition**: Malignant tumor with phenotypical and biological features of skeletal muscle cells.

**Epidemiology**: 20 % of all soft tissue sarcomas. Most frequent under 20 years. Male.

Varieties: (a) Embryonal, (b) alveolar, and (c) pleomorphic. Spindle cell rhabdomyosarcoma is a rare distinctive variant that can occur in children, adolescents, and adults; until recently, it has been classified in the embryonal category.

Type (a): The most common type (60 %), prevalent from birth to 15 years of age. Type (b): Frequent (30 %), prevalent in 10–25 years old. Type (c): Rare, almost exclusively in 40–60 years old.

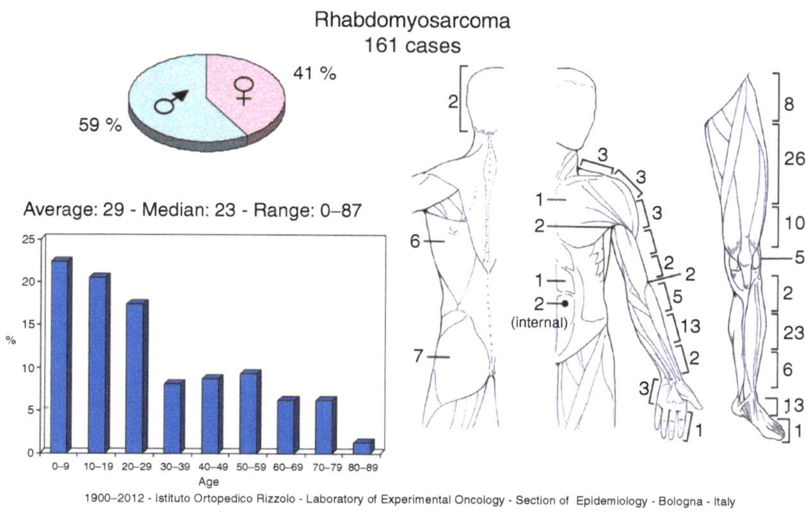


M. Gambarotti, MD
Department of Anatomy and Pathological Histology, Istituto Ortopedico Rizzoli,
Via del Barbiano 1/10, Bologna 40136, Italy
e-mail: marco.gambarotti@ior.it

P. Picci et al. (eds.), *Atlas of Musculoskeletal Tumors and Tumorlike Lesions*,
DOI 10.1007/978-3-319-01748-8_67, © Springer International Publishing Switzerland 2014

**Location**: R. of the limbs constitutes 15 % of all tumors. Type (a): Head, neck, genitourinary tract (botryoid variety), retroperitoneum. Type (b): Limbs (forearm, hand, foot). Type (c): Limbs (prevails in the thigh).
**Clinical**: Type a): Rapidly, invasive, infiltrative growth. Symptoms and signs depend on anatomic location. Types (b–c): Deep, painless unless it compresses a nerve, rapidly enlarging mass in a striated muscle.
**Imaging**: On x-ray erosion of flat bone in 20 % of cases. In types (b–c) is rare, but it occurs mainly in the hand and foot. Periosteal reaction is minimal. Nevertheless, tumor is frequently tightly bound to bone. On arteriography common aspects to all well-vascularized tumors: neovascularity and tumoral blush. On CT large, lobulated mass with central hypodensities, septation, rim-like enhancement surrounding areas of necrosis after contrast administration. On MRI type (a): homogeneous, slight hyperintensity on T1, bright on T2 with some dark internal strands, no enhancement of large parts of the mass on contrast T1, no necrosis. Type (b): Inhomogeneous, intermediate intensity with multiple, central hypointensity areas on T1, heterogeneous, with white central zones on T2 (necrotic areas), strong peripheral enhancement on contrast T1. Type (c): Mixed signal intensities on T1, more inhomogeneous on T2 with central fluid collection near to low-signal-intensity areas, ringlike, peripheral enhancement around these.
**Histopathology**: Firm or soft mass, with myxoid and cystic component, a mottled white-gray to pink-tan, with frequent hemorrhage and necrotic areas. Botryoid variety has a polypoid and myxoid appearance compared to a bunch of grapes. Immunostains for desmin and myoglobin are essential for diagnosis. Type (a): Usually, undifferentiated, small, round cells with round-oval, hyperchromic nucleus. Nuclear membrane is evident. Chromatin granules are distinct, and nucleoli are clearly evident. Thin, perinuclear cytoplasmic ring. Mitotic figures and pleomorphism are frequent. Cells are distributed without any order with scarce collagen stroma among them. At other times, in more differentiated zones, the volume of the cells increases, and the nucleus becomes larger, vesicular, and eccentric, with a large nucleolus. Cytoplasm is eosinophilic, granular, and more abundant, with filaments wrapped around the nucleus. Transverse striation may be recognized. Racket or tadpole or ribbon shapes, angulated like broken straws of hay, are observed. Globose cells may have a vacuolated cytoplasm, and this may be reduced to filaments that from the nucleus irradiate to the periphery: spider-web cells. Type (b): Undifferentiated, small, round-oval cells, collected in solid islands and forming alveolar aspect. Rough bands of dense collagen, convoluted and ramified, containing dilated capillaries, surround these cellular zones. Peripheral cells adhere in a single layer to the fibrous septa. Cells in the center are more loosely arranged or freely floating. Rhabdomyoblasts are much rarer. Type (c): Association of loosely arranged, haphazardly oriented, small and large, round or pleomorphic cells. Larger and more irregular in outline racket-shaped and tadpole-shaped rhabdomyoblasts

with stringy, granular, and vacuolated cytoplasm. Transverse striation is extremely rare. Much less collagen stroma between cells is observed. Immunohistochemistry is positive for vimentin, desmin, pan-muscle actin, MyoD1, and myogenin (diffuse positivity in type (b) but negative for smooth cell actin. There are two main karyotypic aberrations in alveolar rhabdomyosarcoma: t(2;13) that fuses the PAX3 gene on 2q35 with the FOXOA1 gene on 13q14 and is present in 60–70 % of cases and t(1;13) that fuses PAX7 on 1p36 with FOXOA1 on 13q14 and is present in 10–20 % of cases. The latter translocation is associated with a better prognosis, the first with a worse prognosis.

**Course and Staging**: Rapid and aggressive. Frequent recurrence if inadequately treated and mets in lungs and lymph nodes. Usually, stage IIA, within one of the major muscles of the limb, or stage IIB. Stage I is rare. American rhabdomyosarcoma intergroup study distinguishes 4 stages: stage 1: localized tumor, completely resected; stage 2: marginal excision with or without lymph node metastases; stage 3: intralesional excision; and stage 4: distant metastases at time of diagnosis.

**Treatment**: Optimal treatment is preoperative chemotherapy, wide excision with resection of regional lymph nodes, postoperative chemotherapy, plus radiation therapy when excision is not wide. Overall 5-year survival is excellent for stage 1 (83 %) and for stage 2 (75 %), less favorable for stage 3 (52 %), and poor for stage 4 (20 %). Tumors in the head and neck, retroperitoneum, abdomen, and extremities have a worse prognosis than neoplasms in the orbit and in paratesticular region. Alveolar type has a less favorable prognosis than embryonal.

**Immunohistochemical Panel**

| | |
|---|---|
| VIM | + |
| Desmin | + |
| MS act | + |
| Smooth M act | – |
| MyoD 1 | + |
| Myogenin | + |

**Chromosomal Translocations**

| | | | |
|---|---|---|---|
| Alveolar rhabdomyosarcoma | | | |
| t(2;13) (q35;q14) | PAX3-FOXOA1 | 60–70 % | Worse prognosis |
| t(1;13) (p36;q14) | PAX7-FOXOA1 | 10–20 % | Better prognosis |

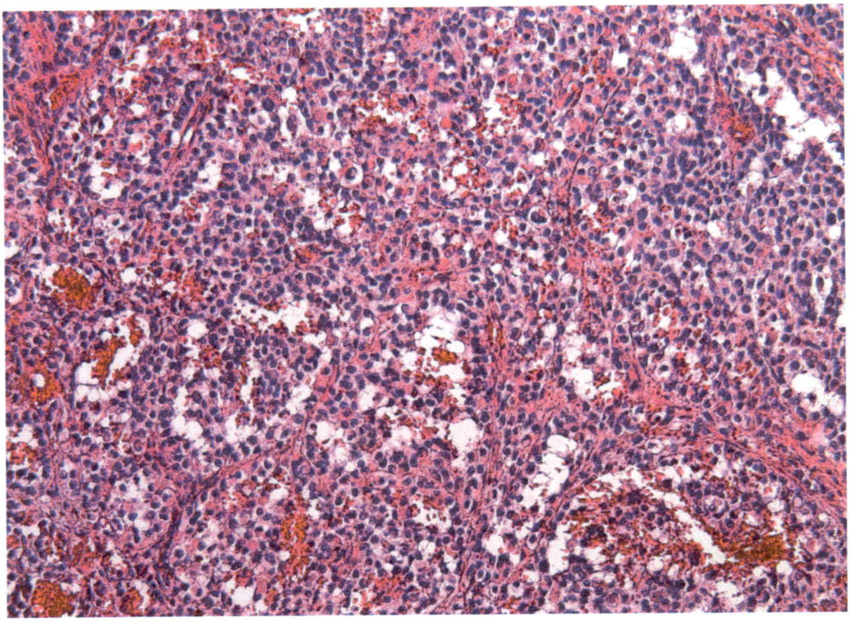

Alveolar rhabdomyosarcoma. Undifferentiated small cells with skeletal muscle differentiation organized in nodules with loosening of cohesiveness of cells in the center of nodules, resulting in an alveolar-like appearance

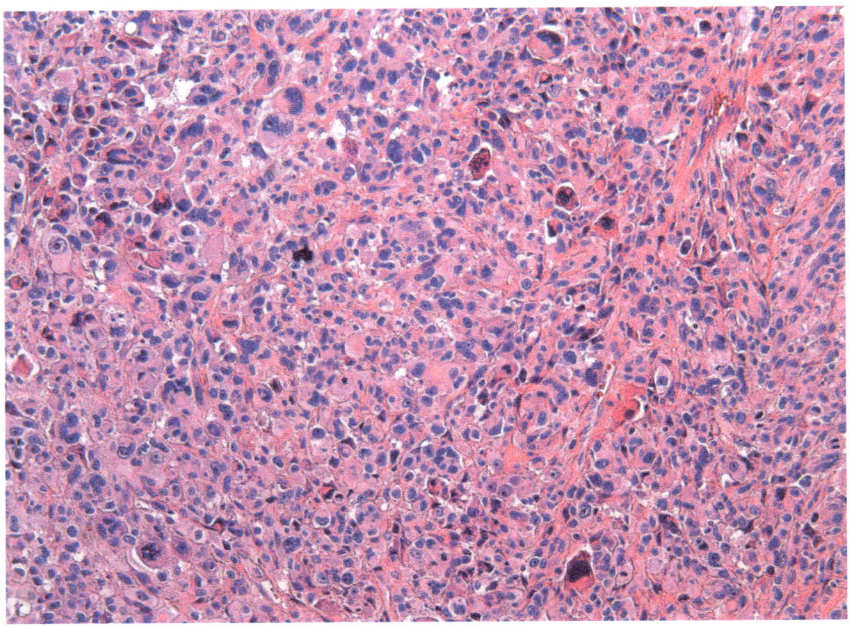

Pleomorphic rhabdomyosarcoma. UPS-like pleomorphic high-grade neoplasm with skeletal muscle differentiation

## Selected Bibliography

Hayes-Jordan A, Andrassy R (2009) Rhabdomyosarcoma in children. Curr Opin Pediatr 21(3):373–378. Review

Morgenstern DA, Rees H, Sebire NJ, Shipley J, Anderson J (2008) Rhabdomyosarcoma subtyping by immunohistochemical assessment of myogenin: tissue array study and review of the literature. Pathol Oncol Res 14(3):233–238. Review

Parham DM (2001) Pathologic classification of rhabdomyosarcomas and correlations with molecular studies. Mod Pathol 14(5):506–514. Review. Erratum in: Mod Pathol 2001 Oct;14(10):1068

Parham DM, Ellison DA (2006) Rhabdomyosarcomas in adults and children: an update. Arch Pathol Lab Med 130(10):1454–1465. Review

Qualman SJ, Coffin CM, Newton WA, Hojo H, Triche TJ, Parham DM, Crist WM (1998) Intergroup Rhabdomyosarcoma Study: update for pathologists. Pediatr Dev Pathol 1(6):550–561. Review

# Chapter 68
# Vascular Tumors: Hemangioma, Epithelioid Hemangioendothelioma, and Angiosarcoma

**Marco Gambarotti**

**Definition**: Hemangiomas are benign tumors that closely resemble normal vessels. It is difficult to distinguish clearly among neoplasm, hamartoma, and malformations. Epithelioid hemangioma is a benign vascular tumor with well-formed but often immature vessels, lined by plump, epithelioid/histiocytoid endothelial cells. Epithelioid hemangioendothelioma is an angiocentric low-grade malignant vascular tumor, composed of epithelioid endothelial cells arranged in short cords and nests set in a distinctive myxohyaline stroma. Angiosarcoma is a malignant tumor composed of cells that variably recapitulate the morphologic features of normal endothelium.

In the previous 2002 WHO classification, hemangioendothelioma and angiosarcoma represent a spectrum of malignant vascular neoplasm. In the 2013 WHO classification, they are listed as separate entities reflecting their different genetic abnormalities.

**Epidemiology**: Benign hemangiomas represent 7 % of all benign soft tissue tumors, the most frequent in infancy and childhood. In this age group, they are generally cutaneous or subcutaneous capillary hemangiomas. They grow until body growth has ended. Intramuscular hemangioma, although relatively uncommon, is one of the most frequent deep-seated soft tissue tumors. Adolescents and young adults are most commonly affected, with an equal sex incidence. Hemangiomas can also arise in a synovium-lined surface, particularly in the knee (synovial hemangiomas), or can affect a large segment of the body (angiomatosis) in a contiguous fashion, either by vertical extension to involve multiple tissue planes or by crossing muscle compartments to involve similar tissue types.

Epithelioid hemangioma affects a wide age range, peaking in the 3rd through 5th decades. Females appear to be more commonly affected than males. Epithelioid

M. Gambarotti, MD
Department of Anatomy and Pathological Histology,
Istituto Ortopedico Rizzoli, Via del Barbiano 1/10, Bologna 40136, Italy
e-mail: marco.gambarotti@ior.it

P. Picci et al. (eds.), *Atlas of Musculoskeletal Tumors and Tumorlike Lesions*,
DOI 10.1007/978-3-319-01748-8_68, © Springer International Publishing Switzerland 2014

hemangioendothelioma is a rare vascular tumor occurring in nearly all age groups with the exception of early childhood; both sexes are equally affected.

Angiosarcomas are rare tumors; the majority develops as cutaneous lesions, particularly in patients suffering from lymph edema or after radiation for a previous malignancy. Less than 25 % are deep-seated soft tissue neoplasms evenly distributed throughout the decades with a peak incidence in the 7th decade. Angiosarcomas are very rare in children.

**Location**: Intramuscular hemangioma most commonly affects the lower limb, particularly the thigh.

Epithelioid hemangioma most commonly affects the head, especially the forehead, preauricular area, and scalp, and distal portions of the extremities, especially digits. Epithelioid hemangioendothelioma develops as a solitary tumor in either superficial or deep soft tissue of the extremities. Nearly 50–60 % originate from a vessel, usually a small vein.

Most soft tissue angiosarcomas occur in deep muscles of the lower extremities, followed by the arm, trunk, head and neck, and the abdominal cavity. Rarely lesions are multifocal.

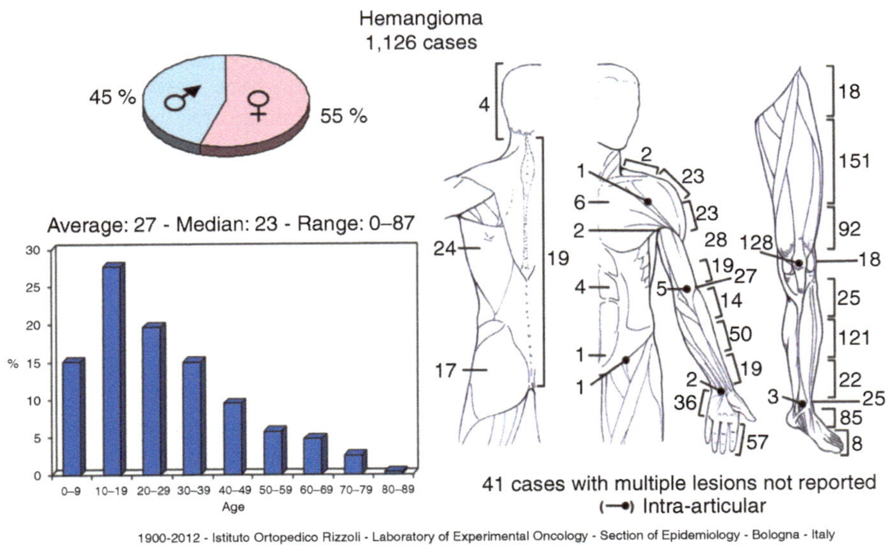

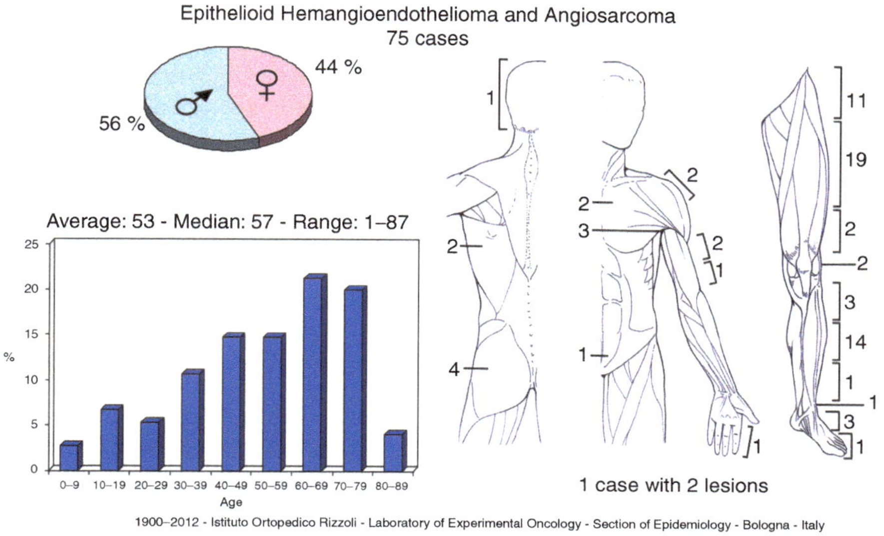


**Clinical**: Superficial (cutaneous/subcutaneous) hemangiomas are reddish-wine-colored painless lesions, generally present at birth. Intramuscular hemangioma arises within the belly of a single muscle; only in the hand and foot it may expand between the fascia, muscles, and tendons. It is possible to observe pain and swelling associated with venous stasis. Pain is sharp and becomes more intense with tension of the muscle. Shortening of muscles causes first joint dysfunction and then joint deformity. In the hand and foot, an increase of skin temperature, of the superficial venous reticulum, telangiectasia, cyanosis, and hyperhidrosis are observed.

The majority of epithelioid hemangioma presents as subcutaneous masses of a year or less in duration. The process is usually uninodular, but multinodularity, generally in contiguous areas, can be present. Dermal examples are less frequent, and deep-seated cases are rare. Epithelioid hemangioendothelioma develops often as a painful nodule in either superficial or deep soft tissue. Deeply situated tumors may be associated with focal ossification that can be detected on plain films.

Soft tissue angiosarcoma develops as enlarging mass, in 1/3 of patients associated with other symptoms such as coagulopathy, anemia, persistent hematoma, or bruisability.

**Imaging**: X-rays are usually negative although small round granular calcifications with a smooth surface and concentric stratifications (phleboliths) can be seen. Vascular tumors present an inhomogeneous pattern on MRI. The lesion often appears as a “bunch of grapes,” occasionally with a serpentine or tubular pattern. On T1, angiomatous tissue has an intermediate intensity between that of muscle and fat, but areas of stagnant blood and hemorrhage can cause high signal intensity. On T2, vascular spaces seen are hyperintense, but fibrous septa and calcified foci are hypointense. Fluid-fluid levels can be appreciated. Malignant vascular lesions often show nonspecific characteristics. The presence of intratumoral necrosis may be demonstrated by the use of contrast agents.

**Histopathology**: Benign hemangiomas can be cavernous, capillary, or pseudovenous. Cavernous hemangiomas are the most frequent. Grossly they are made of a bunch of scattered nodules, from dull red to bluish, with internal cloistering. Histologically, they are composed of extremely dilated vessels with a very thin wall of flattened endothelium and collagen membrane, filled with blood. Capillary hemangiomas are more compact and pinkish. They are composed of many small vessels regularly coated with endothelium spreading into collagen stroma. Pseudovenous hemangiomas appear spongelike with fibrous cloisters and intercommunicating lacunae only partially filled with blood, thrombi, and phleboliths. They are composed of craggy, ramified, labyrinthic cavities with very irregular, thick walls with a pseudovenous fibromuscular structure. Intramuscular hemangiomas have been traditionally classified according to vessel size in small (capillary), large (cavernous), and mixed.

Epithelioid hemangioma is usually 0.5–2.0 cm in size, generally with a rather nonspecific nodular appearance. Subcutaneous examples of epithelioid hemangioma are histologically characterized by a prominent proliferation of small, capillary-sized vessels lined by plump, epithelioid endothelial cells, with a typical immature appearance, sometimes lacking a well-defined lumen. These vessels are rimed by a single cell endothelium layer with an intact myopericytic/smooth muscle layer. The process is usually well demarcated from the surrounding soft tissue, and commonly, it is associated with or centered around a larger vessel, usually a muscular artery. An inflammatory milieu rich in eosinophils and lymphocytes is generally present. Dermal examples of epithelioid hemangioma generally show a more mature appearance with a well-canalized lumen, and endothelial cells are somewhat less plump, frequently more cobblestone, or hobnail-like in appearance. In addition, dermal examples are less circumscribed and are not usually associated with a larger central vein or muscular artery. Epithelioid hemangioendothelioma generally arises as a fusiform intravascular mass that may resemble an organizing thrombus. Histologically, they are composed of short strands, cords, or solid nests of epithelioid eosinophilic endothelial cells, sometimes with intracytoplasmic lumina (vacuoles) containing erythrocytes. Cells appear quite bland with little or no mitotic activity and are embedded in a distinctive matrix that varies from light blue (chondroid-like) to deep pink (hyaline) in color.

Angiosarcomas are high-grade tumors composed of highly malignant cells displaying mitotic activity. These cells may vary in appearance from spindle (resembling a fibrosarcoma) to epithelioid (resembling an undifferentiated carcinoma) and are arranged in sheets, small nests, cords, or rudimentary vascular channels irregular in shape; freely intercommunicate with one another in a sinusoidal fashion; and infiltrate surrounding tissues in a destructive fashion. Malignant endothelial cells can be also arranged in intraluminal buds, projections, or papillae. Extensive hemorrhage is a characteristic feature of most angiosarcomas.

**Course and Treatment**: Complete local excision and eventually follow-up are the optimal management for symptomatic benign hemangiomas (including epithelioid hemangioma). Local recurrence is reported to occur in 1/3 of epithelioid hemangiomas. Metastases do not occur. Behavior of epithelioid hemangioendothelioma is intermediate between hemangiomas and angiosarcomas, with a metastatic rate of 17 % and mortality of 3 %. Soft tissue angiosarcomas are highly aggressive tumors. Local recurrences develop in about 20 % of patients, and 50 % of these may be expected to die within the first year from diagnosis with metastatic disease in the lung, lymph nodes, bone, and soft tissues.

**Immunohistochemical Panel**

| | |
|---|---|
| VIM | + |
| CD31 | + |
| CD34 | ± |
| Fli-1 | + |
| CK | ± |
| ERG | + (Am J Surg Pathol 35(3):432–441) |

**Chromosomal Translocations**

| | | |
|---|---|---|
| Epithelioid hemangioendothelioma | | |
| t(1;3)(p36.3;q25) | WWTR1-CAMTA1 | 100 % |

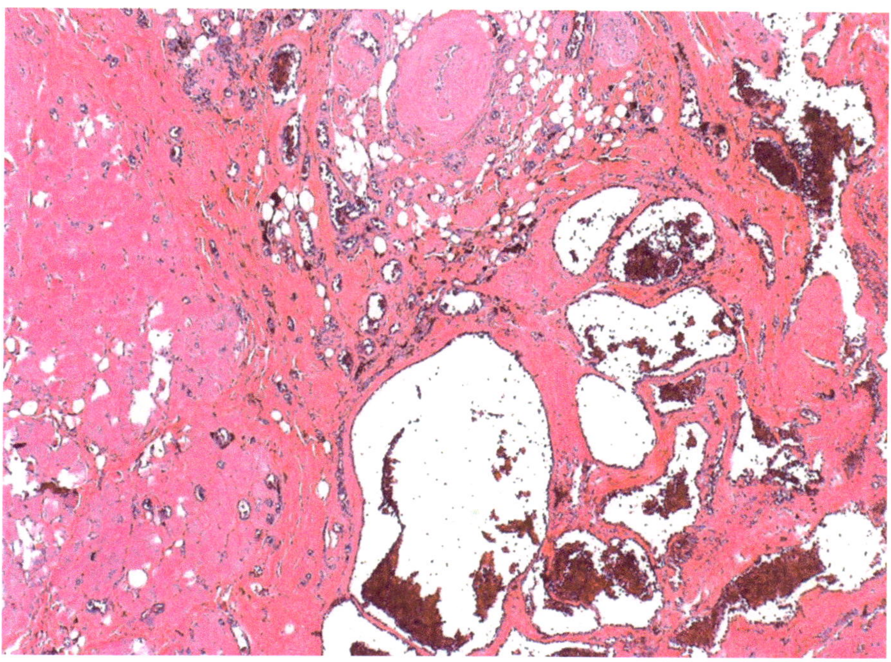

Hemangioma. Blood-filled cavities rimmed by flat endothelium in variable-sized vessels (capillary, cavernous, mixed)

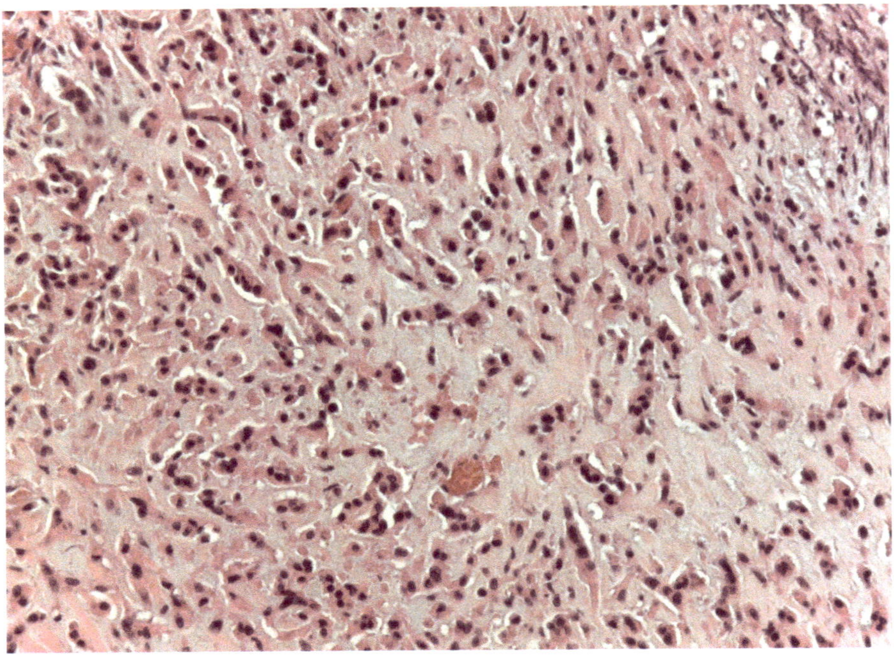

Epithelioid hemangioendothelioma. Short strands and cords of epithelioid cells in a chondroid-like matrix

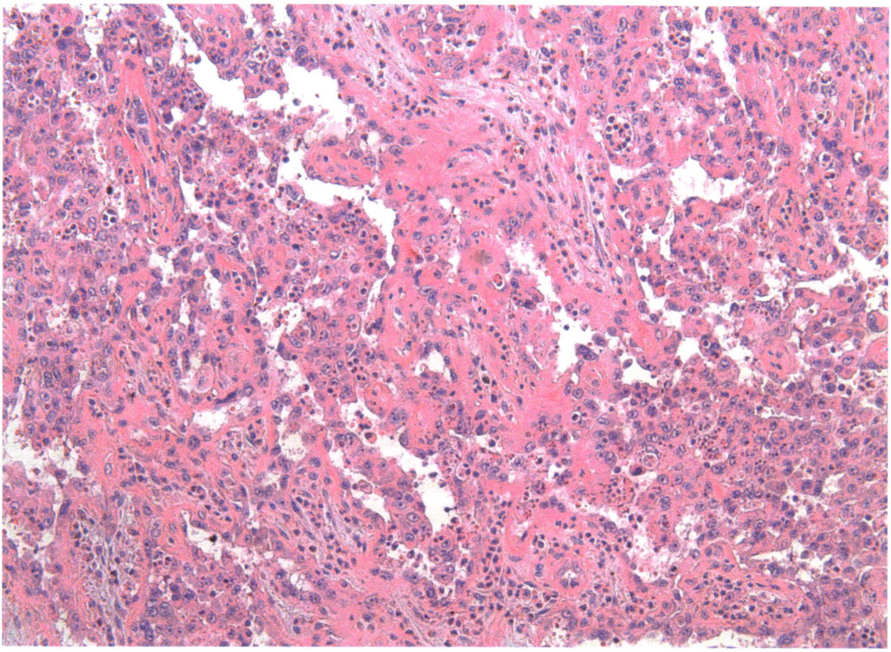

Angiosarcoma. Irregular and haphazard blood-filled cavities rimmed with highly malignant and atypical cells

## Selected Bibliography

Arbiser JL, Bonner MY, Berrios RL (2009) Hemangiomas, angiosarcomas, and vascular malformations represent the signaling abnormalities of pathogenic angiogenesis. Curr Mol Med 9(8):929–934. Review

Errani C, Sung YS, Zhang L, Healey JH, Antonescu CR (2012) Monoclonality of multifocal epithelioid hemangioendothelioma of the liver by analysis of WWTR1-CAMTA1 breakpoints. Cancer Genet 205:12–17

Ganjoo K, Jacobs C (2010) Antiangiogenesis agents in the treatment of soft tissue sarcomas. Cancer 116(5):1177–1183. Review

Moukaddam H, Pollak J, Haims AH (2009) MRI characteristics and classification of peripheral vascular malformations and tumors. Skeletal Radiol 38(6):535–547. Review

Penel N, Marréaud S, Robin YM, Hohenberger P (2011) Angiosarcoma: state of the art and perspectives. Crit Rev Oncol Hematol 80:257–263

Rosai J (2010) Morphologic clues in vascular tumors. Int J Surg Pathol 18(3 Suppl):66S–70S. Review

# Chapter 69
# Solitary Fibrous Tumor

**Marco Gambarotti**

**Definition**: A mesenchymal neoplasm composed of fibroblast with collagen bands and "staghorn" vascularization. According to WHO classification, hemangiopericytoma is currently considered to be closely related, if not identical, to solitary fibrous tumor.

**Epidemiology**: Rare. No sex predilection. Adults.

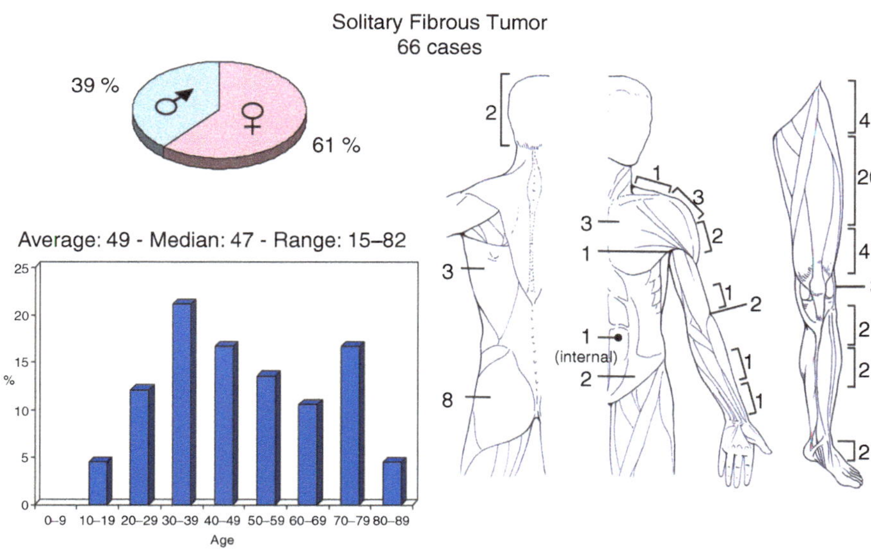


1900–2012 - Istituto Ortopedico Rizzolo - Laboratory of Experimental Oncology - Section of Epidemiology - Bologna - Italy

M. Gambarotti, MD
Department of Anatomy and Pathological Histology, Istituto Ortopedico Rizzoli,
Via del Barbiano 1/10, Bologna 40136, Italy
e-mail: marco.gambarotti@ior.it

P. Picci et al. (eds.), *Atlas of Musculoskeletal Tumors and Tumorlike Lesions*,
DOI 10.1007/978-3-319-01748-8_69, © Springer International Publishing Switzerland 2014 

**Location**: Lower limb (thigh), retroperitoneum, pelvic region, upper limb, trunk, head, and neck. Deeply located.
**Clinical**: Slow enlarging painless mass, increased skin temperature, telangiectasia, unilateral varicose veins, pulsation, audible bruit, urinary retention, hydronephrosis, constipation, vomiting, hypoglycemia.
**Imaging**: On angiography: rich vascularity, encircled by tortuous vessels, arborizing from one pedicle, rapid arteriovenous flow. On CT: heterogeneous mass sharply enhanced after contrast. On MRI: homogeneous intermediate intensity with high-signal hemorrhagic zones on T1, inhomogeneous bright lesion on T2, prominent serpentine vessels with signal intensity depending on the flow within them, strong enhancement with ill-defined necrotic areas that do not become enhanced.
**Histopathology**: Soft, from pale to dull red or brownish red, heavily bleeding, thin pseudocapsule, cystic hemorrhagic necrotic areas. Thick, diffused network of capillaries, totally collapsed or wide-open sinusoids, surrounded by compact proliferation of cells with oval nuclei, distinct nuclear membrane, granular chromatin, small nucleolus, ill-defined cytoplasm, thick reticular fibers all around. Cells are organized in a so-called patternless pattern. Mature flattened endothelium. Typical "staghorn-like" shape of vessels. There is no strict correlation between morphology and behavior. However, histological aspects that could predict an aggressive and malignant behavior are high cellularity, cellular atypia, extensive necrosis, and >4/10 HPF mitotic figures. On immunohistochemistry, SFT/HPC is positive for vimentin, CD34, CD99, and Bcl2.
**Course and Staging**: Slow growth, benign more frequent than malignant. About 10 % behave aggressively with local and distant recurrence. Benign, stage 3; malignant, stage IIB. If malignant, recurrence and metastases are frequent.
**Treatment**: As the biological behavior of SFT is unpredictable, wide excision is the treatment of choice. The role of radio- and chemotherapy is unclear.

**Immunohistochemical Panel**

| | |
|---|---|
| VIM | + |
| CD34 | + |
| CD99 | + |
| Bcl2 | + |

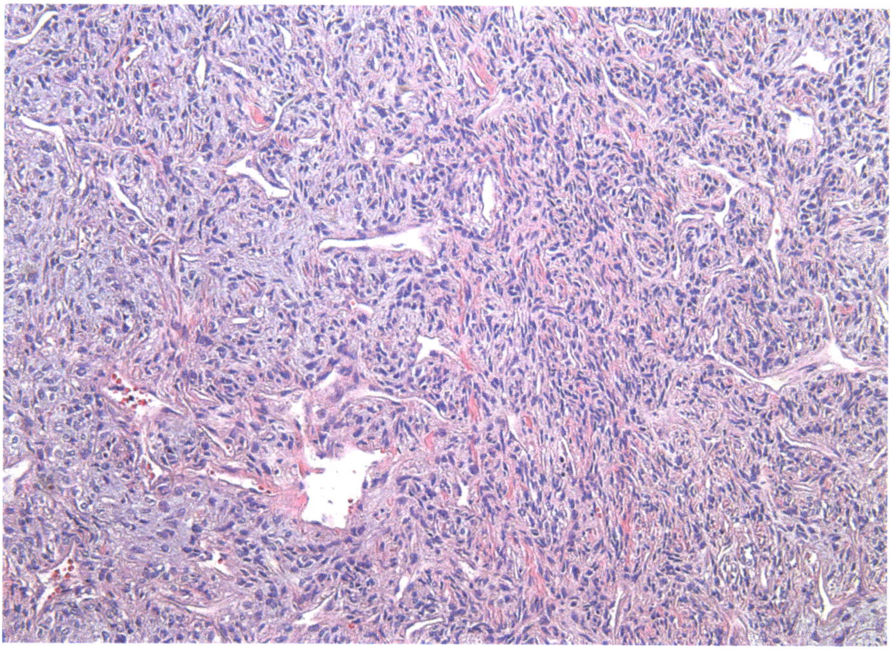

Solitary fibrous tumor. Bland-appearing spindle cells organized around ramified staghorn vessels in the so-called hemangiopericytic pattern or in a patternless pattern

## Selected Bibliography

Anders JO, Aurich M, Lang T, Wagner A (2006) Solitary fibrous tumor in the thigh: review of the literature. J Cancer Res Clin Oncol 132(2):69–75. Review

Carlson JW, Fletcher CD (2007) Immunohistochemistry for beta-catenin in the differential diagnosis of spindle cell lesions: analysis of a series and review of the literature. Histopathology 51(4):509–514. Review

Park MS, Araujo DM (2009) New insights into the hemangiopericytoma/solitary fibrous tumor spectrum of tumors. Curr Opin Oncol 21(4):327–331. Review

Wan S, Ning L, Hong R, Wu W, Fan S, Tsuchiya H, Tomita K (2010) Clinicopathological features of solitary fibrous tumours in the extremities: four case reports and a literature review. J Int Med Res 38(2):694–704. Review

Wignall OJ, Moskovic EC, Thway K, Thomas JM (2010) Solitary fibrous tumors of the soft tissues: review of the imaging and clinical features with histopathologic correlation. AJR Am J Roentgenol 195(1):W55–W62. Review

# Chapter 70
# Neurofibromas

**Marco Gambarotti**

**Definition**: Benign tumor of the peripheral nerve sheath. (a) Solitary and (b) multiple in neurofibromatosis.
**Epidemiology**: (a) More frequent than (b) (90 %). (a): No sex predilection. 20–40 years old. (b): Males, youngsters.

**Solitary Neuarofibroma**
**138 cases**

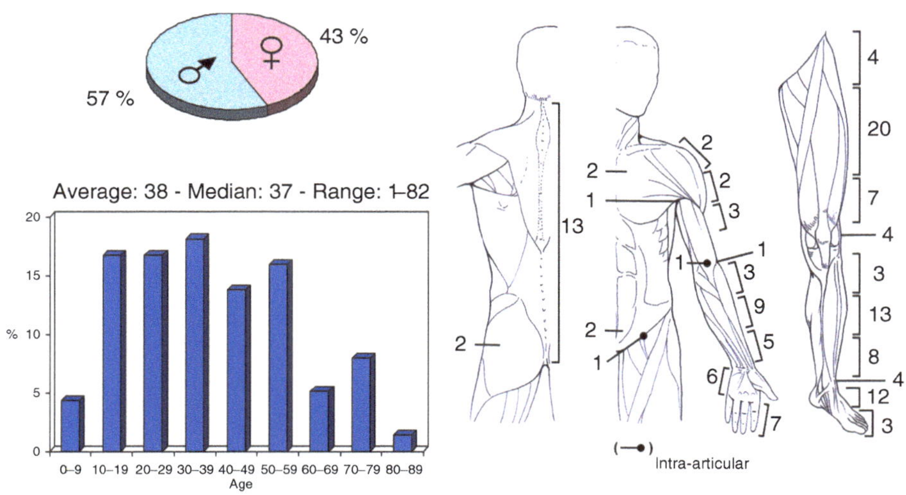


1900–2012 - Istituto Ortopedico Rizzoli - Laboratory of Experimental Oncology - Section of Epidemiology - Bologna - Italy

M. Gambarotti, MD
Department of Anatomy and Pathological Histology, Istituto Ortopedico Rizzoli,
Via del Barbiano 1/10, Bologna 40136, Italy
e-mail: marco.gambarotti@ior.it

P. Picci et al. (eds.), *Atlas of Musculoskeletal Tumors and Tumorlike Lesions*,
DOI 10.1007/978-3-319-01748-8_70, © Springer International Publishing Switzerland 2014

**Location**: (a): Superficial lesion in subcutis or dermis. (b): All sites and all organs.
**Clinical**: (a) Painless nodule, few symptoms. (b) Café au lait spots, typically in the axilla. More than six spots >1.5 cm are diagnostic; symptoms depend on location of neurofibromas: pigmented hamartomas of the iris (90 %), skeletal abnormalities (40 %), disorders of growth, sexual maturation, and abnormalities of the lung.
**Imaging**: On MRI: nerve trapped within or obliterated by the mass. Rarely encapsulated. Much more homogeneous, isointense to muscle with frequent areas of high signal on T1, more inhomogeneous, typical target appearance, higher than fat intensity on T2, centrally higher enhancement never with necrosis on contrast T1.
**Histopathology**: Spindle mass within the nerve, firm, grayish with pink areas. Interlacing bundles of elongated cells with dark wavy nuclei intimately associated with thick collagen fibers and mucoid material among bundles. Rare mitotic figures and necrotic areas. Scattered mast cells, lymphocytes, xanthoma cells. In type (b), there may be plexiform N. In large nerves, cordonal masses of convoluted aspect, a mixture of nerve trunks of different shapes and sizes, crossing each other along different planes or proliferation of Schwann cells producing collagen. Diffused N. in the subcutis of head and neck, a whitish plaque-like thickening of skin, diffusing along connective septa and among adipose tissue cells. Less evident fibrillary collagen matrix, more ovoid, round monomorphous cells, in a finely fibrillary homogeneous matrix or with organoid organization which looks like Meissner's bodies.
**Course and Staging**: Malignant transformation is rare in type (a) but more frequent (5–30 %) in type (b). Usually, stage 1 or 2.
**Treatment**: Marginal excision in (a). In (b), only large, painful, and located in areas where continued growth would compromise organ function are treated by surgery.

**Immunohistochemical Panel**

| | |
|---|---|
| VIM | + |
| S100 (focal) | + |
| Neurofilament proteins (NF) | ± |

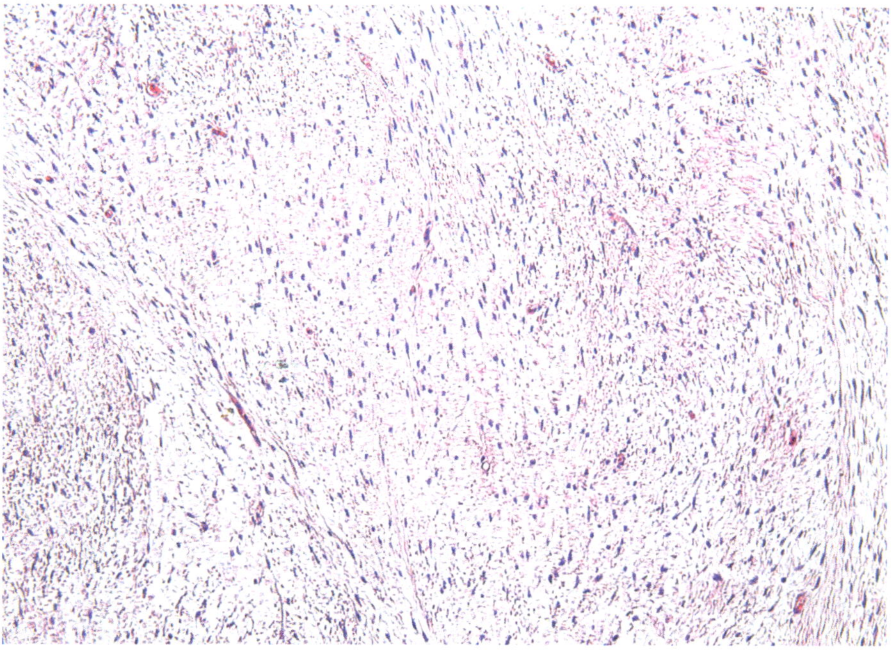

Loose sheets of bland-appearing spindle cells with wavy nuclei and myxoid background

## Selected Bibliography

Goutagny S, Kalamarides M (2010) Meningiomas and neurofibromatosis. J Neurooncol 99(3):341–347. Review

Neville H, Corpron C, Blakely ML, Andrassy R (2003) Pediatric neurofibrosarcoma. J Pediatr Surg 38(3):343–346; discussion 343–346. Review

Riccardi VM (2007) The genetic predisposition to and histogenesis of neurofibromas and neurofibrosarcoma in neurofibromatosis type 1. Neurosurg Focus 22(6):E3. Review

Shah KN (2010) The diagnostic and clinical significance of café-au-lait macules. Pediatr Clin North Am 57(5):1131–1153. Review

Woertler K (2010) Tumors and tumor-like lesions of peripheral nerves. Semin Musculoskelet Radiol 14(5):547–558. Review

# Chapter 71
# Schwannoma (Neurilemoma, Neurinoma)

**Marco Gambarotti**

**Definition**: An encapsulated benign nerve sheath tumor composed of differentiated neoplastic Schwann cells.
**Epidemiology**: 20–50 years old. No sex predilection. Rarely associated with neurofibromatosis.

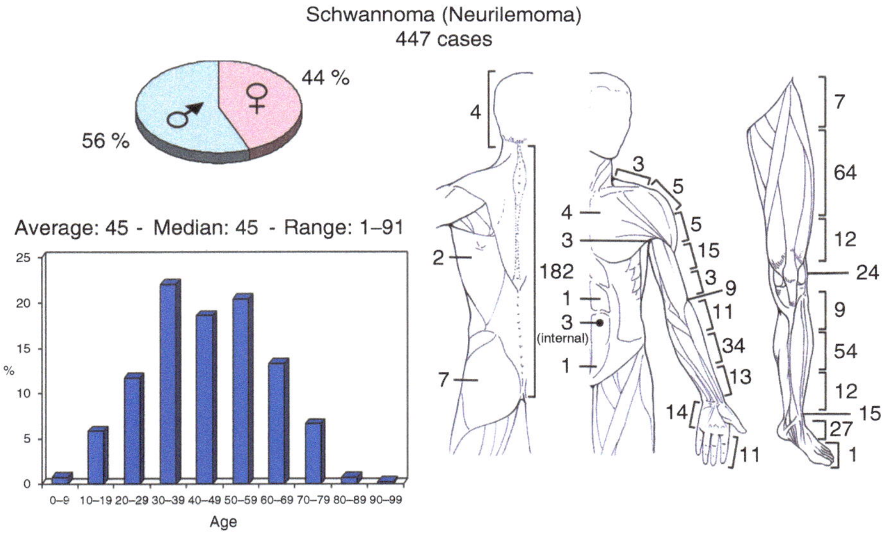


1900–2012 - Istituto Ortopedico Rizzoli - Laboratory of Experimental Oncology - Section of Epidemiology - Bologna - Italy

M. Gambarotti, MD
Department of Anatomy and Pathological Histology, Istituto Ortopedico Rizzoli,
Via del Barbiano 1/10, Bologna 40136, Italy
e-mail: marco.gambarotti@ior.it

P. Picci et al. (eds.), *Atlas of Musculoskeletal Tumors and Tumorlike Lesions*, 337
DOI 10.1007/978-3-319-01748-8_71, © Springer International Publishing Switzerland 2014

**Location**: Spinal roots, nerves of the mediastinum or of the retroperitoneum, peroneal, and ulnar nerve. Single nodule.
**Clinical**: Slow and progressive pain and neurological symptoms. Liquoral block, compression of the other roots or of the spinal cord, rachialgia, nocturnal increase in pain, stiffness, spinal contractures. Sharply painful, irradiated pain, paresthesia, algodystrophic syndrome when in peripheral nerve.
**Imaging**: On x-ray: scalloping of bone. On CT: well-circumscribed, homogeneous lesion, with muscle density. Frequent nonenhancing necrotic and cystic areas that cause an inhomogeneous hyperdense lesion after contrast administration. On MRI: nerve along the site of the mass. Capsulated. Homogeneous, isointense to muscle with frequent areas of low signal on T1, inhomogeneous, sometimes target appearance, higher than fat intensity on T2, diffused or peripheral enhancement with central necrotic unchanged zones on contrast T1.
**Histopathology**: Globose "sausage-like" or "dumbbell mass," elongated along the sac, well delimited by a capsule, soft, pink, white, or yellow, connected to a nerve root. Rarely, it expands like an hourglass inside and out of the intervertebral root foramen. S. of the peripheral nerves appears like an "onion bulb" or a piece of fruit attached to its stem. When very large, it may cause wide osteolyses. Usually encapsulated. Proliferation of spindle cells with ill-defined cytoplasm, in interlacing bundles, nuclear palisading, whorling of the cells, Verocay body formed by compact rows of well-aligned nuclei and cell processes that assume a roughly oval shape. Compact areas of spindle cells (Antoni A) alternating with loosely arranged areas (Antoni B). Rare if any mitotic figures. Large, irregularly spaced vessels with thickened wall and filled with thrombus material.
**Course and Staging**: Slow growth, rare recurrences after even incomplete surgery. Usually, stage 1 or 2.
**Treatment**: Marginal excision is curative. Malignant transformation exceptional.

**Immunohistochemical Panel**

| | |
|---|---|
| VIM | + |
| S100 | + |
| GFAP | + |

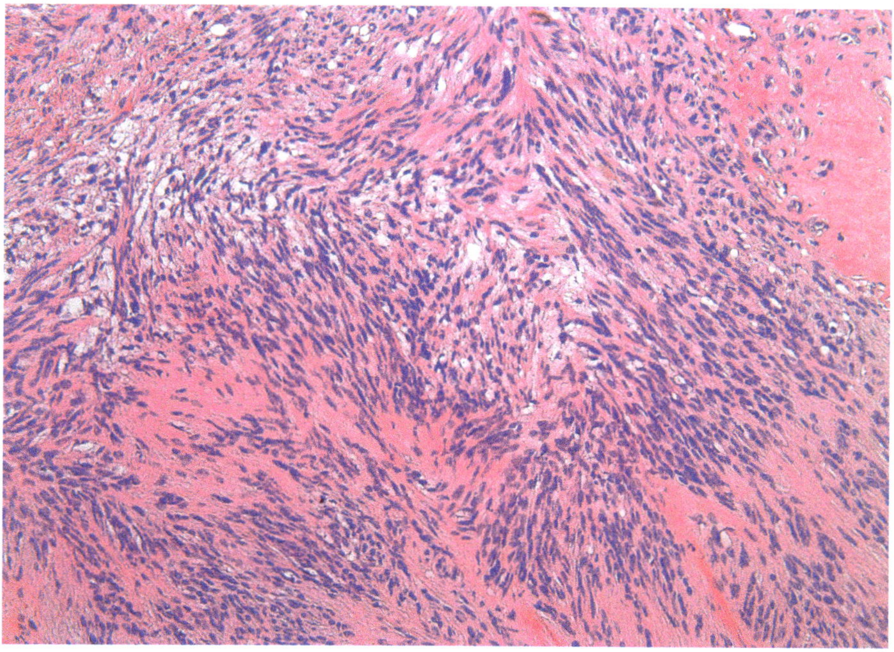

Elongated spindle cells with wavy nuclei organized in palisades (Antoni A areas) or loose sheets of cells (Antoni B areas)

## Selected Bibliography

Asthagiri AR, Helm GA, Sheehan JP (2007) Current concepts in management of meningiomas and schwannomas. Neurol Clin 25(4):1209–30. Review

Benz MR, Czernin J, Dry SM, Tap WD, Allen-Auerbach MS, Elashoff D, Phelps ME, Weber WA, Eilber FC (2010) Quantitative F18-fluorodeoxyglucose positron emission tomography accurately characterizes peripheral nerve sheath tumors as malignant or benign. Cancer 116(2):451–8. Erratum in: Cancer. 2010 Feb 1;116(3):775

Coulon A, Milin S, Laban E, Debiais C, Jamet C, Goujon JM (2009) Pathologic characteristics of the most frequent peripheral nerve tumors. Neurochirurgie 55(4–5):454–8. Review. French

Levi AD, Ross AL, Cuartas E, Qadir R, Temple HT (2010) The surgical management of symptomatic peripheral nerve sheath tumors. Neurosurgery 66(4):833–40

Woertler K (2010) Tumors and tumor-like lesions of peripheral nerves. Semin Musculoskelet Radiol 14(5):547–58. Review

# Chapter 72
# Malignant Schwannoma: Malignant Peripheral Nerve Sheath Tumor

**Marco Gambarotti**

**Definition**: Malignant tumor that rises from a peripheral nerve, from a preexisting benign nerve sheath tumor, in a patient with neurofibromatosis type 1 or with features of Schwann cell differentiation.
**Epidemiology**: 5 % of all soft tissue sarcomas. Fifty percent in neurofibromatosis. Males prevail in patients with neurofibromatosis with a medium age of 30 years. Females prevail in sporadic cases in the adult age (>40 years).

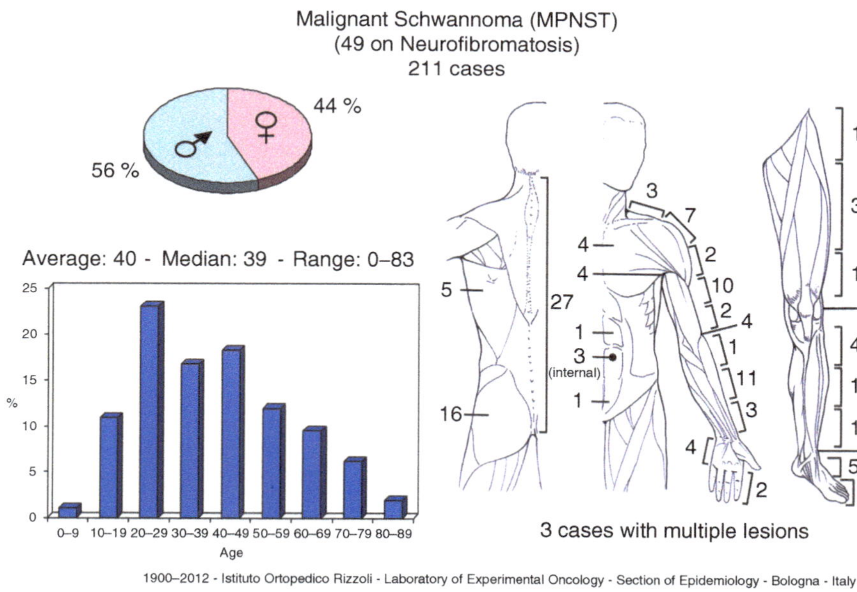


M. Gambarotti, MD
Department of Anatomy and Pathological Histology, Istituto Ortopedico Rizzoli, Via del Barbiano 1/10, Bologna 40136, Italy
e-mail: marco.gambarotti@ior.it

P. Picci et al. (eds.), *Atlas of Musculoskeletal Tumors and Tumorlike Lesions*,
DOI 10.1007/978-3-319-01748-8_72, © Springer International Publishing Switzerland 2014

**Location**: In association with major nerve trunks: sciatic nerve, brachial plexus, sacral plexus. Gluteal region and pelvis, thigh region, shoulder region, and axilla are the most frequent sites.
**Clinical**: Rapidly enlarging mass with pain. Sensory and motor symptoms including projected pain, paresthesias, and weakness.
**Imaging**: Neither CT nor MRI can establish a definite diagnosis. On CT: mild enhancement with some nonenhancing areas of necrosis. On MRI: inhomogeneous and higher intensity than muscle on T1, markedly inhomogeneous and bright on T2, common necrosis in the center of the mass with peripheral capsule-like or irregular enhancement on contrast T1, a wide halo of edema unusual in benign lesions.
**Histopathology**: Large fusiform or eccentric mass within a major nerve. When the tumor spreads along the epi- and perineurium, it has a rosary bead aspect proximal or distal to the principal mass. Tumor is usually >5 cm, deep-seated, but rarely superficial, and fleshy, with hemorrhagic and necrotic diffused areas. Spindle cells with markedly irregular contours with wavy-, buckled-, or comma-shaped nuclei, lightly stained and usually indistinct cytoplasm. Densely cellular sweeping fascicules alternate with hypocellular, myxoid zones where the parallel orientation of cells is lacking. A peculiar nodular or whorled arrangement of cells can be present. Focal nuclear palisading is present in only 10 % of cases. Hemangiopericytoma-like vascular pattern can be present. Heterologous elements such as glandular differentiation, skeletal muscle (malignant triton tumor), bone, cartilage, and angiosarcomatous areas are present in about 15 % of cases. On immunohistochemistry, focal S-100 protein expression in spindled tumors; diffused strong S-100 protein expression in epithelioid MPNST.
**Course and Staging**: Frequent local recurrence when surgery is inadequate. Metastases frequently develop in the lungs, liver, subcutaneous, and bone, rarely in the lymph nodes. Usually, stage IIB.
**Treatment**: Very wide or radical local excision. Often, multicentric origin and diffusion along the nerve make even radical surgery ineffective. Overall 5-year survival is 30 % in patients with neurofibromatosis, 75 % in sporadic cases. In neurofibromatosis, prognosis is worse because the tumor involves the trunk and proximal extremities with late diagnosis, it is larger and of higher grade, and multiple sarcomas may occur. Usefulness of radiotherapy and chemotherapy is uncertain.

**Immunohistochemical Panel**

| | |
|---|---|
| VIM | + |
| S100 | + (30 % of cells in 30 % of cases) |
| GFAP | + (30 % of cells in 30 % of cases) |

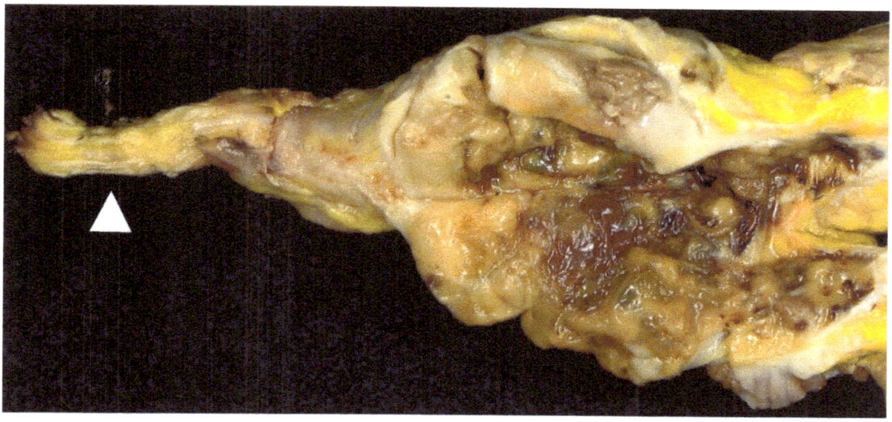

The tumor arises from a peripheral nerve (*arrowhead*)

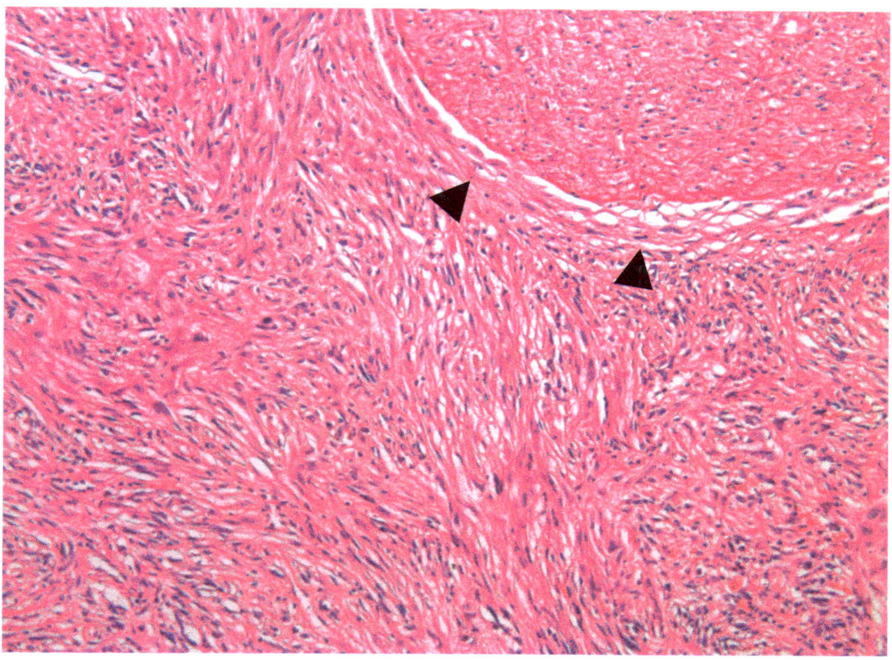

Highly atypical spindle cells with wavy or plump nuclei organized in storiform or fascicular pattern, frequently close to a peripheral nerve (*arrowheads*)

## Selected Bibliography

Grimm S, Chamberlain MC (2009) Adult primary spinal cord tumors. Expert Rev Neurother 9(10):1487–1495. Review

Gupta G, Mammis A, Maniker A (2008) Malignant peripheral nerve sheath tumors. Neurosurg Clin N Am 19(4):533–543, v. Review

Katz D, Lazar A, Lev D (2009) Malignant peripheral nerve sheath tumour (MPNST): the clinical implications of cellular signalling pathways. Expert Rev Mol Med 11:e30. Review

Kim DH, Murovic JA, Tiel RL, Moes G, Kline DG (2005) A series of 397 peripheral neural sheath tumors: 30-year experience at Louisiana State University Health Sciences Center. J Neurosurg 102(2):246–255

Ogawa BK, Skaggs DL, Kay RM (2009) Malignant peripheral nerve sheath tumor of the lumbar spine. Am J Orthop (Belle Mead NJ) 38(5):E89–E92. Review

# Chapter 73
# Myofibroblastic Sarcoma

**Marco Gambarotti**

**Definition**: Sarcomas composed of myofibroblasts.

Myofibroblastic sarcomas display a range of differentiation. Low-grade myofibroblastic sarcoma is identified as a specific entity in the WHO 2013 classification, while the definition of high-grade myofibroblastic sarcoma is not well established. There is evidence that myofibroblastic differentiation in pleomorphic sarcomas is associated with a more aggressive behavior.

**Epidemiology**: Low-grade myofibroblastic sarcomas occur predominantly in adult patients (age range: 4–75 years, mean 38), while high-grade myofibroblastic sarcomas can also occur in children. There is a slight male predominance.

M. Gambarotti, MD
Department of Anatomy and Pathological Histology, Istituto Ortopedico Rizzoli,
Via del Barbiano 1/10, Bologna 40136, Italy
e-mail: marco.gambarotti@ior.it

P. Picci et al. (eds.), *Atlas of Musculoskeletal Tumors and Tumorlike Lesions*,
DOI 10.1007/978-3-319-01748-8_73, © Springer International Publishing Switzerland 2014

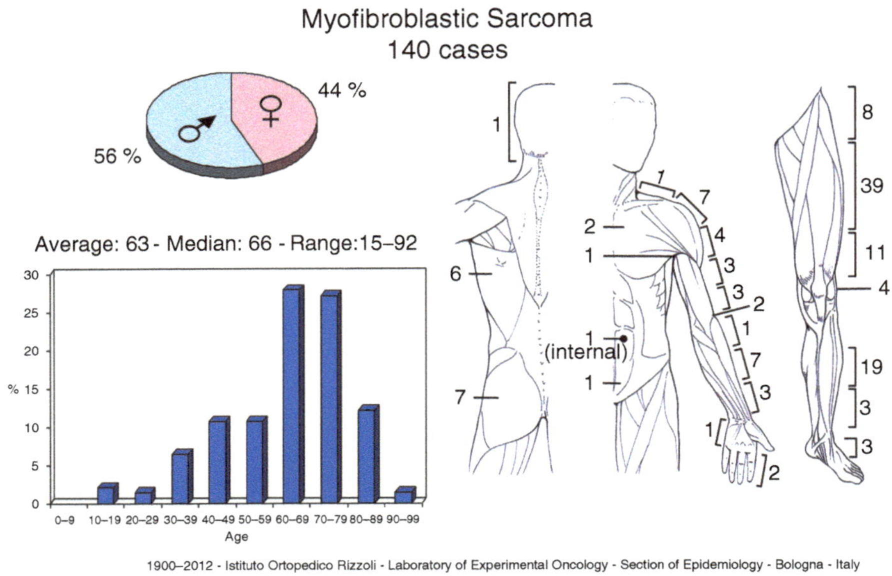


**Location**: Low-grade myofibroblastic sarcomas most commonly occur in the head and neck region, including the oral cavity, pharynx and parapharyngeal regions, and proximal extremities and trunk; occasional cases can occur in the abdomen or pelvis. They usually arise in the deep soft tissues, but cases have been seen in the subcutis and submucosa. Visceral lesions are rare. Cases have also been described in bone, including maxilla, mandible, femur, and ilium. High-grade myofibroblastic sarcomas arise in deep soft tissues, predominantly in lower limbs and trunk, with occasional cases in head and neck.

**Clinical and Imaging**: Enlarging painless mass, very often with infiltrative margins. High-grade sarcomas with hemorrhage and necrosis are heterogeneous at MRI.

**Histopathology**: Grossly, low-grade myofibroblastic sarcomas are firm with a pale fibrous-appearing cut surface, ill-defined infiltrative margins, or sometimes with pushing margins. High-grade myofibroblastic sarcomas are large solid tumors with hemorrhage and necrosis. Histologically, low-grade myofibroblastic sarcomas are characterized by a proliferation of spindle cells arranged in a fascicular or in a storiform pattern. The neoplastic cells show tapered fusiform elongated to wavy nuclei with discernible eosinophilic cytoplasm. Sometimes the nuclei are rounded and vesicular with small punctuated nucleoli. There is at least focally moderate nuclear atypia. The margins are predominantly infiltrative, with separation rather than destruction of skeletal muscle bundles. Mitotic activity is variable but atypical mitoses are rare. Stroma is often minimal and can be variably collagenous. High-grade myofibroblastic sarcomas are composed of pleomorphic, spindle, or epithelioid cells arranged in a fascicular or in a storiform growth pattern, with scattered atypical mitotic figures. Both low- and high-grade myofibroblastic sarcomas show variable positivity for actins and/or desmin. Calponin and CD34 can be positive. EMA, S100, beta-catenin, and caldesmon are negative. Ultrastructure is the gold standard in defining normal myofibroblasts which are distinguished from smooth muscle cells by fibronectin fibril and fibronexus junctions, which are considered the most characteristic ultrastructural markers of myofibroblasts. Fibronexus, a cell-to-matrix junction, first described and designated by Singer in 1979, is a unit composed of fibronectin and 5 nm microfilaments that converge at the cell surface. It is not perfectly clear whether these features are always present in myofibroblastic sarcomas.

**Course and Staging**: About 33 % of low-grade myofibroblastic sarcomas locally recur, especially after incomplete excision. Metastases have been reported in approximately 10 % of cases. Progression to high-grade sarcoma has been documented. High-grade myofibroblastic sarcomas recur in 33 % of cases with metastases in over 70 %. These tumors, like the other pleomorphic sarcomas with myogenic differentiation, have a worse outcome than undifferentiated sarcomas.

**Treatment**: Low-grade myofibroblastic sarcomas are best managed by wide surgical excision and long-term follow-up to detect possible late metastases. High-grade myofibroblastic sarcomas should be managed by excision with wide margins and adjuvant radiation therapy and/or systemic chemotherapy.

**Immunohistochemical Panel**

| | |
|---|---|
| VIM | + |
| MS act | ± |
| Smooth M act | ± |
| Desmin | ± |
| Caldesmon | − |
| Calponin | + |

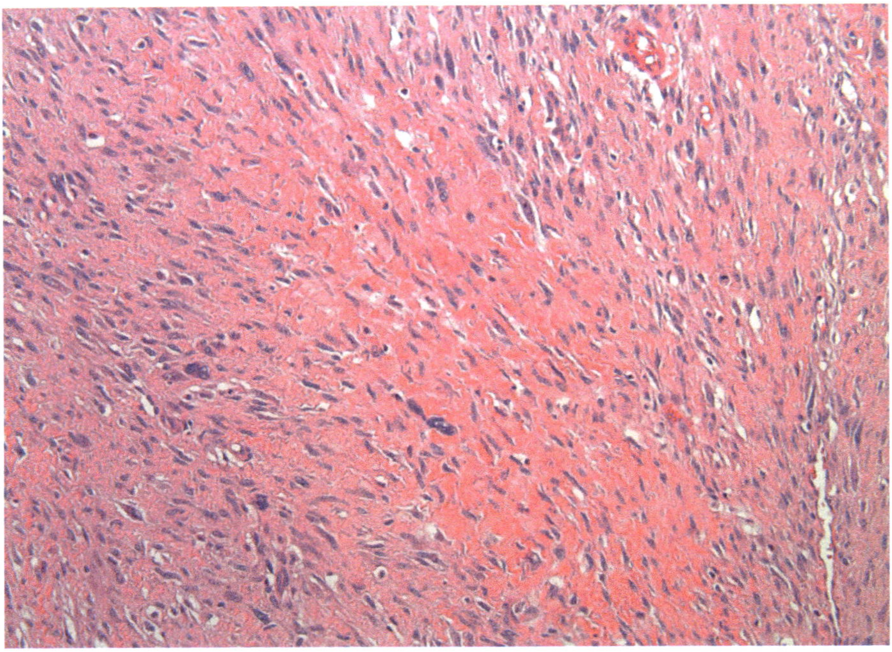

Fascicular to storiform arrangement of spindle cells with short tapered wavy nuclei. Stroma is variably collagenous. Some tumors show a tissue culture-like growth pattern that closely resembles nodular fasciitis

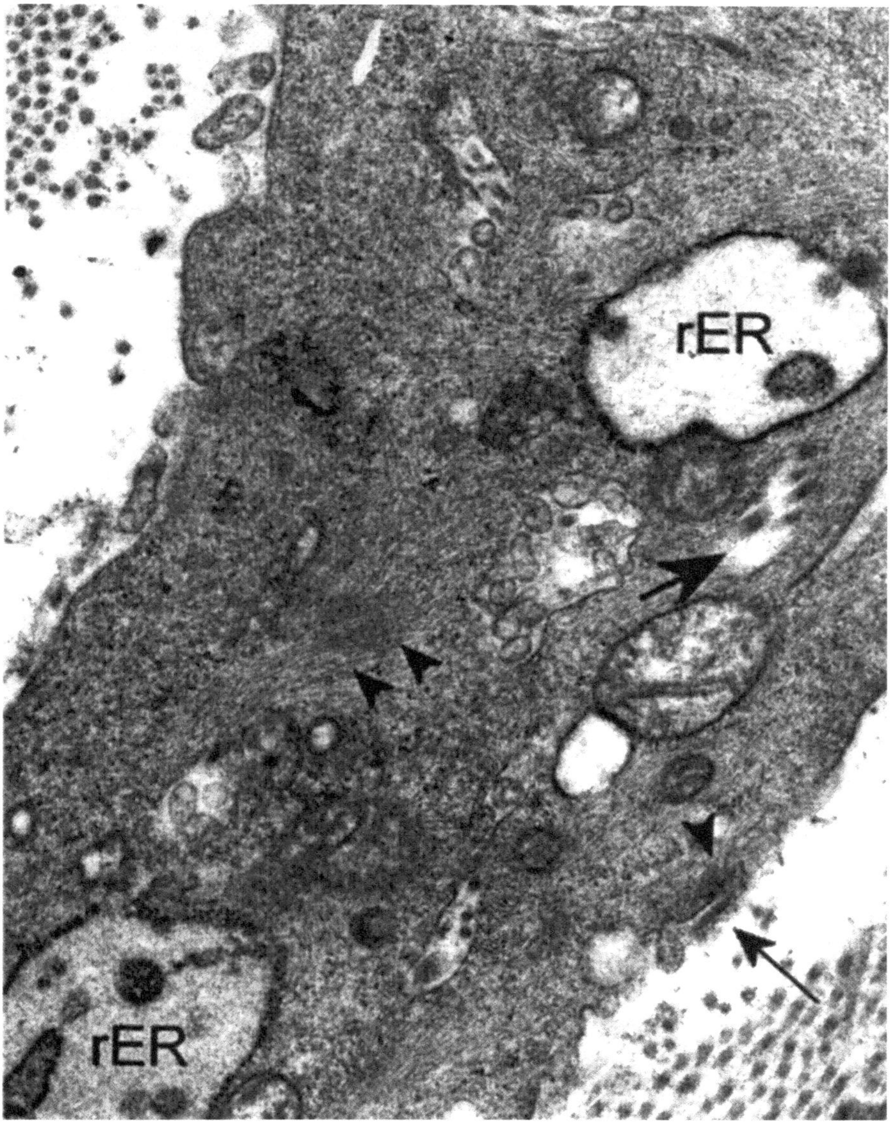


Portion of a myofibroblast showing intermediate filaments, actin filaments (*arrowheads*), dilated rough endoplasmic reticulum cisternae (rER), collagen inclusions (*large* and *short arrows*), and fibronexus (*thin* and *long arrow*), ×9,000

## Selected Bibliography

Eyden B, Banerjee SS, Shenjere P, Fisher C (2009) The myofibroblast and its tumours. J Clin Pathol 62(3):236–249. Review

Fisher C (2004) Myofibroblastic malignancies. Adv Anat Pathol 11(4):190–201. Review

Fletcher CD (1998) Myofibroblastic tumours: an update. Verh Dtsch Ges Pathol 82:75–82. Review

Fletcher CDM, Bridge JA, Hogendoorn PCW, Mertens F (eds) (2013) WHO classification of tumors of soft tissue and bone, 4th edn. International Agency for Research on Cancer, Lyon

Meng GZ, Zhang HY, Zhang Z, Wei B, Bu H (2009) Myofibroblastic sarcoma vs nodular fasciitis: a comparative study of chromosomal imbalances. Am J Clin Pathol 131(5):701–709

Perez-Montiel MD, Plaza JA, Dominguez-Malagon H, Suster S (2006) Differential expression of smooth muscle myosin, smooth muscle actin, h-caldesmon, and calponin in the diagnosis of myofibroblastic and smooth muscle lesions of skin and soft tissue. Am J Dermatopathol 28(2):105–111

# Chapter 74
# Undifferentiated Pleomorphic Sarcoma (UPS)

**Marco Gambarotti**

**Definition**: A soft tissue sarcoma characterized by tumor cells with diffuse pleomorphism in the absence of a specific line of differentiation. The old term malignant fibrous histiocytoma (MFH) is obsolete as immunohistochemistry demonstrated that the phenotype of the neoplastic cells is closely aligned with a fibroblast than a histiocyte. In the 2013 WHO classification, UPS is classified in the group of undifferentiated/unclassified sarcomas that are divided into pleomorphic (UPS), round cell, spindle cells, and epithelioid subsets. Together they account for up to 20 % of all soft tissue sarcomas.
**Epidemiology**: Males. Most frequent in late adult life (50–70 years).

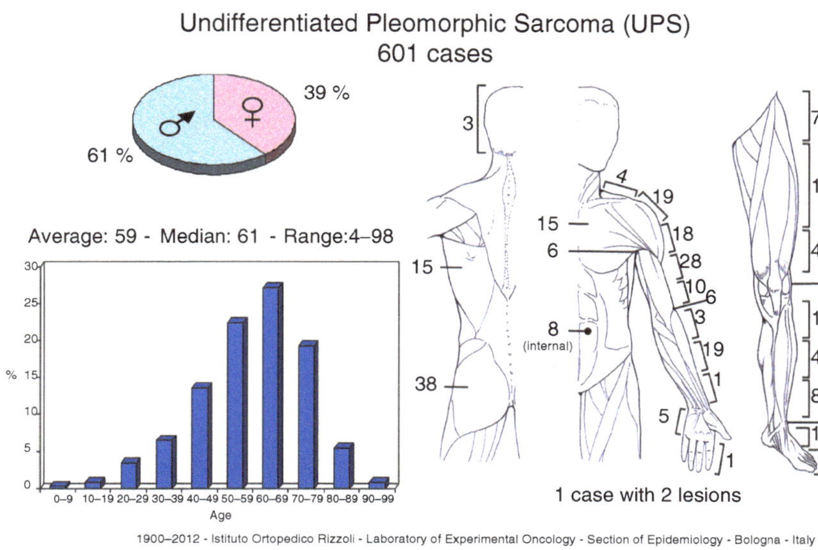


M. Gambarotti, MD
Department of Anatomy and Pathological Histology, Istituto Ortopedico Rizzoli,
Via del Barbiano 1/10, Bologna 40136, Italy
e-mail: marco.gambarotti@ior.it

P. Picci et al. (eds.), *Atlas of Musculoskeletal Tumors and Tumorlike Lesions*,
DOI 10.1007/978-3-319-01748-8_74, © Springer International Publishing Switzerland 2014

**Localization**: Skeletal muscles of extremities and retroperitoneum. Tumor is deep in 90 % of cases.
**Clinical**: Globose and painless swelling with no characteristic clinical features other than a frequently rapid growth rate.
**Imaging**: On X-ray: nonspecific STT displacing the adjacent fat. Peripheral calcifications are rare (9 %). Florid periosteal reaction and smooth cortical erosion may be observed. In these cases, bone scan is always very hot. On angiography: typical changes occurring in sarcomas, with very large avascular areas of tumor due to necrosis or hemorrhage. Major vessels are almost never infiltrated. On CT: inhomogeneous, similar to or lower density than that of muscle, strong enhancement of the solid component, central hypodense area of necrosis, hemorrhage, myxomatous tissue, with large cavities with fluid contents, and a thick wall that are mistaken with a hematoma. On MRI: poorly defined margins, homogeneous, muscular intensity on T1 and heterogeneous high signal intensity on T2, dark central necrotic zones and strong enhancement at the periphery on contrast T1, internal low signal intensity septa of collagen bands on T1 and T2. In MFH, central myxoid area is black on T1 and white on T2. Hematoma is white on T1; fluid levels show low signal intensity for hemosiderin deposits and high for supernate on both sequences.
**Histopathology**: White-gray mass with no distinctive macroscopic features other than the frequent presence of necrosis. Pleomorphic aspect, cellular atypia with hyperchromic nuclei, coarse chromatin, large nucleoli, and numerous typical and atypical mitotic figures. Histologically, UPS resembles other specific type of pleomorphic sarcoma, with frequent multinucleated tumor giant cells and a frequent patternless pattern. In some areas, there is a distinctive orientation around eosinophilic areas or vessels that gives the appearance of a cartwheel: "the storiform pattern." Collagen production may produce an accentuation of this feature. Clusters of histiocytes, foam cells, and inflammatory cells are sprinkled. In the 2013 WHO classification, UPS is classified in the group of undifferentiated/unclassified sarcomas that are divided into pleomorphic (UPS), round cell (similar to other specific types of round cell sarcoma, especially Ewing's sarcoma), spindle cells, and epithelioid (similar to a metastatic carcinoma or melanoma) subsets. On immunohistochemistry, undifferentiated pleomorphic sarcomas are positive for vimentin, and by definition, no pattern of protein expression that would identify a specific line of differentiation can be identified.
**Course and Staging**: The majority of UPS are high-grade sarcomas with a metastatic incidence that varies between 30 and 50 %, with the common metastatic sites being the lung, bone, and liver; regional lymph node metastases are decidedly uncommon. Usually, this tumor is stage IIB. IIIB presentation due to lymphatic or pulmonary metastases is not uncommon. UPS may be secondary to benign processes or radiation therapy. Several studies have suggested that pleomorphic sarcomas with myogenic differentiation are clinically more aggressive than those without myogenic differentiation.
**Treatment**: Wide excision or yet better radical surgery. Radiotherapy is effective in 50 % of cases and is used as primary procedure to very well delimit the mass and to reduce the lesion making the operation possible and easier. Patients treated with adjuvant chemotherapy have a better survival.

**Immunohistochemical Panel**

| | |
|---|---|
| VIM | + |

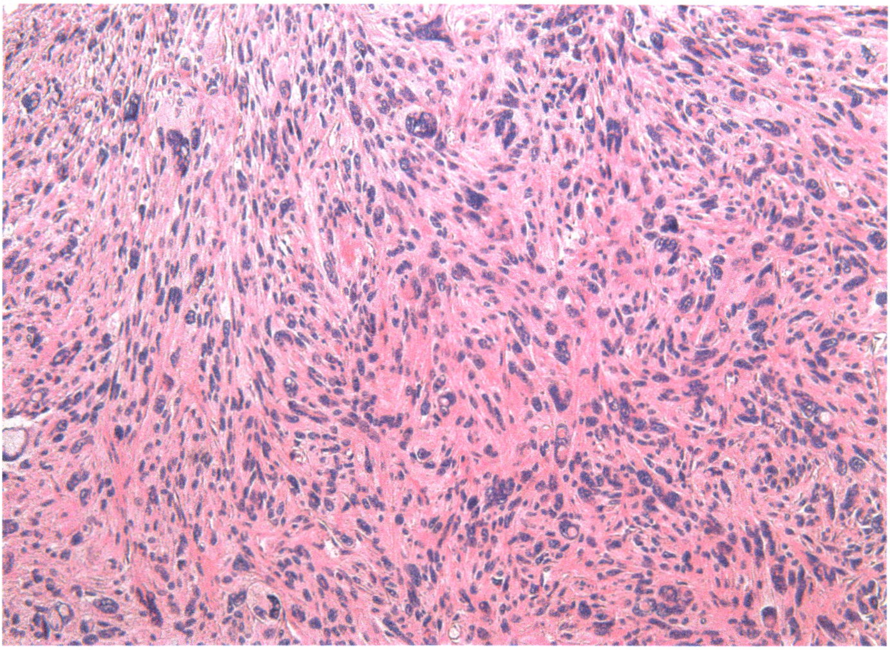

High-grade pleomorphic undifferentiated neoplasm with focal storiform pattern

## Selected Bibliography


Al-Agha OM, Igbokwe AA (2008) Malignant fibrous histiocytoma: between the past and the present. Arch Pathol Lab Med 132(6):1030–1035. Review

Erlandson RA, Antonescu CR (2004) The rise and fall of malignant fibrous histiocytoma. Ultrastruct Pathol 28(5–6):283–289. Review. Erratum in: Ultrastruct Pathol. 2005; 29(2):157

Fletcher CDM, Bridge JA, Hogendoorn PCW, Mertens F (eds) (2013) WHO classification of tumors of soft tissue and bone, 4th edn. International Agency for Research on Cancer, Lyon

Matushansky I, Charytonowicz E, Mills J, Siddiqi S, Hricik T, Cordon-Cardo C (2009) MFH classification: differentiating undifferentiated pleomorphic sarcoma in the 21st century. Expert Rev Anticancer Ther 9(8):1135–1144. Review

Nascimento AF, Raut CP (2008) Diagnosis and management of pleomorphic sarcomas (so-called "MFH") in adults. J Surg Oncol 97(4):330–339. Review

Randall RL, Albritton KH, Ferney BJ, Layfield L (2004) Malignant fibrous histiocytoma of soft tissue: an abandoned diagnosis. Am J Orthop (Belle Mead NJ) 33(12):602–608. Review

# Chapter 75
# Myxofibrosarcoma

**Marco Gambarotti**

**Definition**: Malignant fibroblastic lesions with myxoid stroma, pleomorphism, and a distinctive curvilinear vascular pattern. The designation of myxofibrosarcomas and myxoid malignant fibrous histiocytoma has been considered almost synonymous in the 2013 WHO classification of soft tissue tumors. This classification does not mention a minimal amount of myxoid matrix for the definition of myxofibrosarcoma.
**Epidemiology**: Among the most common sarcomas in elderly patients. Overall age range is wide, but they are more frequent in the sixth to eighth decade, whereas they are exceptional under 20 years of age. There is a slight male predominance.

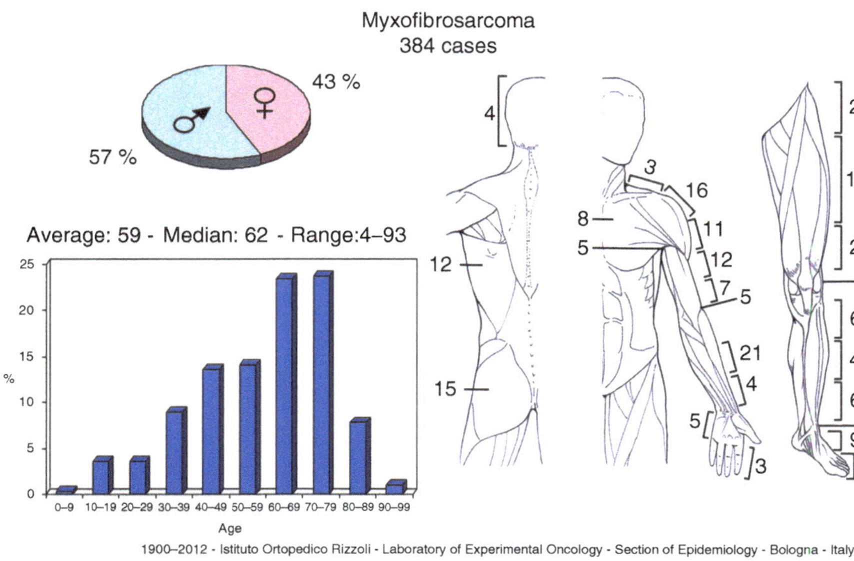


M. Gambarotti, MD
Department of Anatomy and Pathological Histology, Istituto Ortopedico Rizzoli,
Via del Barbiano 1/10, Bologna 40136, Italy
e-mail: marco.gambarotti@ior.it

P. Picci et al. (eds.), *Atlas of Musculoskeletal Tumors and Tumorlike Lesions*,
DOI 10.1007/978-3-319-01748-8_75, © Springer International Publishing Switzerland 2014

**Location**: The most frequent sites are the lower limbs, followed by upper limbs and limb girdles. Rare in trunk, head and neck, and hands and feet. Most cases described in the retroperitoneum and in the abdominal cavity represent dedifferentiated liposarcomas. About half of cases arise in dermal/subcutaneous tissue, with the remaining arising in the underlying fascia and skeletal muscle.
**Clinical and Imaging**: Enlarging painless mass, with infiltrative margins. Very often heterogeneous at MRI.
**Histopathology**: Grossly, they appear as multiple gelatinous to firm nodules with infiltrative margins. Histologically, all cases share distinct morphological features: multinodular growth with incomplete fibrous septa, a myxoid stroma composed of hyaluronic acid, and prominent elongated, curvilinear, thin-walled vessels with a perivascular condensation of tumor cells. Frequently, the so-called pseudolipoblasts (vacuolated neoplastic fibroblastic cells with cytoplasmic acid mucin) are seen. Low-grade lesions are hypocellular, composed of few non-cohesive, plump spindled or stellate cells with ill-defined cytoplasm and hyperchromatic nuclei; mitosis is infrequent. High-grade lesions are composed in large part of solid sheets and cellular fascicles of spindle and pleomorphic tumor cells with numerous, often atypical, mitoses and areas of necrosis. Immunohistochemically, tumor cells stain positive for vimentin.
**Course and Staging**: In up to 50–60 % of cases, local recurrences unrelated to histological grade repeatedly occur. In contrast, metastases and death from tumor are closely related to tumor grade: low-grade tumors do not metastasize, while metastases develop in 20–35 % of intermediate- and high-grade neoplasms. Metastases occur in the lung, bone, and lymph nodes. Low-grade lesions that recur may subsequently increase in grade. The depth of the lesion does not influence the rate of local recurrence, while deep-seated neoplasms have a higher percentage of metastases and tumor-associated mortality. Overall 5-year survival rate is 60–70 %.
**Treatment**: Excision with wide margins and adjuvant radiation therapy and/or systemic chemotherapy.

**Immunohistochemical Panel**

| | |
|---|---|
| VIM | + |

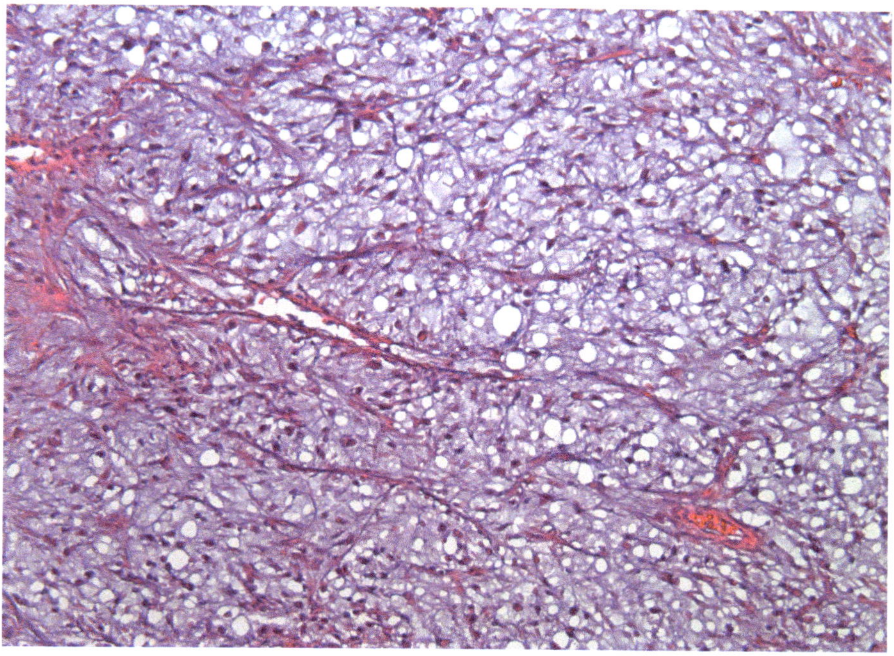

Bland-appearing spindle cells in a loose myxoid extracellular matrix. Cells are arranged in lobules, with elongated curvilinear blood vessels

## Selected Bibliography


Antonescu CR, Baren A (2004) Spectrum of low-grade fibrosarcomas: a comparative ultrastructural analysis of low-grade myxofibrosarcoma and fibromyxoid sarcoma. Ultrastruct Pathol 28(5–6):321–332

Arnaoutoglou C, Lykissas MG, Gelalis ID, Batistatou A, Goussia A, Doukas M, Xenakis TA (2010) Low grade fibromyxoid sarcoma: a case report and review of the literature. J Orthop Surg Res 5:49

Fletcher CDM, Bridge JA, Hogendoorn PCW, Mertens F (eds) (2013) WHO classification of tumors of soft tissue and bone, 4th edn. International Agency for Research on Cancer, Lyon

Mertens F, Fletcher CD, Antonescu CR, Coindre JM, Colecchia M, Domanski HA, Downs-Kelly E, Fisher C, Goldblum JR, Guillou L, Reid R, Rosai J, Sciot R, Mandahl N, Panagopoulos I (2005) Clinicopathologic and molecular genetic characterization of low-grade fibromyxoid sarcoma, and cloning of a novel FUS/CREB3L1 fusion gene. Lab Invest 85(3):408–415

Monson E, Vancourt R, Dawson J (2010) Myxoinflammatory fibroblastic sarcoma: a case report and review of the literature. J Foot Ankle Surg 49(1):86.e1–3. Review

Shidham VB, Ayala GE, Lahaniatis JE, Garcia FU (1999) Low-grade fibromyxoid sarcoma: clinicopathologic case report with review of the literature. Am J Clin Oncol 22(2):150–155. Review

# Chapter 76
# Synovial Sarcoma (SS)

**Marco Gambarotti**

**Definition**: Malignant mesenchymal tumor which displays a variable degree of epithelial differentiation and a specific translocation t(X;18).
**Epidemiology**: 5–10 % of all sarcomas of soft tissues. Males. 15–40 years old.

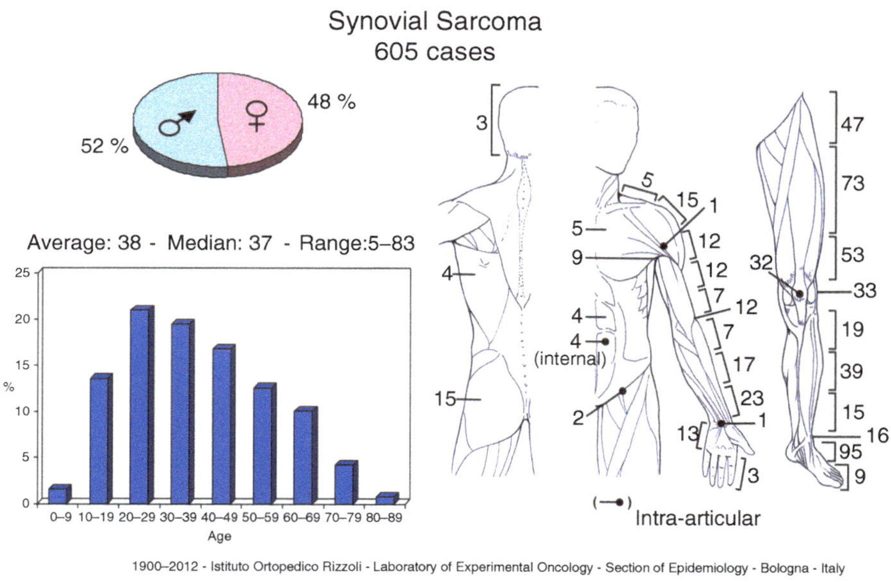


M. Gambarotti, MD
Department of Anatomy and Pathological Histology, Istituto Ortopedico Rizzoli,
Via del Barbiano 1/10, Bologna 40136, Italy
e-mail: marco.gambarotti@ior.it

P. Picci et al. (eds.), *Atlas of Musculoskeletal Tumors and Tumorlike Lesions*,
DOI 10.1007/978-3-319-01748-8_76, 

**Location**: 80 % extremities. Lower limb (60 %: thigh, knee, foot, ankle) and upper limb (23 %: forearm, wrist, shoulder). Only 10 % within a joint. Usually, close to a major joint, intimately related to tendons, tendon sheaths, bursae, and beyond the confines of the joint capsule.
**Clinical**: A palpable deep-seated mass with pain. Pain may be the first symptom of the disease. Tumor grows slowly and insidiously, and the duration of symptoms ranges from 2 to 4 years, although in some cases they have been noted for as long as 20 years.
**Imaging**: On x-ray: a round or oval, lobulated swelling of moderate density near a joint. Usually, bone is uninvolved. Periosteal reaction, bone erosion, or bone invasion (15–20 %). Multiple, small and spotty calcifications, or cloudy and faded shadow or massive and dense radiopacities of bone formation (40 %). Radiopacities more frequent in the periphery. On angiography: richly hypervascularized. On CT: infiltrating ST mass with slightly higher density than muscle, markedly enhanced, with easily detected calcifications, cortical erosion, and joint invasion when tendons or ligaments are involved. On MRI: 90 % hypointense on T1, hyperintense on T2. Fluid levels (15 %). Marked inhomogeneity, enhancement, and septation on T2. A triple-signal pattern on T2 (30 %): white-like fluid, gray-like fat, and dark-like fibrous tissue. Small high signal areas on T1 (45 %) are foci of hemorrhage. This and the triple pattern are suggestive of SS.
**Histopathology**: 1–20 cm. Slow-growing lesions are a firm, lobular mass fairly well circumscribed, attached to surrounding tissues, from yellow to gray-white. More rapidly growing lesions are a soft, globose mass poorly encapsulated, with a variegated, friable, or shaggy appearance, from mottled pink-yellow to light brown, with frequent necrosis, hemorrhage, and cystic areas. Calcifications can be seen. Two different types of cells: epithelial and spindle cells. (1) Biphasic type: two cellular components; (2) monophasic fibrous type: only spindle cells; (3) rare monophasic epithelial type: only epithelioid cells; and (4) poorly differentiated type (20 %). In type (1), there is a balanced distribution of the two components. Epithelioid cells are globose, cubic, or cylindrical, with large vesicular nuclei, abundant pale cytoplasm, and with well-defined limits. They are disposed in solid cords, whorls, nests, islands, and border irregular pseudoglandular, cleft-like or cyst-like spaces with granular or homogeneous secretions. Fibrous cells are well-oriented, plump, and spindle-shaped and have a uniform appearance, with scant and indistinct cytoplasm and oval, dark nuclei. They form solid, compact masses with irregular nodular arrangement. A fine net of collagen fibers surrounds the single cells of the fibrous component. Mitotic figure infrequent. In type (2), spindle cells are often arranged in a herringbone or HPC-like pattern. In type (3), a pseudoglandular feature with a poorly developed spindle cells pattern prevails. In type (4), hypercellular areas with round (similar to Ewing's sarcoma) or spindle or epithelioid cells with severe nuclear atypia and high mitotic activity.

Synovial sarcoma shows cytokeratin and/or EMA expression, although it can be quite focal and confined to only rare cells in an entire section. There is nuclear expression of TLE-1 protein. The balanced reciprocal translocation t(X;18)(p11.2;q11.2) is found in more than 95 % resulting in fusion of the SYT gene on chromosome 18 with the SSX1, SSX2, or rarely SSX4 gene on the X chromosome.

**Course and Staging**: Usually, it grows between and adheres to the tendons, joint capsules, bursae mucosae, fasciae, skeleton, muscles, and interosseous membranes. It pushes its growth along these multiple planes, infiltrating these structures and creating intravascular plugs. Local recurrence is common also 10 years after inadequate surgery. Metastases develop in about 50 % of cases. Usually, stage IIB.
**Treatment**: Surgery aims at obtaining wide margins, also sacrificing functionally important structures or amputating the limb. Wide conservative surgery may be difficult for the characteristic growth of the tumor. It may be useful to associate adjuvant chemotherapy and pre- or postoperative radiation therapy. Excision of the regional lymph nodes is appropriate for the high rate of metastases in this site. Overall 10-year survival varies from 15 to 35 %. Favorable prognostic factors are <5 cm, youth, women, distal extremities, calcifications, <15 mits/10HPF, degree of glandular differentiation, and no intravascular plugs.

**Immunohistochemical Panel**

| | |
|---|---|
| VIM | + |
| CK | +/– |
| EMA | +/– |
| TLE1 | + |
| CD99 | +/– |
| S100 | +/– |
| Bcl2 | +/– |

**Chromosomal Translocations**

| | | |
|---|---|---|
| t(X;18) (p11;q11) | SYT-SSX (ssx1 e ssx2) | 95 % |
| | SYT-ssx4 | <1 % |

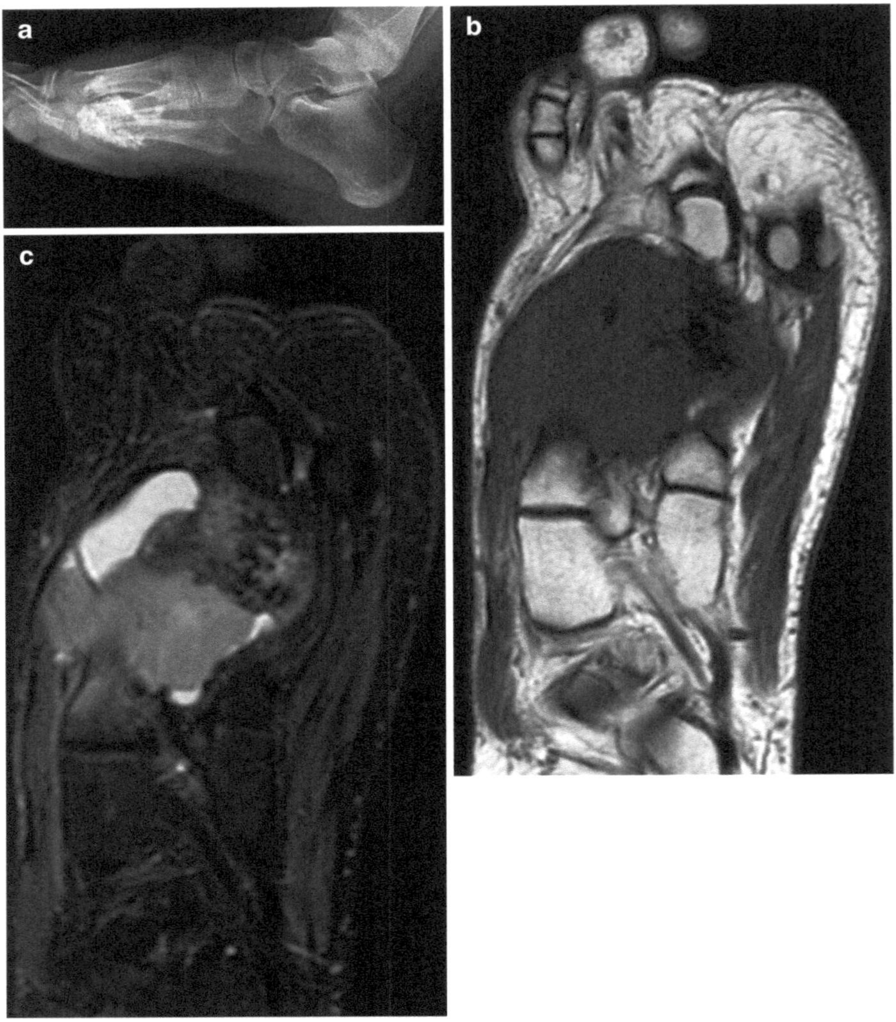


Radiograph, axial and sagittal T1 and T2 FS MR images. The mass is calcified, displaces the bones, and has an heterogeneous signal on MR

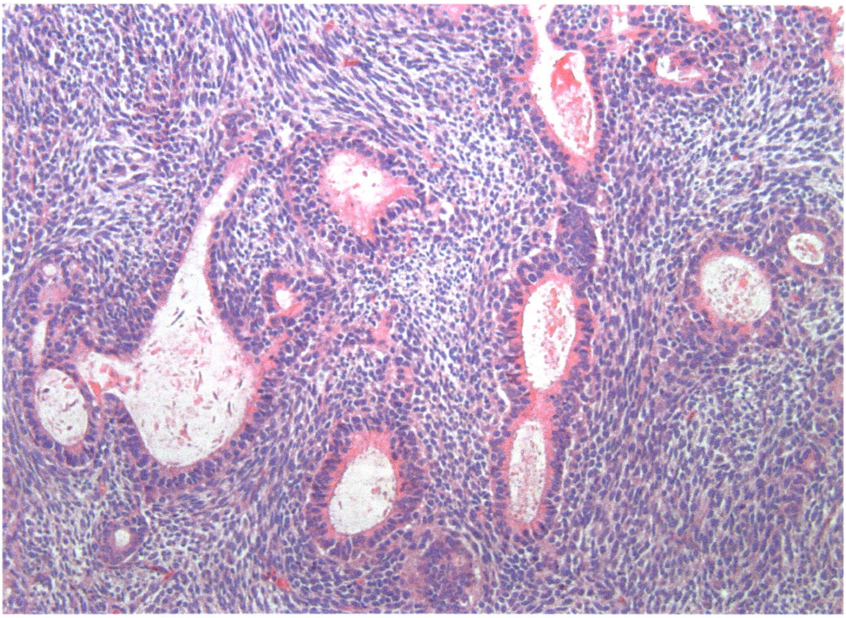

Biphasic synovial sarcoma: glandular component and spindle cell component intermixed in the same tumor

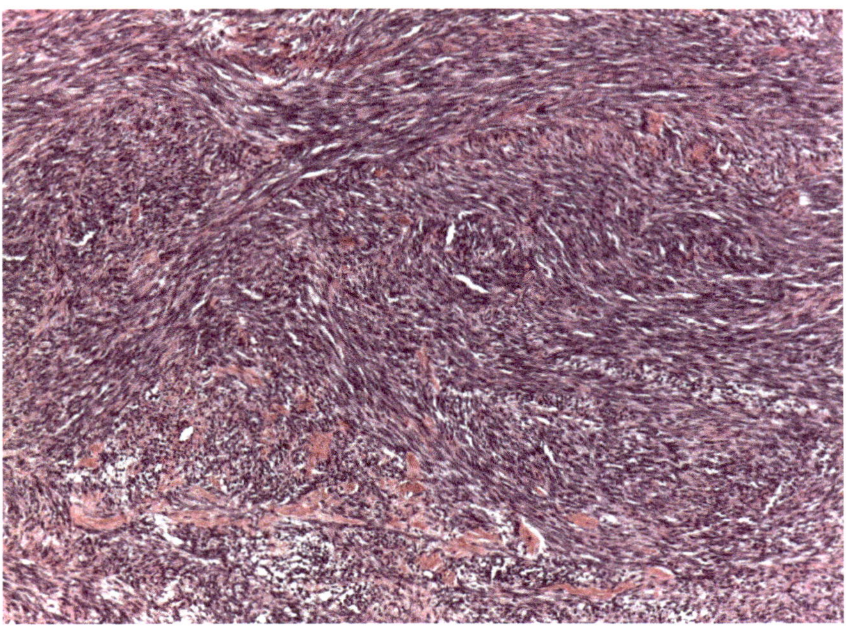

Monophasic type: spindle cell sarcoma with variable grade of malignancy organized in a herringbone pattern or HPC-like pattern

## Selected Bibliography

Bixby SD, Hettmer S, Taylor GA, Voss SD (2010) Synovial sarcoma in children: imaging features and common benign mimics. AJR Am J Roentgenol 195(4):1026–1032

Cai W, Sun Y, Wang W, Han C, Ouchida M, Xia W, Zhao X, Sun B (2011) The effect of SYT-SSX and extracellular signal-regulated kinase (ERK) on cell proliferation in synovial sarcoma. Pathol Oncol Res 17:357–367

O'Sullivan PJ, Harris AC, Munk PL (2008) Radiological features of synovial cell sarcoma. Br J Radiol 81(964):346–356

Palmerini E, Staals EL, Alberghini M, Zanella L, Ferrari C, Benassi MS, Picci P, Mercuri M, Bacci G, Ferrari S (2009) Synovial sarcoma: retrospective analysis of 250 patients treated at a single institution. Cancer 115(13):2988–2998

West RB (2010) Expression profiling in soft tissue sarcomas with emphasis on synovial sarcoma, gastrointestinal stromal tumor, and leiomyosarcoma. Adv Anat Pathol 17(5):366–373. Review

# Chapter 77
# Alveolar Soft Part Sarcoma

**Marco Gambarotti**

**Definition**: Sarcoma with alveolar, pseudo-endocrine, and organoid structure.
**Epidemiology**: Rare, females, 15–35 years old.

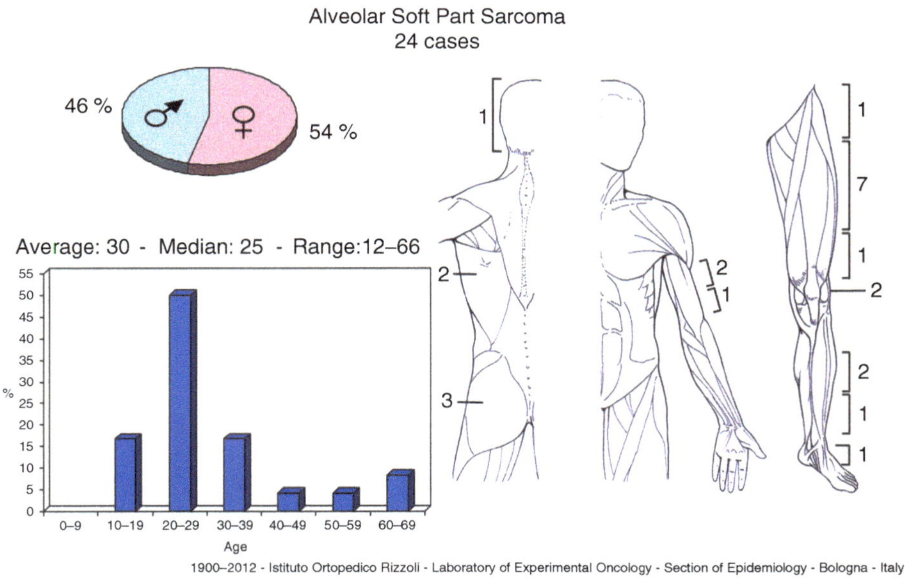


M. Gambarotti, MD
Department of Anatomy and Pathological Histology, Istituto Ortopedico Rizzoli,
Via del Barbiano 1/10, Bologna 40136, Italy
e-mail: marco.gambarotti@ior.it

P. Picci et al. (eds.), *Atlas of Musculoskeletal Tumors and Tumorlike Lesions*,
DOI 10.1007/978-3-319-01748-8_77, © Springer International Publishing Switzerland 2014

**Location**: Deep-seated, lower limb within the muscle, thigh.
**Clinical**: Globose or multinodular mass, slowly increasing, and painless. Tumor may be large, present from many years, and richly vascularized with pulsation on the mass.
**Imaging**: On CT: an intense uptake of contrast medium. On angiography: tumoral neoangiogenesis and indirect signs of arteriovenous shunt. On MRI: bright on both T1 and T2, homogeneous on the first and inhomogeneous with sharp peripheral demarcation of the mass on the second.
**Histopathology**: Soft, yellowish-white, or gray-red-purple, with hemorrhagic and necrotic areas, poorly delimited. Typical very glandular, acinar, and alveolar appearance. Nests of tumor cells are separated by thin-walled vascular spaces. Cells are large, epithelioid, globose-polyhedric, with eosinophilic granular to clear cytoplasm. The nucleus is vesicular, the nucleolus evident, mitotic figures rare, pleomorphism and hyperchromia, and sporadic giant cells with monstrous nuclei. Intense neoangiogenesis and frequent vascular invasion. At genetic level, this tumor is characterized by an unbalanced translocation: der(17)t(X;17)(p11;p25), resulting in the fusion of a gene of unknown function ASPL on chromosome 17 to the TFE3 transcription factor gene on the X chromosome.
**Course and Staging**: Recurrences are frequent when treated inadequately. Metastases may be present at diagnosis and at the first signs of the lesion. At times, they occur very late, even 10 years after the diagnosis. Usually, stage II B or III B.
**Treatment**: Wide or radical excision associated to adjuvant radio- and/or chemotherapy.

**Chromosomal Translocations**

| | | |
|---|---|---|
| t(X;17) (p11;q25) | ASPL-TFE3 (type 1, type 2) | 100 % |

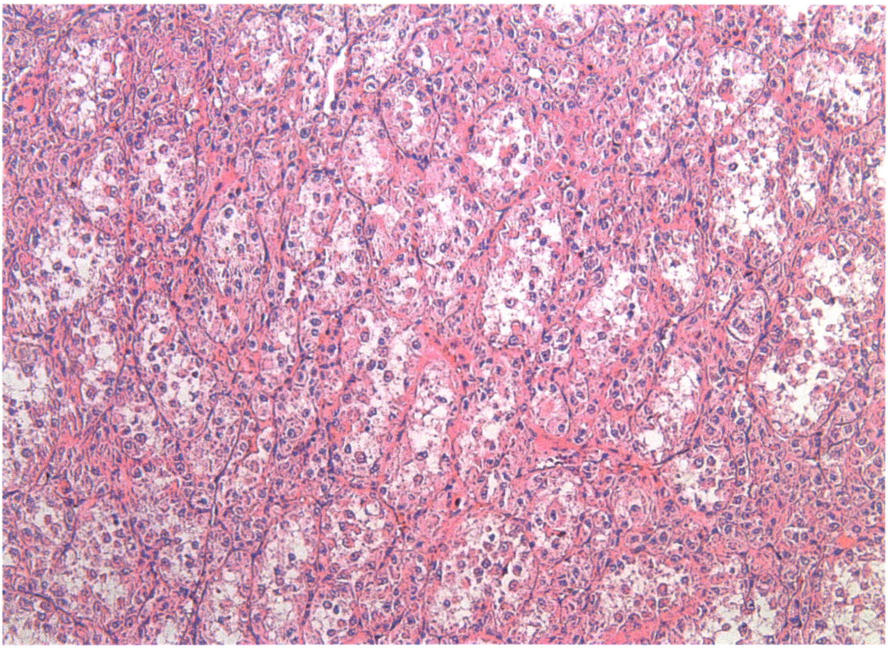

Eosinophilic wide cells losing cohesion and giving the alveolar appearance

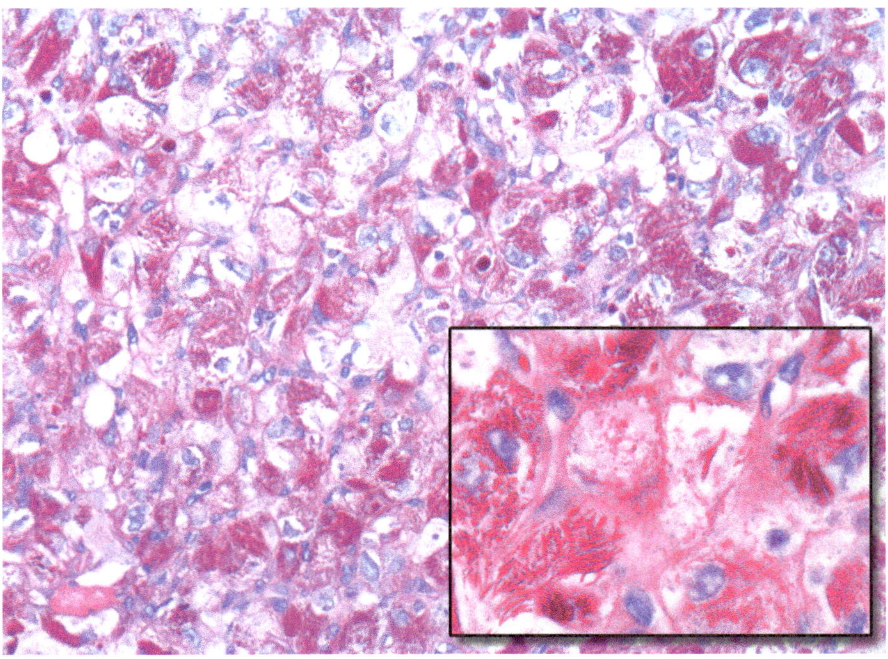

PAS stain with diastase digestion demonstrates rod-shaped crystals (inset: higher magnification)

## Selected Bibliography

Agarwal S, Gupta R, Iyer VK, Mathur SR, Ray R (2011) Cytopathological diagnosis of alveolar soft part sarcoma, a rare soft tissue neoplasm. Cytopathology 22:318–322

Folpe AL, Deyrup AT (2006) Alveolar soft-part sarcoma: a review and update. J Clin Pathol 59(11):1127–1132. Review

Khanna P, Paidas CN, Gilbert-Barness E (2008) Alveolar soft part sarcoma: clinical, histopathological, molecular, and ultrastructural aspects. Fetal Pediatr Pathol 27(1):31–40

Weiss SW (2002) Alveolar soft part sarcoma: are we at the end or just the beginning of our quest? Am J Pathol 160(4):1197–1199. Review

Zarrin-Khameh N, Kaye KS (2007) Alveolar soft part sarcoma. Arch Pathol Lab Med 131(3):488–491. Review

# Chapter 78
# Epithelioid Sarcoma

**Marco Gambarotti**

**Definition**: Malignant mesenchymal neoplasm that exhibits a predominantly epithelial phenotype.
**Epidemiology**: Rare, males, 20–30 years old.

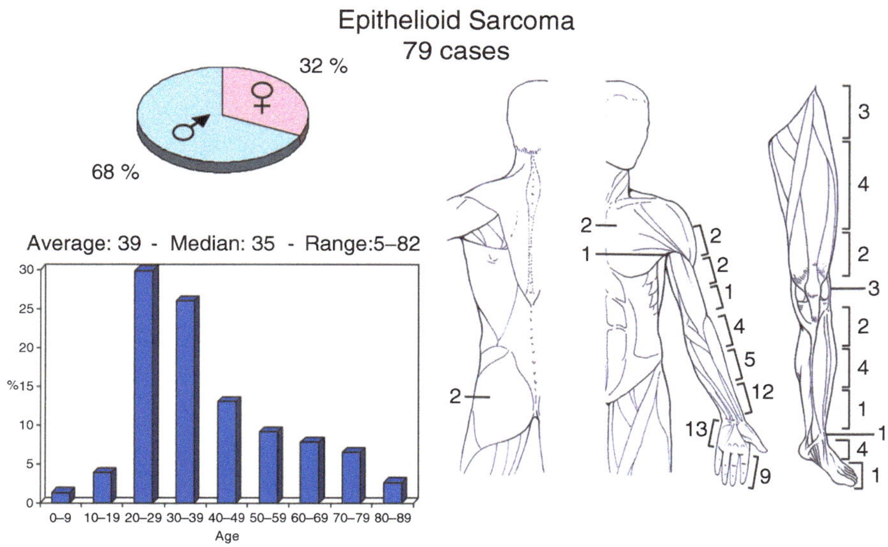


1900–2012 - Istituto Ortopedico Rizzoli - Laboratory of Experimental Oncology - Section of Epidemiology - Bologna - Italy

M. Gambarotti, MD
Department of Anatomy and Pathological Histology, Istituto Ortopedico Rizzoli,
Via del Barbiano 1/10, Bologna 40136, Italy
e-mail: marco.gambarotti@ior.it

P. Picci et al. (eds.), *Atlas of Musculoskeletal Tumors and Tumorlike Lesions*,
DOI 10.1007/978-3-319-01748-8_78, © Springer International Publishing Switzerland 2014

**Location**: Among the most frequent sarcomas occurring in the hand and forearm. Palmar region of the hand, dorsal region of the forearm, plantar aspect of the foot, anterior region of the leg, and in the subcutaneous tissue.
**Clinical**: One or more small, hard, superficial nodules, adherent to the derma, slightly elevated on the skin. They grow slowly ulcerating the skin surface. Some nodules are deep. Painless. Almost woody, nodules adhere and fixed to the surrounding tissues.
**Imaging**: On x-ray: calcifications (20 %), bone erosion rare. On MRI: isointense with muscle on T1, inhomogeneous, hyperintense, with an irregularly outlined lesion on T2.
**Histopathology**: Multinodular, hard, whitish, infiltrating, and adhering to the skin, fascia, tendon sheaths, muscles, vessels, nerves, and periosteum. Pseudo-granulomatous pattern with a central area of necrosis surrounded by proliferation of large polygonal cells, with eosinophilic cytoplasm, vesicular nucleus, and evident nucleolus. At the periphery spindle cells arranged in bundles. In the proximal-type variant (that generally arise in the deeper soft tissue of pelvo-perineal, genital, and inguinal region), the cells are larger with pleomorphic carcinoma-like or rhabdoid features, and the pseudo-granulomatous pattern is mostly absent. Immunohistochemistry shows positivity for vimentin, cytokeratin, and EMA. These tumors express CD34 in about 50 % of cases. There is loss of INI-1 protein in more than 90 % of cases.
**Course and Staging**: Metastases in regional lymph nodes and lungs. It diffuses in the limb in a proximal direction along the tendons, muscles, neurovascular bundles, and the lymphatic network of the derma as it creates multiple superficial ulcerated nodules. Usually, stage IIB.
**Treatment**: Radical excision with dissection of the regional lymph nodes. Often, even after wide excision, local recurrence is frequent. Poor prognosis.

**Immunohistochemical Panel**

| | |
|---|---|
| VIM | + |
| CK | + |
| EMA | + |
| CD34 | ± |
| INI-1 | − |

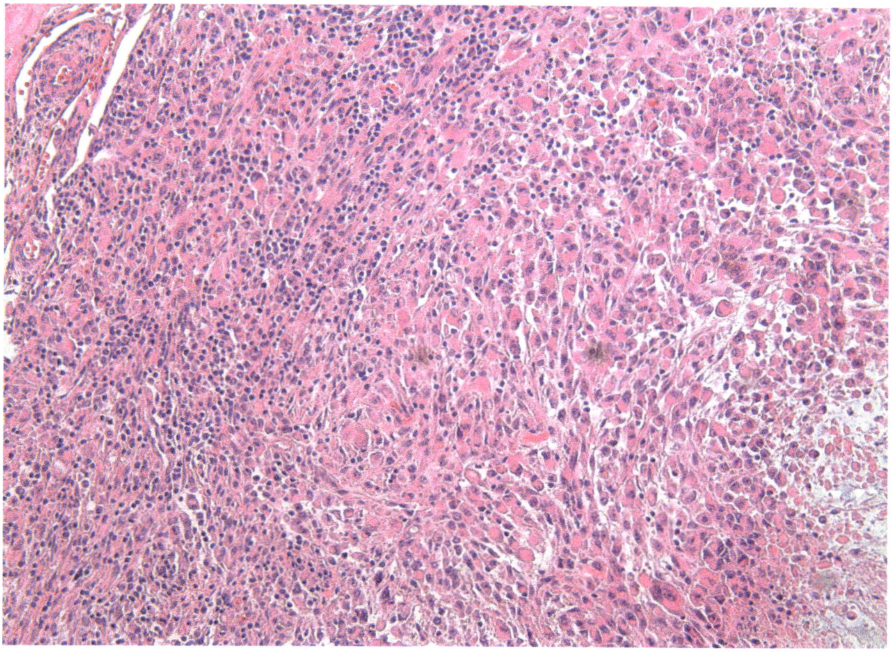

Malignant-appearing spindle to epithelioid cells organized in pseudo-granulomatous structures with necrotic centers

## Selected Bibliography


Armah HB, Parwani AV (2009) Epithelioid sarcoma. Arch Pathol Lab Med 133(5):814–819. Review

Dei Tos AP, Wagner AJ, Modena P, Comandone A, Leyvraz S (2009) Epithelioid soft tissue tumors. Semin Oncol 36(4):347–357. Review

Fisher C (2006) Epithelioid sarcoma of Enzinger. Adv Anat Pathol 13(3):114–121. Review

Gasparini P, Facchinetti F, Boeri M, Lorenzetto E, Livio A, Gronchi A, Ferrari A, Massimino M, Spreafico F, Giangaspero F, Forni M, Maestro R, Alaggio R, Pilotti S, Collini P, Modena P, Sozzi G (2011) Prognostic determinants in epithelioid sarcoma. Eur J Cancer 47(2):287–295

López-Ríos F, Rodríguez-Peralto JL, Castaño E (2000) Epithelioid sarcoma and its variants. Hum Pathol 31(4):520. Review

# Chapter 79
# Clear Cell Sarcoma

**Marco Gambarotti**

**Definition**: Sarcoma with melanocytic differentiation of tendons and aponeuroses.
**Epidemiology**: Very rare, females, average age of 25 years.

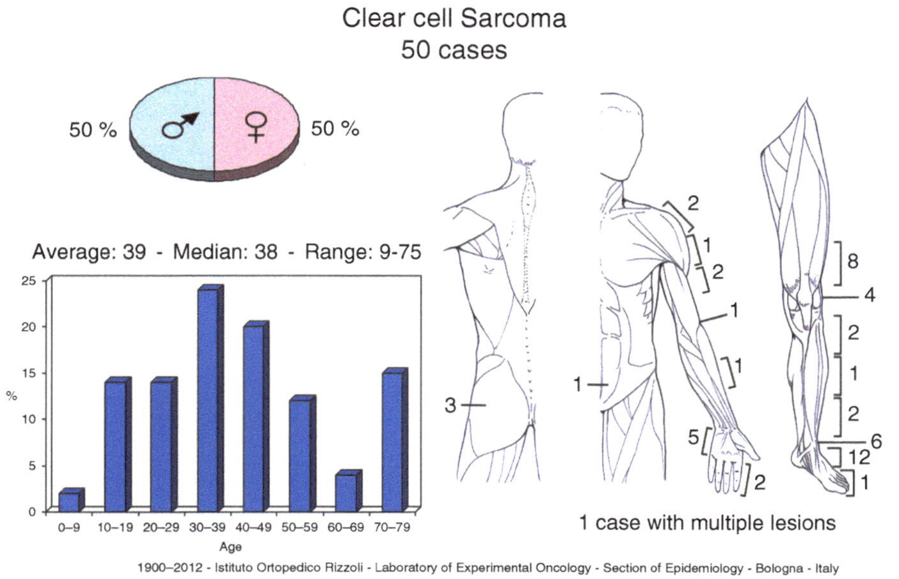


M. Gambarotti, MD
Department of Anatomy and Pathological Histology, Istituto Ortopedico Rizzoli,
Via del Barbiano 1/10, Bologna 40136, Italy
e-mail: marco.gambarotti@ior.it

P. Picci et al. (eds.), *Atlas of Musculoskeletal Tumors and Tumorlike Lesions*,
DOI 10.1007/978-3-319-01748-8_79, © Springer International Publishing Switzerland 2014

**Location**: Foot, ankle, knee, and upper limb.
**Clinical**: Slowly growing, moderate size, painless, globose mass.
**Diagnosis**: On MRI, elliptical, smoothly outlined mass, slightly or markedly hyperintense on T1, white lesion with multiple low-intensity septations on T2, homogeneous enhancement after gadolinium.
**Histopathology**: Adherent to a tendon or to an aponeurosis, but not to the subcutaneous or to the skin. Firm, globose, well-defined boundaries, capsulated appearance. Round, oval, spindle cells with clear cytoplasm. Nuclei are round, ovoid or elongated, vesicular, with a large nucleolus. Mitotic figures are rare. Cells are arranged in nests surrounded by a fine reticular stroma or rough collagen bands that may give an almost acinar or glandular impression at first glance. Immunohistochemistry shows consistent positivity for S-100, HMB45, MART-1, and MiTF. Molecular findings are characterized by a specific translocation: t(12;22)(q13;q12), with an EWS-ATF1 fusion transcript.
**Course and Staging**: Local recurrences are frequent when treatment is inadequate due to the not very aggressive clinical presentation. Metastases are late (>6 years) but very frequent in lymph nodes and lungs. Usually, stage II B.
**Treatment**: Very wide or radical excision with regional lymph nodes.

**Immunohistochemical Panel**

| | |
|---|---|
| S100 | + |
| HMB45 | + |
| MART-1 | + |
| Mitf | + |

**Chromosomal Translocations**

| | | |
|---|---|---|
| t(12;22) (q13;q12) | EWSR1-ATF1 (type 1, type2) | >90 % |

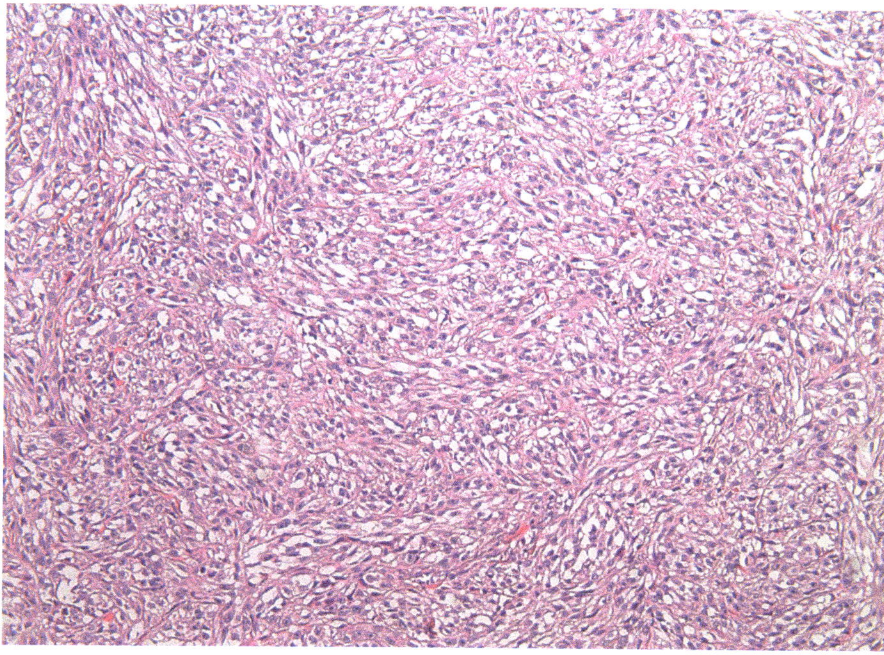

Spindle cell neoplasm with optically clear cytoplasms

## Selected Bibliography

Dim DC, Cooley LD, Miranda RN (2007) Clear cell sarcoma of tendons and aponeuroses: a review. Arch Pathol Lab Med 131(1):152–156. Review

Kosemehmetoglu K, Folpe AL (2010) Clear cell sarcoma of tendons and aponeuroses, and osteoclast-rich tumour of the gastrointestinal tract with features resembling clear cell sarcoma of soft parts: a review and update. J Clin Pathol 63(5):416–423. Review

Malchau SS, Hayden J, Hornicek F, Mankin HJ (2007) Clear cell sarcoma of soft tissues. J Surg Oncol 95(6):519–522. Review

Meis-Kindblom JM (2006) Clear cell sarcoma of tendons and aponeuroses: a historical perspective and tribute to the man behind the entity. Adv Anat Pathol 13(6):286–292. Review

Sandberg AA, Bridge JA (2001) Updates on the cytogenetics and molecular genetics of bone and soft tissue tumors: clear cell sarcoma (malignant melanoma of soft parts). Cancer Genet Cytogenet 130(1):1–7. Review

# Chapter 80
# Extraskeletal Myxoid Chondrosarcoma

**Marco Gambarotti**

**Definition**: Soft tissue sarcoma composed of spindle and epithelioid tumor cells associated with extracellular myxoid matrix. Extraskeletal myxoid chondrosarcomas are genetically distinct from osseous chondrosarcoma.
**Epidemiology**: Rare, male/female ratio 2:1, adults (40–70 years).

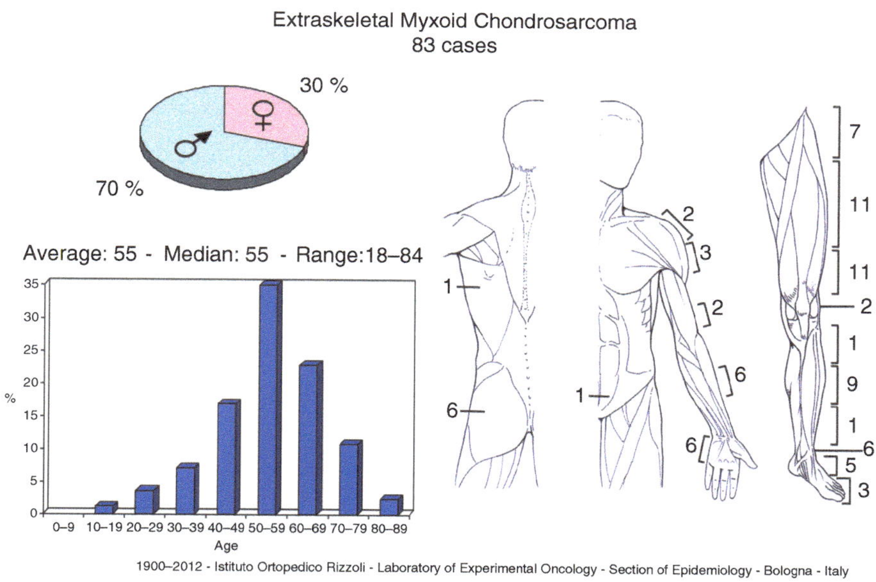


M. Gambarotti, MD
Department of Anatomy and Pathological Histology, Istituto Ortopedico Rizzoli,
Via del Barbiano 1/10, Bologna 40136, Italy
e-mail: marco.gambarotti@ior.it

P. Picci et al. (eds.), *Atlas of Musculoskeletal Tumors and Tumorlike Lesions*,
DOI 10.1007/978-3-319-01748-8_80, © Springer International Publishing Switzerland 2014

**Location**: Deep-seated, proximal extremities and trunk.
**Clinical**: Enlarging painless soft tissue mass.
**Imaging**: Although imaging characteristics are nonspecific, most tumors appear lobulated, and highly myxoid tumors have a homogeneous high intensity signal on T2-weighted MRI images. Tumors with necrosis or hemorrhage have a more heterogeneous appearance.
**Histopathology**: There are cytologically uniform oval and spindle cells embedded in flocculent myxoid ground substance, with fibrous bands dividing individual lobules of tumor. Cells may be arranged as cords, strands, nests, and sheets and may have typically uniform dark-staining nuclei with indistinct nucleoli. In hypervascular matrix-poor regions, tumor cells often display larger, more vesicular nuclei with visible nucleoli; some tumor cells may have distinctly rhabdoid morphology. Tumor is hypovascular and the matrix stains alcian blue positive at pH 4.0 and 1.0 (chondroitin sulfate positive). At the genetic level, the majority of tumors exhibit the t(9;22)(q22;q12) translocation involving EWS and CHN genes.
**Course and Staging**: Five-year survival rates are high (>80 %); however, late lung metastases occur, and 10- and 15-year disease-free survival rates are considerably lower.
**Treatment**: Wide surgical excision.

**Chromosomal Translocations**

| | | |
|---|---|---|
| t(9;22) (q22;q12) | EWS-NR4A3 (CHN, TEC, NOR1) | 72 % |
| t(9;17) (q22;q11) | TAF2N-NR4A3 (CHN, TEC, NOR1) | 16 % |
| t(9;15) (q22;q21) | TCF12-NR4A3 (CHN, TEC, NOR1) | <1 % |

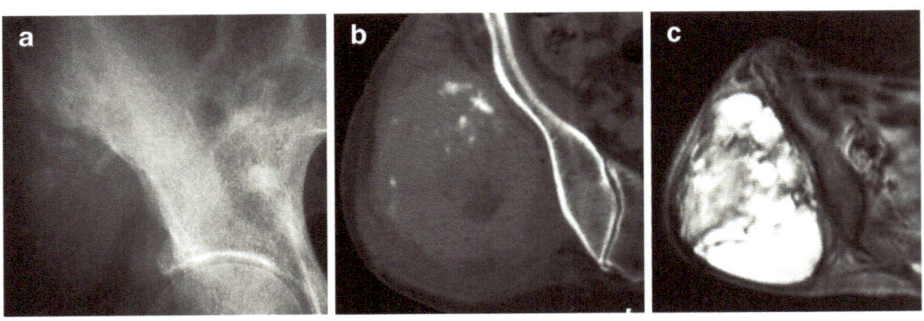


Radiograph, CT, and T2 axial MR image. The soft tissue mass contains cartilaginous calcifications, is lobulated, and has a high signal on T2 MR image

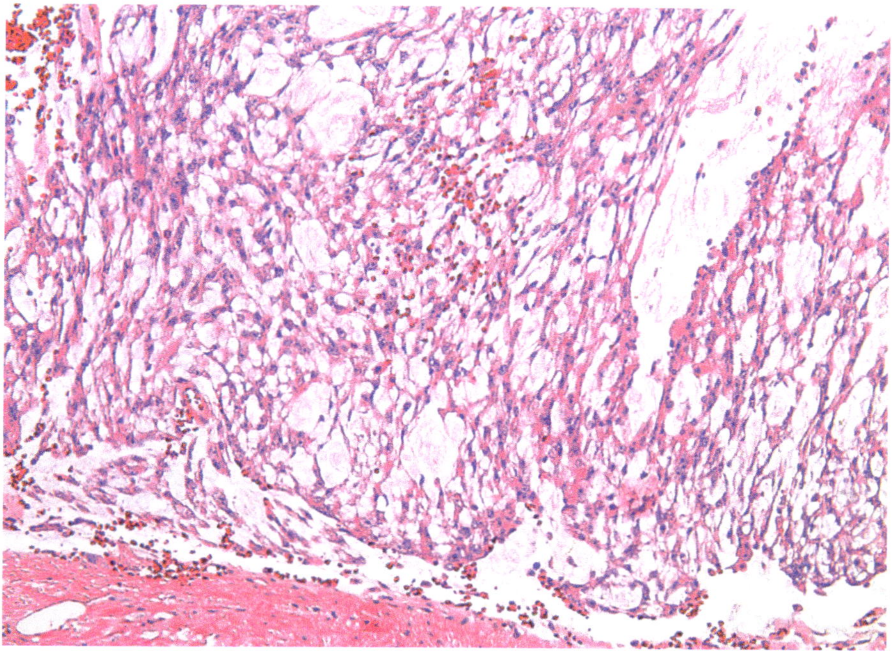

Cords and nests of eosinophilic cells embedded in a myxoid matrix

## Selected Bibliography

Filion C, Motoi T, Olshen AB, Laé M, Emnett RJ, Gutmann DH, Perry A, Ladanyi M, Labelle Y (2009) The EWSR1/NR4A3 fusion protein of extraskeletal myxoid chondrosarcoma activates the PPARG nuclear receptor gene. J Pathol 217(1):83–93

Hisaoka M, Hashimoto H (2005) Extraskeletal myxoid chondrosarcoma: updated clinicopathological and molecular genetic characteristics. Pathol Int 55(8):453–463. Review

McGrory JE, Rock MG, Nascimento AG, Oliveira AM (2001) Extraskeletal myxoid chondrosarcoma. Clin Orthop Relat Res 382:185–190

Tateishi U, Hasegawa T, Nojima T, Takegami T, Arai Y (2006) MRI features of extraskeletal myxoid chondrosarcoma. Skeletal Radiol 35(1):27–33

Wang WL, Mayordomo E, Czerniak BA, Abruzzo LV, Dal Cin P, Araujo DM, Lev DC, López-Terrada D, Lazar AJ (2008) Fluorescence in situ hybridization is a useful ancillary diagnostic tool for extraskeletal myxoid chondrosarcoma. Mod Pathol 21(11):1303–1310

# Chapter 81
# Extraskeletal Osteosarcoma

**Marco Gambarotti**

**Definition**: A sarcoma arising in the extraskeletal somatic soft tissue in which neoplastic cells produce osteoid or bone matrix or both.
**Epidemiology**: Rare, males, 50–60 years.

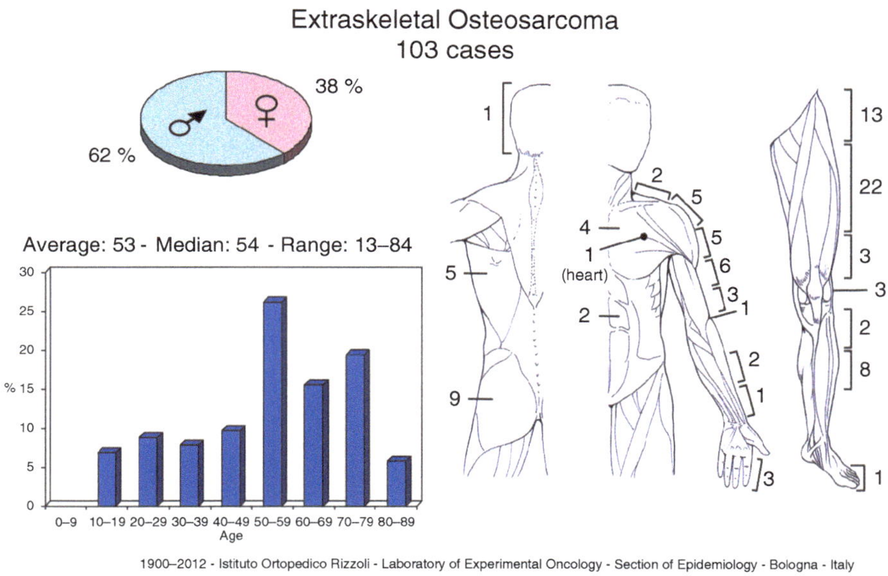


M. Gambarotti, MD
Department of Anatomy and Pathological Histology, Istituto Ortopedico Rizzoli,
Via del Barbiano 1/10, Bologna 40136, Italy
e-mail: marco.gambarotti@ior.it

P. Picci et al. (eds.), *Atlas of Musculoskeletal Tumors and Tumorlike Lesions*, 381
DOI 10.1007/978-3-319-01748-8_81, © Springer International Publishing Switzerland 2014

**Location**: Deep soft tissues of the thigh and buttocks.
**Clinical**: Progressively enlarging painless mass.
**Imaging**: X-rays, CT, and MRI reveal a large deep-seated soft tissue mass with variable mineralization. By definition, these lesions do not arise from bone but secondarily involve the periosteum, cortex, or medullary canal.
**Histopathology**: Highly cellular, mitotically active tumors. There is usually considerable nuclear pleomorphism with necrosis, and lacelike eosinophilic osteoid outlines individual cells or clusters of cells. Mineralized osteoid (bone) is relatively uncommon. Lobules of cellular atypical hyaline cartilage may be present. All histological patterns of osseous osteosarcoma may be seen.
**Course and Staging**: High-grade tumors, poor prognosis, high rate of metastases.
**Treatment**: Wide or radical excision with systemic chemotherapy.

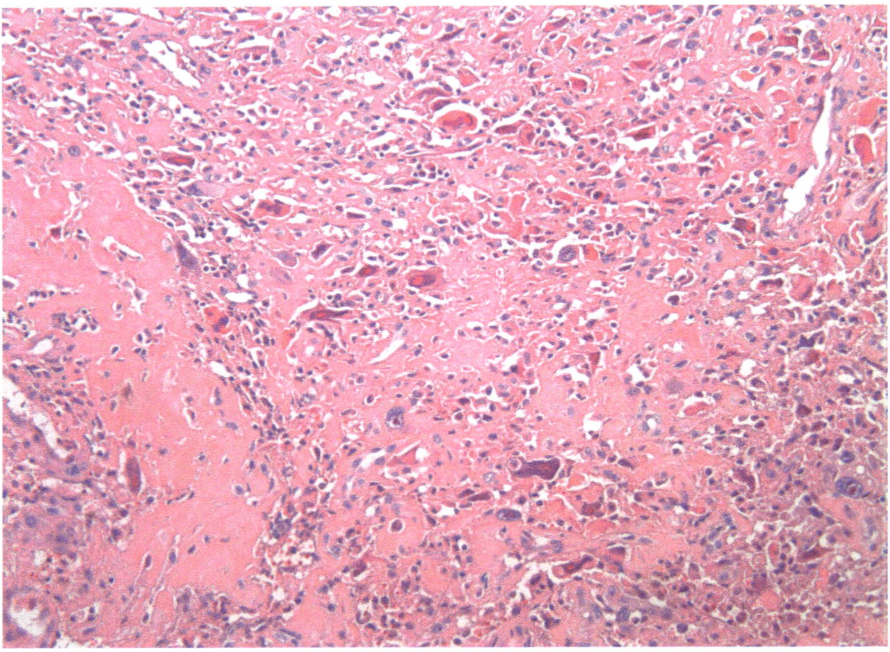

Osteoid-producing cells haphazardly organized, growing in the soft tissue

## Selected Bibliography

Goldstein-Jackson SY, Gosheger G, Delling G, Berdel WE, Exner GU, Jundt G, Machatschek JN, Zoubek A, Jürgens H, Bielack SS, Cooperative Osteosarcoma Study Group COSS (2005) Extraskeletal osteosarcoma has a favourable prognosis when treated like conventional osteosarcoma. J Cancer Res Clin Oncol 131(8):520–526

Lee JS, Fetsch JF, Wasdhal DA, Lee BP, Pritchard DJ, Nascimento AG (1995) A review of 40 patients with extraskeletal osteosarcoma. Cancer 76(11):2253–2259

Lidang Jensen M, Schumacher B, Myhre Jensen O, Steen Nielsen O, Keller J (1998) Extraskeletal osteosarcomas: a clinicopathologic study of 25 cases. Am J Surg Pathol 22(5):588–594

McCarter MD, Lewis JJ, Antonescu CR, Brennan MF (2000) Extraskeletal osteosarcoma: analysis of outcome of a rare neoplasm. Sarcoma 4(3):119–123

Patel SR, Benjamin RS (1995) Primary extraskeletal osteosarcoma–experience with chemotherapy. J Natl Cancer Inst 87(17):1331–1333

# Chapter 82
# Extraskeletal Ewing's Sarcoma

**Marco Gambarotti**

**Definition**: Highly malignant poorly differentiated tumor composed of small round cells. It is the soft tissue counterpart of the more frequent Ewing's sarcoma of the bone.
**Epidemiology**: Soft tissue Ewing's sarcoma is by far less frequent than the bone variant. Most common in adults, but it may occur at any age. There is predominance in the male gender.

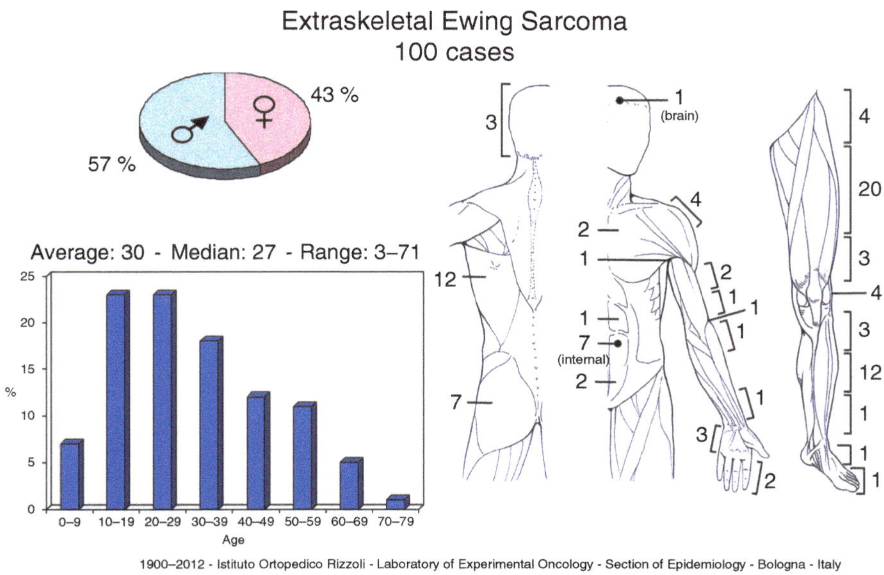


M. Gambarotti, MD
Department of Anatomy and Pathological Histology, Istituto Ortopedico Rizzoli,
Via del Barbiano 1/10, Bologna 40136, Italy
e-mail: marco.gambarotti@ior.it

P. Picci et al. (eds.), *Atlas of Musculoskeletal Tumors and Tumorlike Lesions*,
DOI 10.1007/978-3-319-01748-8_82, © Springer International Publishing Switzerland 2014

**Location**: Any soft tissue site can be involved, but most common are the paravertebral region, retroperitoneum, and chest wall, followed by the extremities.
**Clinical**: Large, destructive mass.
**Histopathology**: Grossly it is a large, destructive mass, frequently with necrosis and hemorrhage. Histologically, Ewing's sarcoma shows a sheetlike to vaguely lobular growth pattern, a well-developed capillary vascularization, and a uniform cell population of round blue cells with small amounts of clear to light eosinophilic cytoplasm, regular nuclear contours, finely dispersed chromatin, and small nucleoli. Geographic necrosis and individual apoptotic cells are frequently seen. Pseudorosettes are a consistent clue for diagnosis of PNET. Immunohistochemical and molecular findings are the same as for the bone counterpart.
**Course and Staging**: 10-year survival rate of 60 % with current treatment regimens.
**Treatment**: Current multimodal therapy (surgery, chemotherapy, and radiotherapy).

**Immunohistochemical Panel**

| | |
|---|---|
| VIM | + |
| CD99 | + |
| Caveolin-1 | + |
| Fli1 | + |
| CAM 5.2 | ± |
| S100 | ± |
| LAC | – |
| TdT | – |

**Chromosomal Translocations**

| | | |
|---|---|---|
| t (11,22)(q24;q12) | EWSR1-FLI1 | 90–95 % |
| t (21;22)(q22;q12) | EWSR1-ERG | 5–10 % |
| t (2;22)(q35;q12) | EWSR1-FEV | Rare |
| t (7;22)(p22;q12) | EWSR1-ETV1 | Rare |
| t (17;22)(q21;q12) | EWSR1-E1AF | Rare |
| t (2;16)(q35;p11) | FUS-FEV | Rare |
| t (16;21)(q11;p22) | FUS-ERG | Rare |
| t (20;22)(q13;q12) | EWSR1-NFACT2 | Rare |
| t (4;22)(q31;q12) | EWSR1-SMARCA5 | Rare |
| t (6;22)(p21;q12) | EWSR1-POU5F1 | Rare |
| t (2;22)(q31;q12) | EWSR1-SP3 | Rare |
| t (4;19)(q35;q13) | EWSR1-POU5F1 | Rare |
| Submicroscopic inv(22) in t(1;22)(p36.1;q12) | EWSR1-PATZ | Rare |

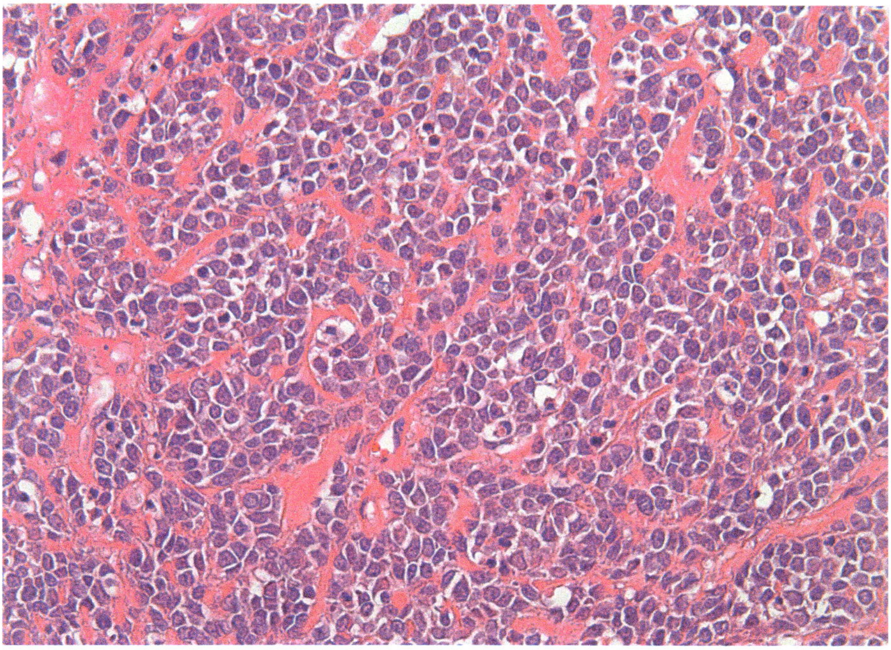

Uniform cell population of small round to ovoid undifferentiated blue cells with intensely stained nuclei and small amounts of clear to light eosinophilic cytoplasm

## Selected Bibliography


Applebaum MA, Worch J, Matthay KK, Goldsby R, Neuhaus J, West DC, Dubois SG (2011) Clinical features and outcomes in patients with extraskeletal Ewing sarcoma. Cancer 117:3027–3032

Gamberi G, Cocchi S, Benini S, Magagnoli G, Morandi L, Kreshak J, Gambarotti M, Picci P, Zanella L, Alberghini M (2011) Molecular diagnosis in Ewing family tumors: the Rizzoli experience–222 consecutive cases in four years. J Mol Diagn 13(3):313–324. doi:10.1016/j.jmoldx.2011.01.004. Epub 2011 Mar 31

Guiter GE, Gamboni MM, Zakowski MF (1999) The cytology of extraskeletal Ewing sarcoma. Cancer 87(3):141–148

Koscielniak E, Morgan M, Treuner J (2002) Soft tissue sarcoma in children: prognosis and management. Paediatr Drugs 4(1):21–28

O'Keeffe F, Lorigan JG, Wallace S (1990) Radiological features of extraskeletal Ewing sarcoma. Br J Radiol 63(750):456–460

Sumegi J, Nishio J, Nelson M, Frayer RW, Perry D, Bridge JA (2011) A novel t(4;22)(q31;q12) produces an EWSR1-SMARCA5 fusion in extraskeletal Ewing sarcoma/primitive neuroectodermal tumor. Mod Pathol 24(3):333–342

Kolibri 360 GmbH
Ferdinand-Jühlke-Straße 7
99095 Erfurt, Deutschland
produktsicherheit@kolibri360.de

# Atlas of Musculoskeletal Tumors and Tumorlike Lesions